Springer Proceedings in Materials

Volume 45

Springer Proceedings in Materials publishes the latest research in Materials Science and Engineering presented at high standard academic conferences and scientific meetings. It provides a platform for researchers, professionals and students to present their scientific findings and stay up-to-date with the development in Materials Science and Engineering. The scope is multidisciplinary and ranges from fundamental to applied research, including, but not limited to:

- Structural Materials
- Metallic Materials
- Magnetic, Optical and Electronic Materials
- Ceramics, Glass, Composites, Natural Materials
- Biomaterials
- Nanotechnology
- Characterization and Evaluation of Materials
- Energy Materials
- Materials Processing

To submit a proposal or request further information, please contact one of the following Springer Publishing Editors according to your affiliation location:

European countries: **Mayra Castro** (mayra.castro@springer.com)
India, South Asia and Middle East: **Priya Vyas** (priya.vyas@springer.com)
South Korea: **Smith Chae** (smith.chae@springer.com)
Southeast Asia, Australia and New Zealand: **Ramesh Nath Premnath** (ramesh.premnath@springer.com)
The Americas: **Michael Luby** (michael.luby@springer.com)
China and all the other countries or regions, as well as topics in materials chemistry: **Maggie Guo** (maggie.guo@cn.springernature.com)

This book series is indexed in **SCOPUS** and EI Compendex database.

Smail Hamamda · Abdelmalek Zahaf ·
Yurii Sementsov · Serhii Nedilko ·
Kateryna Ivanenko
Editors

Proceedings of the 2nd International Conference of Nanotechnology for Environmental Protection and Clean Energy Production

ICNEP 2023, 9–10 October, Constantine, Algeria

Springer

Editors
Smail Hamamda
Université Frères Mentouri Constantine 1
Constantine, Algeria

Abdelmalek Zahaf
Université Frères Mentouri Constantine 1
Constantine, Algeria

Yurii Sementsov
Chuiko Institute Surface Chemistry
Kyiv, Ukraine

Serhii Nedilko
Taras Shevchenko National University
of Kyiv
Kyiv, Ukraine

Kateryna Ivanenko
Taras Shevchenko National University
of Kyiv
Kyiv, Ukraine

ISSN 2662-3161 ISSN 2662-317X (electronic)
Springer Proceedings in Materials
ISBN 978-981-97-1915-0 ISBN 978-981-97-1916-7 (eBook)
https://doi.org/10.1007/978-981-97-1916-7

This Springer imprint is published by the registered company Springer Nature Singapore Pte Ltd.
The registered company address is: 152 Beach Road, #21-01/04 Gateway East, Singapore 189721, Singapore

Paper in this product is recyclable.

Organization

Conference

The International Conference of Nanotechnology for Environmental Protection and Clean Energy Production (ICNEP) provides a forum for specialists and practitioners to present and discuss their research results in the several areas of the conference and also state-of-the-art findings in using the applied nanotechnology and renewable energy to solve national problems that face developing countries. The conference ICNEP publishes papers on experimental studies and applications in the domain of nanoparticles and materials characterization. The objective of the conference is not only the exchange of knowledge and experience, since the conference is an open door to students, but also provides opportunities for researchers to target future collaboration on current issues.

Honorary Chair

Pr. A. Bouras, Constantine 1—Frères Mentouri University, Algeria

Conference Chair

Pr. S. Hamamda, Constantine 1—Frères Mentouri University, Algeria

Program Chairs

Pr. S. Hamamda, Constantine 1—Frères Mentouri University, Algeria
Dr. A. Zahaf, Constantine 1—Frères Mentouri University, Algeria

Invited Lecturers

Pr. Yu. I. Sementsov, Chuiko Institute Surface Chemistry,Kiev, Ukraine
Pr. S. G. Nedilko, Taras Shevchenko National University, Kiev, Ukraine
Pr. K. O. Ivanenko, Taras Shevchenko National University, Kiev, Ukraine
Dr. A. Settar, INSA centre Val de Loire Compus de Bourge, France

Local Committee

ICNEP established a local committee from the Laboratory of Thermodynamic and Treatment of Materials Surface as an organizer, Department of Physics, Faculty of Exact Sciences, Constantine 1—Frères Mentouri University, Algeria. The list below comprises the Local Committee members
Pr. S. Hamamda, Constantine 1—Frères Mentouri University, Algeria
Pr. A. Karaali, Constantine 1—Frères Mentouri University, Algeria
Dr. T. Dorbani, Constantine 1—Frères Mentouri University, Algeria
Dr. A. Zahaf, Constantine 1—Frères Mentouri University, Algeria
Mr. N. Barama, Constantine 1—Frères Mentouri University, Algeria
Dr. M. C. Bouleklab, Constantine 1—Frères Mentouri University, Algeria
Dr. H. Filali, Constantine 1—Frères Mentouri University, Algeria
Dr. S. Touahri, Constantine 1—Frères Mentouri University, Algeria
Dr. A. Bounemeur, National Polytechnic School of Constantine (ENPC), Algeria

Scientific Program Committee

ICNEP established an international committee of selected well-known experts in nanotechnology and renewable energy who are willing to be mentioned in the program and to review a set of papers each year. The list below comprises the scientific program committee members
Pr. S. Hamamda, Constantine 1—Frères Mentouri University, Algeria
Pr. A. Karaali, Constantine 1—Frères Mentouri University, Algeria
Pr. B. Boudine, Constantine 1—Frères Mentouri University, Algeria
Pr. Z. Ouili, Constantine 1—Frères Mentouri University, Algeria

Pr. N. Bellel, Constantine 1—Frères Mentouri University, Algeria
Dr. T. Dorbani, Constantine 1—Frères Mentouri University, Algeria
Pr. H. Belkheir, Annaba University, Algeria
Pr. A. Saker, Annaba University, Algeria
Pr. A. Mefteh, Tunis University, Tunisia
Pr. F. Bouchear, Constantine 1—Frères Mentouri University, Algeria
Pr. A. Boumaza, Khenchela University, Algeria
Pr. A. Boudjada, Constantine 1—Frères Mentouri University, Algeria
Pr. A. Chaker, Constantine 1—Frères Mentouri University, Algeria
Pr. R. Guemini, O. E. B. University, Algeria
Pr. M. Zaabat, O. E. B. University, Algeria
Pr. A. Nouiri, O. E. B. University, Algeria
Pr. S. L. Revo, Taras Shevchenko National University, Kiev, Ukraine
Pr. Yu. I. Sementsov, Chuiko Institute Surface Chemistry, Kiev, Ukraine
Pr. L. Kisterska, Bakul institute Super Hard Materials, National Academy of Sciences, Ukraine
Pr. S. G. Nedilko, Taras Shevchenko National University, Kiev, Ukraine
Pr. Yu. A. Kunitsky, Technical Center NASU, Kiev, Ukraine
Academicien Pr. V. a. Makara, Research Center NASU, Kiev, Ukraine
Pr. W. Paulus, Montpellier, France
Dr. S. Touahri, Constantine 1—Frères Mentouri University, Algeria

Preface

This proceeding book about the Nanotechnology for Environmental Protection and Clean Energy Production conference covers a broad spectrum of topics, the conference focused on nanocomposites and nanomaterials applications, physic-chemical nanomaterials science, nanoscale physics, solids and structures, renewable energy, nanoparticles, metallic and polymer materials, as well as eco-friendly materials and nanocomposites. These diverse topics represent the multidimensional nature of nanotechnology and its profound impact on environmental protection and clean energy production. The main aim of the conference is to create periodical possibility for students, academics, and researchers to exchange their ideas, experimental results, and novel techniques. We try to explore the frontiers of nanotechnology, where cutting-edge research converges with environmental protection and clean energy production. This proceedings book is a testament to the exceptional work showcased at the conference, embodying the spirit of innovation and collaboration that defines our community.

With the support of organizing universities, we ensured that only the most compelling and innovative research was included in this collection. We believe that this conference represents a good standard conference in the domain of Nanotechnology for Environmental Protection and Clean Energy Production. Our sincerest thanks goes to the program committees and reviewers for their meticulous efforts in maintaining the conference's excellence. The contributions of the authors, alongside the dedication of the local organizing committee, have significantly enriched the content and success of the conference.

This event is supported by the Constantine 1—Frères Mentouri University, Constantine, Algeria, and mainly financed by Laboratory of Thermodynamic and Treatment of Materials Surface and Directorate General for Scientific Research and Technological Development (DGRSDT).

Constantine, Algeria — Smail Hamamda
Constantine, Algeria — Abdelmalek Zahaf
Kyiv, Ukraine — Yurii Sementsov
Kyiv, Ukraine — Serhii Nedilko
Kyiv, Ukraine — Kateryna Ivanenko

Contents

Nanocomposites and Nanomaterials Applications

Multiwall Carbon Nanotubes Effects on the Coefficient of Thermal Expansion of High Density Polyethylene

H. Filali, M. Bouleklab, S. Saadallah, T. Dorbani, N. Barama, and S. Hamamda

Abstract The aim of this work is studying the influence of multiwall nanotube additions in high-density polyethylene using different complementary techniques. We were able to monitor the degradation of PE + X%MWCNTs using dilatometry, a tool that is highly sensitive to the structural behavior of materials. The small quantities of nanotubes inserted did not reveal any significant changes in the nanocomposites change-of-state temperature. Experimental methods using both calorimetry and thermogravimetry have shown that, overall, degradation temperatures tend to be approximately the same order of magnitude. According to Raman spectroscopy and x-ray diffraction shows that crystallinity elevates with increasing concentration.

Keywords Dilatation · Calorimetry · Raman · Nanotube · Carbon · Anisotropy · PE

1 Introduction

Polymer composites are a particularly interesting class of materials in the socio-economic field. The advantages of using polymers are ease of handling, light weight, and low production costs. However, polymers are less rigid than metals and, above all, non-conductive, which limits their applications in certain technical fields.

The incorporation of nano-objects, and carbon nanotubes in particular, into polymer structures present a possibility to alter their mechanical, thermal and electrical properties, thus broadening their range of applications [1]. Nanofillers have been widely used as reinforcements in the manufacturing of high-performance nanocomposites [2], such properties make them highly regarded in various fields of technology.

H. Filali (✉) · M. Bouleklab · S. Saadallah · T. Dorbani · N. Barama · S. Hamamda
Laboratory of Thermodynamic and Treatment of Materials Surface, Constantine 1 University, Route Ain el Bey, Constantine, Algeria
e-mail: hichemfilali5@yahoo.com

S. Hamamda et al. (eds.), *Proceedings of the 2nd International Conference of Nanotechnology for Environmental Protection and Clean Energy Production*, Springer Proceedings in Materials 45, https://doi.org/10.1007/978-981-97-1916-7_1

Nanomaterials have also become essential for energy and hydrogen storage, energy conversion, transport and industry [3–5].

In particular, the introduction of CNT into polymers provides not only reinforcement, but also an improvement in other physical and chemical properties, such as electrical and thermal conductivity, thermal, fire and corrosion resistance [6].

High-density polyethylene was chosen as the material. It is a commonly utilized thermoplastic material. With its easy availability, low cost and good processability, polyethylene is also a material of choice as a matrix for composites [1]. All these changing characteristics make polyethylene a much sought-after material, especially since the insertion of carbon nanotubes as a reinforcing element. Polyethylene is used in a variety of fields, including industry, medicine and environmental protection.

The discovery of fullerenes and the subsequent development of carbon nanotubes have revolutionized and contributed to the improvement and replacement of heavy and costly materials with polymer-based nanocomposites containing carbon nanotubes. The major interest in these materials stems from the fact that carbon nanotubes have excellent mechanical properties and good electrical and thermal conductivity [7–10].

They have a very high form factor. They accelerate crystallization by acting as a nucleating agent in a polymer, responsible for increasing the thermal stability of nanomaterials [11]. They have a low and/or negative coefficient of thermal expansion. Their addition to metal or polymer matrices is responsible for the drastic reduction in expansion [12, 13]. Their addition to metal or polymer matrices is responsible for the drastic reduction in expansion [14, 15].

Today, the search for polymer-based nanocomposites with thermodynamic properties grows considerably owing to its multifunctional uses in advanced engineering and the development of new devices sensitive to thermal expansion.

Several groups have studied polyethylene (PE)-based composites reinforced with CNTs. This combination makes it possible to obtain new combinations of nanocomposites.

Authors in [16] Reported a significant improvement in the mechanical performance of PE containing 1, 3 and 5%MWCNTS. The authors [17] noted a negligible influence on the viscoelastic behavior of PE + X%MWCNTs where (X = 0.2, 0.4, 0.6, 0.8, 1.0%MWCNTs).

References [18, 19] Studied the thermal expansion and diffusion coefficients of PE-CNT and discovered the improvement in the TG, diffusion and thermal expansion coefficients of the system studied.

This work focuses on the dilatometry of the influence of the insertion of multiwall carbon nanotubes in high-density polyethylene.

Thermal expansion behavior is mainly important when PE nanomaterials are used in conjunction with other materials. However, it is preferable to use materials with compatible CTEs. Therefore, understanding the coefficient of thermal expansion (CTE) of PE nanocomposites reinforced with MWCNTs is an important issue. According to the literature, most research has focused on thermomechanical properties. The dilatometric behavior of carbon nanotube-reinforced polyethylene-based nanocomposites has been little studied. This is probably due to the fragility of the material and the difficulty of obtaining samples long enough for accurate dilatation measurements.

2 Methods and Preparations

The high-density polyethylene (HDPE) manufactured by "Ufa OrgSynthesis" (Russia) with 918.5 $\pm$ 1.5 kg/m^3 density and multi-walled carbon nanotubes (MWCNTs) were used for the sample preparation. The nanotubes were made by the CVD method in a rotating reactor at the O. Chuiko Institute of Surface Chemistry, NASU (Kyiv, Ukraine). Powder of the Al_2O_3–MoO_3–Fe_2O_3 mixture was applied as a catalyst. Propylene was just used as a carbon source [20], and to avoid the formation of agglomerations of MWCNTs.

During synthesis, highly dispersed silica (VDK brand A 300) was added as a catalyst as well. The average size of the nanotubes was nearly 15 nm, and the average length was between 2 and 5 μm. The specific surface area, calculated through argon desorption, was about 300 m^2/g, while the apparent density was near 30 g/dm^3 [17].

The deagglomeration of MWCNTs was accomplished through the use of an ultrasonic disperser (UZD-N) or a universal rotary hydrodynamic homogenizer in aqueous solutions with varying compositions (Fig. 1).

The samples were prepared via a multistage procedure. First, starting components, the HDPE and MWCNTs powders, were mixed in the revolver-type rotational mixer. The rotation rate and time were 120 rev/min and 4 h, respectively. The isotropic mixtures were being taken after mixing. The sample mixture was loaded into a vacuum extruder cylinder and heated to 120 °C for 40 min at a pressure of 103 Pa. After that, the mixture was pushed through a die. The cylinders of the composites were 3 mm in diameter, and their high was near 5 mm. Different fractions of MWCNTs (0.01, 0.04, 0.07 and 0.1%) were used to prepare the starting mixtures.

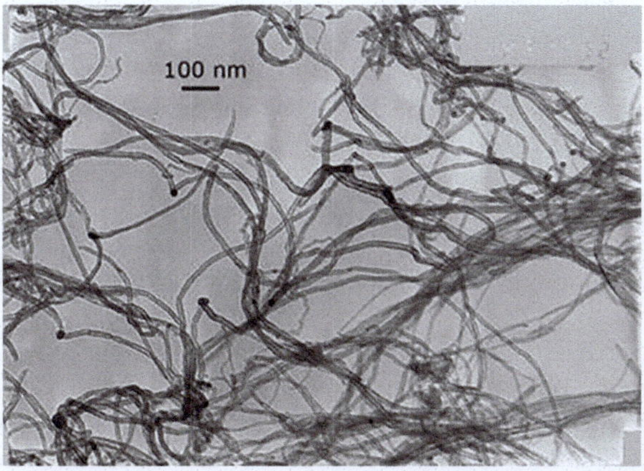

Fig. 1 TEM image of MWCNTs by SELMI at an accelerating voltage of 100 kV

For the characterization of the samples, we used the following experimental techniques: X-ray diffraction (XRD), Raman spectroscopy, Infrared spectroscopy, differential scanning calorimetry (DSC), thermogravimetry (TG), and dilatometry. The XRD was performed using an XPERTPRO diffractometer, with a CuKα radiation (copper anode, λCuKα = 1.541838 Å) at 45 kV and 40 mA, and a temperature of 25 °C. The samples were scanned over the 2θ range of 10–90°. The Raman spectra were recorded using a Bruker Senterra spectrometer (Bruker Optics Inc., Billerica, MA, USA) with an operating wavelength of incident light of 785 nm. The DSC and TG curves were recorded using a Netzsch Jupiter STA 449 F3 calorimeter (Netzsch, Selb, Germany), and the samples were analyzed in an Al_2O_3.

sample crucible over the temperature range of 30–145 °C with a heating rate of 5 °C/min. The dilatometry measurements were monitored over the same temperature range. Relative dimensional variations ΔL/L, and the thermal expansion coefficient α(T) (CTE) were measured by using NETZSCH 402C type dilatometer (NETZSCH, Selb, Germany). Each measurement was repeated three times, and the average data was used for further analysis. The precision of the α(T) measurements was approximately 10^{-7} $°C^{-1}$.

The starting mixtures of the MWCNTs different fractions (0.01, 0.04, 0.07 and 0.1%) were taken, so four types of the samples were prepared for measurements.

Multiwall carbon nanotubes obtained by Chemical vapor deposition method in a rotating reactor [20]. The mean diameter of the carbon nanotubes was 10–20 nm; their specific surface area, which was determined by argon desorption, was 200–400 m^2/g; and their bulk density varied from 20 to 40 g/dm^3.

3 Results and Discussion

The dilatometry results for the four samples are shown in Fig. 2. Determination of the expansion α as a function of temperature in the Z direction for all the samples reveals similar curves.

In the Z direction, the dilatometric curves of the 4 samples have the same shape. They show three domains. In the first, from ambient to 130 °C, they merge. This implies that the effect of the nanotubes is the same. The interaction forces are of the same order of magnitude. The second interval is around 140–145 °C, where each curve contains an anomaly whose intensity changes as a function of concentration. It is probably linked to the degradation phase. The PE + 0.1%MWCNTs interval lags behind the other three. The anomaly is extensive and of very low intensity compared with the others. In the third area, we return to the initial situation where all curves merged.

All in all, the four samples each contained an anomaly whose onset temperatures were a function of concentrations. Its origin is linked to the degradation of the matrix. The temperatures of appearance are a function of the concentrations. So, the greater the quantity of nanotubes, the better the protection of the samples. Their degradation is delayed (Fig. 3).

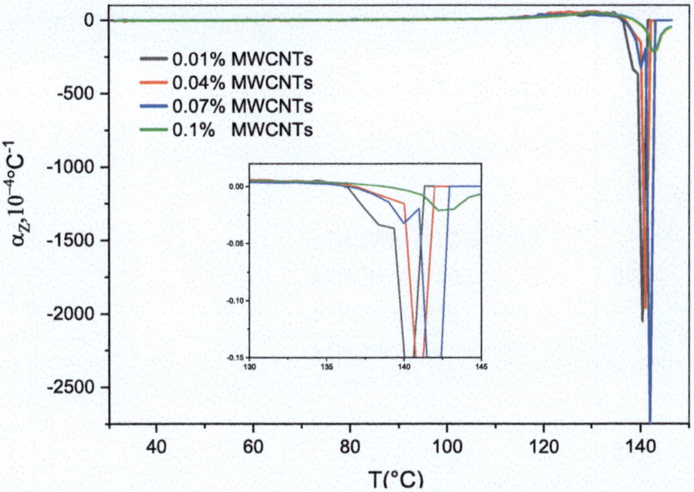

Fig. 2 The coefficient of thermal expansion measured in the longitudinal direction of PE + X%MWCNTS

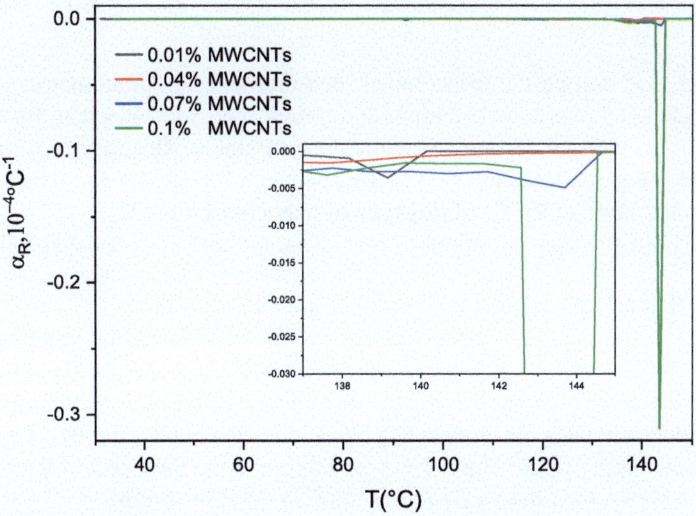

Fig. 3 The coefficient of thermal expansion measured in the radial direction of PE + X%MWCNTS

α of the PE + 0.1%MWCNTs shows a very intense peak, followed by PE + 0.07%MWCNTS. The anomaly disappeared in PE + 0.01MWCNTs and PE + 0.04NT. Beyond the peaks, the curves overlap and behave in a similar way to that observed in the first domain.

In the radial direction, the degradation phenomenon is practically comparable with that in the Z direction.

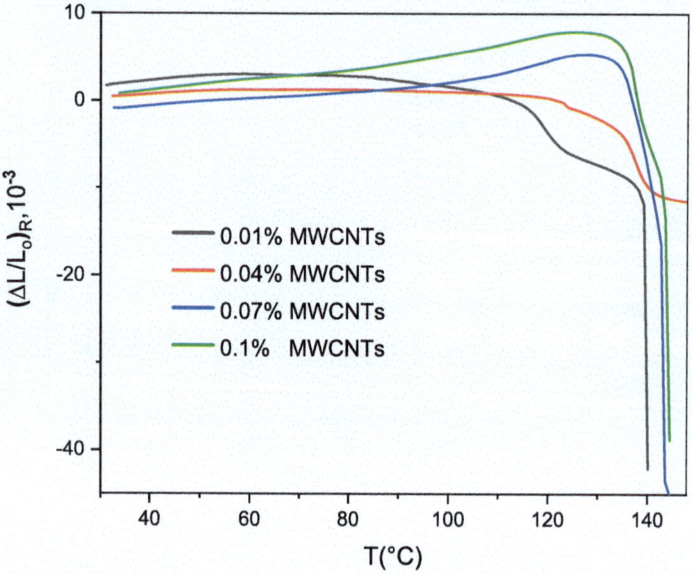

Fig. 4 Relative variation ΔL/L measured in the radial direction of PE + X%MWCNTS

We note that despite the direction of measurement, the dilatometric variations remain unchanged over a wide temperature range. This indicates that, for very low concentrations, thermal expansion coefficients are stable. This may be attributed to the strengthening of interatomic bonds. This behavior is consistent with the increased melting temperature (140 °C) of this type of nanocomposite [21].

Probably, the low concentrations of carbon nanotubes have good dispersion in PE matrices.

We see that the degradation temperatures are all above 140 °C.

Figure 4 shows the relative dimensional variations in the radial direction.

At low temperatures, the four curves are nearly identical and overlap. From 80 °C onwards, the curves separate from each other and the differences become significant as the temperature rises. At around 125 °C, all the curves pass through a maximum and then fall sharply. This behavior is probably linked to the effect of degradation.

Figure 5 shows the relative ΔL/L variation in the Z direction.

The four samples are behaved in the same way. The curves are merged over the entire temperature range. They each contain a dilatometric anomaly appearing at the same temperature.

As before, dimensional changes in both directions are relatively stable over the 30–100 °C range, which is not as wide as in the case of expansion.

With regard to differential scanning calorimetry, inserting different concentrations of multiwall carbon nanotubes into the polyethylene did not change the shape of the DSC curves (Fig. 6).

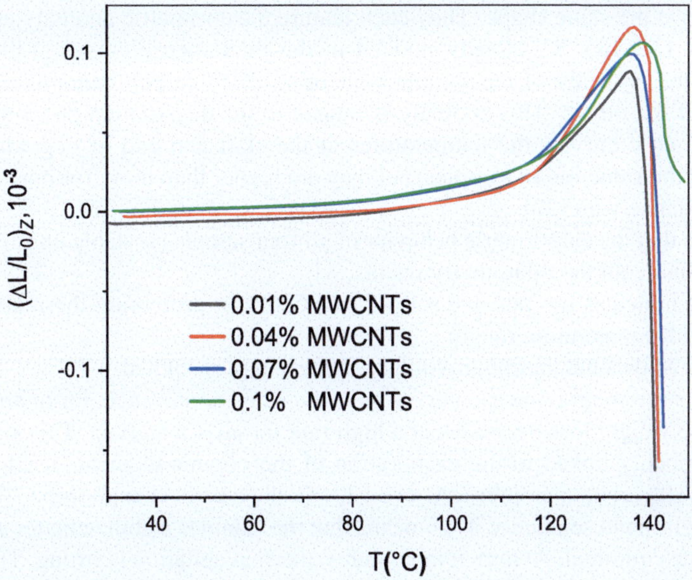

Fig. 5 Relative variation ΔL/L measured in the longitudinal direction of PE + X%MWCNTS

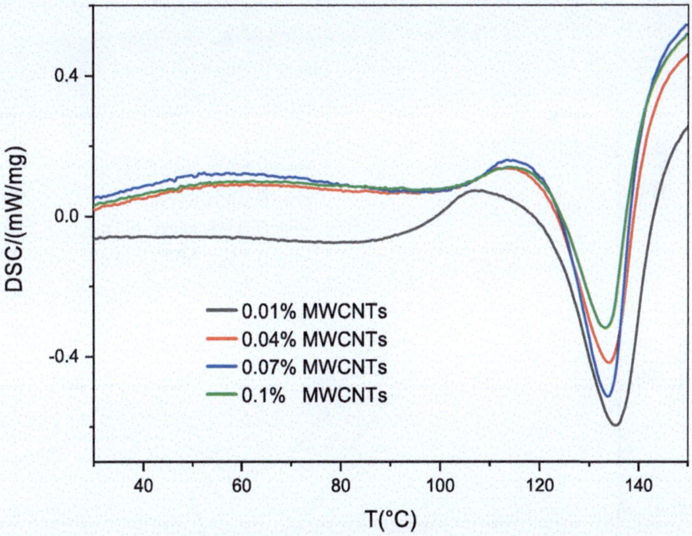

Fig. 6 Differential scanning calorimetry of PE + X%MWCNTS samples

They have the same shape. They each feature a calorimetric anomaly appearing at roughly 130–135 °C, already noticed in the dilatometric curves. 130–135 °C However, the anomaly of the sample containing 0.1% carbon nanotubes is larger than that of the others. This anomaly is related to the degradation phenomena. On the calorimetric curves, the temperatures at the start and end of degradation are practically the same for all samples, but they are higher than those for pure PE. This is in accordance with [22].

We note that the calorimetric behavior of all four samples is stable up to a 100 °C. This is in line with the dilatometry results.

All this indicates the positive role of MWCNTs in promoting thermal stability across a wide temperature range.

Figure 7 shows the thermogravimetry of the samples studied.

The four thermogravimetric curves have the same appearance. From ambient to 100 °C, the weight loss decreases at a high rate for all 4 materials. This significant drop is probably linked to the evaporation of the organic elements used to make the multiwall carbon nanotubes. Above 120 °C, the curves change slope. They tend towards right-hand segments. This means that the samples stabilize better and mass loss becomes minimal. At high temperatures, thermal agitation is strong. This led to good dispersion of the MWCNTSs, which in turn implies a more stable material.

We are interested in two aspects of the Raman technique. One is related to the D and G bands of carbon and the other is related to the crystallinity of the nanomaterials studied.

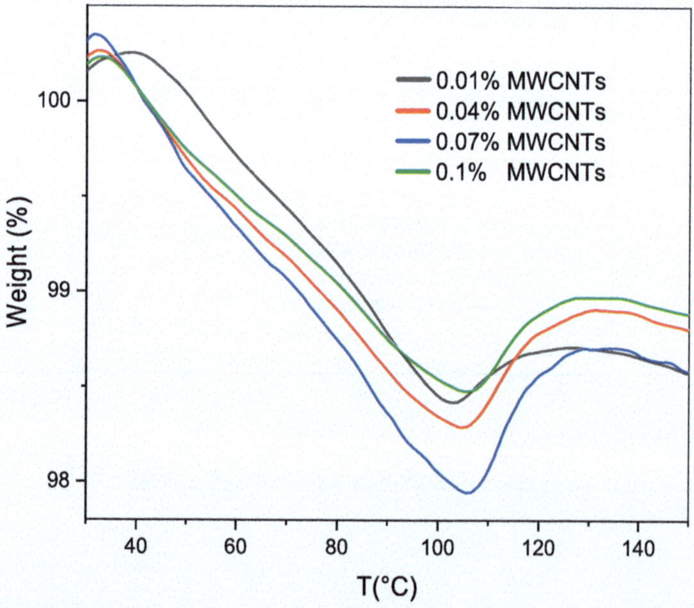

Fig. 7 Thermogravimetry of PE + X%MWCNTS samples

In general, the G band reveals highly crystalline graphite layers and the D band reveals the presence of defects in the graphite layer.

The band in the vicinity of 1420 cm^{-1} is absent in the PE/0.01%MWCNTs nanocomposite, where the concentration is very low. This implies that concentration plays a significant role. So, the behavior of this new class of nanomaterials depends on the amount of nanoscale carbon inserted.

Figure 8 shows the Raman spectra of all 4 samples containing varying percentages of multi-walled carbon nanotubes. All spectra exhibit the characteristic vibrational modes expected for pure HDPE, notably the intense Raman shift at 1297 cm^{-1} associated with CH_2 twisting. Additionally, evident are the orthorhombic crystalline phase vibrations, in particular the peak at 1420 cm^{-1} attributed to methylene bending [23]. Tracking the relative intensity of this 1420 cm^{-1} peak enables semi-quantitative evaluation of the crystalline fraction within the composite samples. For the HDPE composite containing the lowest 0.01% MWCNTs loading, the 1420 cm^{-1} peak is nearly imperceptible even upon background subtraction. This implies an almost entirely amorphous morphology for the HDPE matrix at this MWCNTs concentration. As MWCNT content is increased to 0.04%, a discernible 1420 cm^{-1} peak emerges, indicating initial formation of HDPE crystalline domains with the addition of higher MWCNT loadings. The intensity continues to grow for 0.07% MWCNTs, further confirming the nucleating influence of MWCNTs in facilitating crystallization of the HDPE polymer chains. Finally, the composite containing the maximum 10% MWCNTs exhibits both the highest absolute intensity of the 1420 cm^{-1} peak and the highest overall degree of crystallinity. Additionally, the characteristic D band Raman shift of MWCNTs becomes slightly visible for the HDPE composite containing 0.07% MWCNTs and continues to increase in intensity with higher MWCNTs loadings due to the greater concentration of nanotubes. In summary, the evolution of the1420 cm^{-1}peak provides evidence that MWCNTs substantially enhance HDPE crystallization in a concentration-dependent manner (Fig. 9).

All four spectra are comparable. They contain virtually the same peaks, but they are very intense compared with those of pure polyethylene. So, nanotubes have an important influence on structural behavior. There are also differences between the intensities of the different X-ray diffraction peaks of the 5 nanocomposites materials studied. This suggests that concentration plays a significant role. The angles of appearance of intense anomalies also depend on concentration. The obtained result can be merged with some techniques in control theory [24–30].

4 Conclusion

Dilatometric measurements enabled us to highlight the degradation phenomenon. The different concentrations did not significantly change the degradation temperatures. The dilatometric measurements in the radial and longitudinal directions were not far apart. Very low concentrations probably lead to relatively weak effects. This

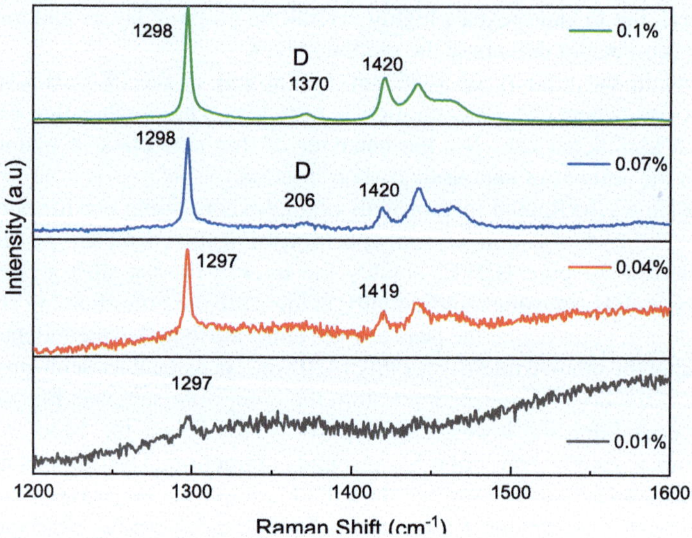

Fig. 8 Raman shift of PE + X%MWCNTS samples

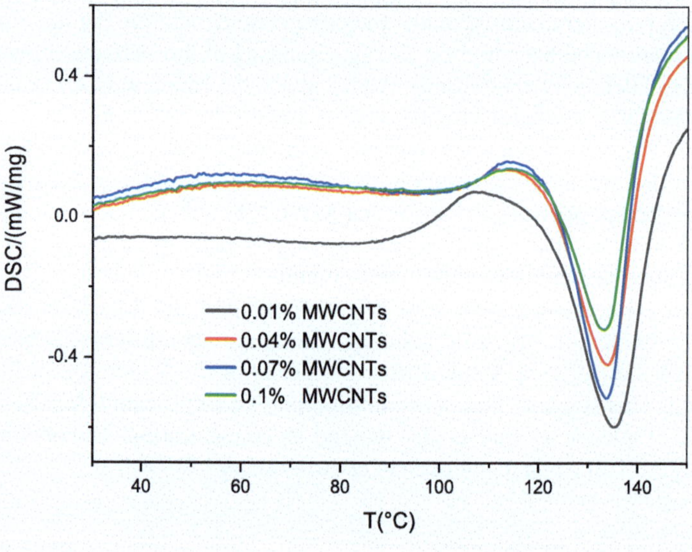

Fig. 9 X-ray diffraction pattern of PE + X%MWCNTS samples

may be due to insignificant anisotropy. Calorimetric and thermogravimetric measurements provide better information on the degradation phenomenon. The Raman shows that crystallinity changes with nanotubes concentration, which is in agreement with the dilatometric results.

In conclusion, all the results obtained in this work are of definite importance (improvement of thermostructural characteristics) and testify to the positive role of the introduction of multiwall carbon nanotubes. They have become an essential part of materials science. Their introduction means that heavy, expensive materials can be replaced by polyethylene-based nanocomposites reinforced with multiwall carbon nanotubes. This advantage has led to considerable gains in various areas of industry.

References

1. Avramenko TG, Khutoryanskaya NV, Naumenko SM, Ivanenko KO, Hamamda S, Revo SL (2019) Effect of carbon nanofillers on processes of structural relaxation in the polymer matrixes. In: Fesenko O, Yatsenko L (eds) Nanocomposites, nanostructures, and their applications. NANO 2018. Springer Proceedings in Physics, vol 221. Springer, Cham
2. Lozovyi F, Ivanenko K, Nedilko S, Revo S, Hamamda S (2016) Thermal analysis of polyethylene + X% carbon nanotubes. Nanoscale Res Lett 11:97
3. Naskar A, Keum J, Boeman R (2016) Polymer matrix nanocomposites for automotive structural components. Nature Nanotech 11:1026–1030
4. Kayser LV, Lipomi DJ (2019) Stretchable conductive polymers and composites based on PEDOT and PEDOT: PSS. Adv Mater 31:1806133
5. Ferreira F, Pinheiro I, de Souza S, Mei L, Lona L (2019) Polymer composites reinforced with natural fibers and nanocellulose in the automotive industry: a short review. Journal of Composites Science 3:51
6. Kazakova MA, Selyutin AG, Semikolenova NV, Ishchenko AV, Moseenkov SI, Matsko MA, Zakharov VA, Kuznetsov VL (2018) Structure of the in situ produced polyethylene based composites modified with multi-walled carbon nanotubes: In situ synchrotron X-ray diffraction and differential scanning calorimetry study. Compos Sci Technol 167:148–154
7. Abishera R, Velmurugan R, Gopal KVN (2016) Reversible plasticity shape memory effect in carbon nanotubes reinforced epoxy nanocomposites. Compos Sci Technol 137:148–158
8. Han Z, Fina A (2011) Thermal conductivity of carbon nanotubes and their polymer nanocomposites: a review. Prog Polym Sci 36:914–944
9. Khan W, Sharma R, Saini P (2016) Carbon nanotube-based polymer composites: synthesis, properties and applications. In: Carbon nanotubes-current progress of their polymer composites. InTech Rijeka, Croatia
10. Abdalla M, Dean D, Theodore M, Fielding J, Nyairo E, Price G (2010) Magnetically processed carbon nanotube/epoxy nanocomposites: morphology, thermal, and mechanical properties. Polymer 51:1614–1620
11. Song L, Qiu Z (2009) Crystallization behavior and thermal property of biodegradable poly(butylene succinate)/functional multi-walled carbon nanotubes nanocomposite. Polym Degrad Stab 94:632–637
12. Alamusi N, Hu N, Jia B, Arai M, Yan C, Li J, Liu Y, Atobe S, Fukunaga H (2012) Prediction of thermal expansion properties of carbon nanotubes using molecular dynamics simulations. Comput Mater Sci 54:249–254
13. Shirasu K, Nakamura A, Yamamoto G, Ogasawara T, Shimamura Y, Inoue Y, Hashida T (2016) Temperature dependence of axial thermal expansion coefficient of multi-walled carbon nanotubes (a procedure based on measurements of coefficient of thermal expansion for aligned carbon nanotube/epoxy composites). Trans JSME 82(844) (in Japanese)
14. Bouleklab MC, Hamamda S, Naoui Y, Nedilko S, Avramenko T, Ivanenko K, Revo S, Teselko P, Strelchuk V, Nikolenko A (2020) Influence of the multiwall carbon nanotubes on the thermal properties of the Fe–Cu nanocomposites. J Alloys Compounds 816:152525

15. Hamamda S, Jari A, Revo S, Ivanenko K, Jari Y, Avramenko T (2017) Thermal analysis of copper-titanium-multiwall carbon nanotube composites. Nanoscale Res Lett 12:251
16. Malekie S, Ziaie F (2015) Study on a novel dosimeter based on polyethylene-carbon nanotube composite. Nucl Instrum Methods Phys Res Sect A Accel Spectrometers Detect Assoc Equip 791:1–5
17. Guedes RM, Pereira CMC, Fonseca A, Oliveira MSA (2013) The effect of carbon nanotubes on viscoelastic behaviour of biomedical grade ultra-high molecular weight polyethylene. Compos Struct 105:263–268
18. Naoui Y, Zerrouk I, Zahaf A, Revo S, Boubertakh A, Hamamda S (2015) Thermal expansion coefficient polyethylene containing nanotubes carbon. In: XLIII international conference advanced problems in mechanics. St. Petersburg Polytechnic University, St. Petersburg, Russia, pp 66–67
19. Wei C, Srivastava D, Cho K (2002) Thermal expansion and diffusion coefficients of carbon nanotube-polymer composites. Nano Lett 2:647–650
20. Revo SL, Sementsov YI, Lozovii FV, Ivanenko EA, Druga L (2008) Structure and resistance of Al-C nanocomposite material. Heat Treat Surf Eng 8:3
21. Napp KE, Völker-Pop L, Fleissner F, Wezisla B (2019) Combined Raman spectroscopy and rheology for characterizing. Spectroscopy 34(9):38–43
22. Ferreira FV, Menezes BRC, Franceschi W, Ferreira EV, Lozano K, Cividanes LS, Coutinho AR, Thim GP (2017) Influence of carbon nanotube concentration and sonication temperature on mechanical properties of HDPE/CNT nanocomposites. Fullerenes Nanotub Carbon Nanostruct 25:531–539
23. Furukawa T, Sato H, Kita Y, Matsukawa K, Yamaguchi H, Ochiai S, Siesler HW, Ozaki Y (2006) Molecular structure, crystallinity and morphology of polyethylene/polypropylene blends studied by Raman mapping, scanning electron microscopy, wide angle X-Ray diffraction, and differential scanning calorimetry. Polym J 38:1127–1136
24. Bounemeur A, Chemachema M, Essounbouli N (2018) Indirect adaptive fuzzy fault-tolerant tracking control for MIMO nonlinear systems with actuator and sensor failures. ISA Trans 1(79):45–61
25. Bounemeur A, Chemachema M (2021) Adaptive fuzzy fault-tolerant control using Nussbaum-type function with state-dependent actuator failures. Neural Comput Appl 33:191–208
26. Bounemeur A, Chemachema M (2023) General fuzzy adaptive fault-tolerant control based on Nussbaum-type function with additive and multiplicative sensor and state-dependent actuator faults. Fuzzy Sets Syst 7:108616
27. Abdelhamid B, Chemachema M, Bouzina S (2023) Fuzzy fault-tolerant control applied on two inverted pendulums with nonaffine nonlinear actuator failures. Int J Robot Control Syst 3(2)
28. Zahaf A, Bououden S, Chadli M, Chemachema M (2022) Robust fault tolerant optimal predictive control of hybrid actuators with time-varying delay for industrial robot arm. Asian J Control 24(1):1–5
29. Zahaf A, Bououden S, Chadli M, Boulkaibet I, Neji B, Khezami N (2023) Dynamic sensorless control approach for Markovian switching systems applied to PWM DC–DC converters with time-delay and partial input saturation. Sensors 23(15):6936
30. Abdelmalek Z, Abdelhamid B, Sofiane B, Boulkaibet I (2019) Fault diagnosis of uncertain hybrid actuators based model predictive control. In: Proceedings of the 4th international conference on electrical engineering and control applications: ICEECA 2019, 17–19 Dec 2019, Constantine, Algeria 2021. Springer, Singapore, pp 961–971

The Influence of Milling Time and Multiwall Carbon Nanotubes Concentrations on the Thermal-Structural Behavior of Iron–Copper Nanocomposites

Mohamed Cherif Bouleklab, Sorror Saadallah, Hichem Filali, Tahar Dorbani, Serhii Nedilko, and Smail Hamamda

Abstract In this work we aim to study the effect of incorporating different concentrations of multiwalled carbon nanotubes into Iron–Copper—Fe–Cu with a ratio of 4:1; MWCNTs of 0.5, 1 and 2% in volume—nanocomposites—prepared via high-energy ball milling of different grinding times of 1 and 2 h, respectively—on the thermal and structural properties. The characterization conducted was heat flow, weight change, relative linear expansion (ΔL/L0) and coefficient of thermal expansion (CTE) for the thermal behavior, whereas for the structural changes and assessments, x-ray diffraction, and Raman spectroscopy were applied.

Keywords Carbon nanotube · Fe–Cu nanocomposite · Thermal expansion · Mechanical milling · Heat flow

1 Introduction

Carbon never ceased to attract the attention of the scientific community, A short time ago, it had been regarded only in three allotropic forms, nowadays their numbers only increased, from graphite foam, fullerenes followed by carbon nanotubes (CNTs) arrival which created a revolution in many fields, especially in materials science, medicine and environmental protection. New types of nanocomposites are obtained by incorporating CNTs into polymers, metals and ceramics. The motivation behind this work is the unique characteristics and exceptional properties of MWCNTs, which

M. C. Bouleklab (✉) · S. Saadallah · H. Filali · T. Dorbani · S. Hamamda
Laboratory of Thermodynamic and Treatment of Materials Surface, Constantine 1 University, 325 Route Ain el Bey, P.O. Box, Constantine, Algeria
e-mail: bouleklab.mcherif@umc.edu.dz

S. Nedilko
Taras Shevchenko National University of Kyiv, Kyiv 01601, Ukraine

S. Hamamda et al. (eds.), *Proceedings of the 2nd International Conference of Nanotechnology for Environmental Protection and Clean Energy Production*, Springer Proceedings in Materials 45, https://doi.org/10.1007/978-981-97-1916-7_2

were revealed only a few decades ago. They possess an advantageous form factor and high specific surface area (up to 1300 $m^2 g^{-1}$) [1]. The strong C–C covalent bonds and the geometry of the MWCNT walls give them outstanding mechanical properties translating into very high Young's modulus (greater than 1 TPa) [2] associated with high resistance to rupture (over 100 GPa) [3]. The space industry is also one of the many fields of application, in particular, the construction of light spacecraft thanks to the use of materials containing these nanofillers. This weight loss is a major asset in the protection of the environment and a significant reduction in greenhouse emissions. Due to the captivating properties of carbon nanotubes, a growing interest in the development of novel metal matrix nanocomposites containing MWCNTs.

Due to their high strength, availability and affordability, iron alloys with carbon reinforcement are among the most commonly used metallic materials in engineering applications and as structural material where mechanical properties are of a key concern. Iron as a transition metal is crucial for electromagnetics and permanent magnets. Because numerous transition metals are used as structural materials in a variety of industries, substantial emphasis has been devoted to the development of techniques for producing novel transition metal-MWCNTs matrix nanocomposites. The transition metals' vacant 3d orbitals enable for 3d–2p hybridization between the metal and graphitic structure MWCNTs [4, 5]. In which result in high interfacial adhesion [6–8].

The Fe–Cu intermetallic nanocomposite systems have been an underlying subject of attention for a long time. despite problems with their manufacturing. This interest was related to the attractiveness of Fe–Cu for many applications as well as their well-known high strength, thermal, and electrical properties [9, 10].

Mechanical alloying (milling) (MA) of the components of the intermetallic system allows for an increase in their solubility compared to equilibrium values and, at the same time, is a rather simple and effective way to produce a large amount of the composite. That is why manufacturing nanocomposite materials by the MA method is today a recognized method of the "cold" treatment of solid powder composites [11, 12]. The materials thus obtained reveal new, interesting properties,

This approach has shown to improve and have favorable impacts on material properties. Nevertheless, the manufacturing of metal matrix-MWCNTs composite has remained problematic due to the difficulty of achieving randomly distribution of CNTs with good interfacial bonding. High-energy ball milling provides high dispersion of CNTs in the matrix. The ball milling procedure, consisting of alternating welding, fracturing, and resoldering of iron powders, can provide iron-based nanocomposites with a homogeneous MWCNTs distribution [13]. It also produces homogeneous and very fine microstructures also it refines the grain size [14–17]. Due to the fascinating properties of carbon nanotubes, there is a growing interest in the development of new metal matrix composites reinforced with multiwall carbon nanotubes.

2 Materials and Methods

2.1 *Specimen Preparation*

Multiwall carbon nanotubes have been synthesized by chemical vapor deposition (CVD) [18]. Their parameters are:

- Average diameter varies between 10 and 20 nm,
- specific surface (determined by argon desorption) is 200–400 m^2/g,
- apparent density is between 20 and 40 g/dm^3.

The matrix comprises of the iron IP-1 and copper PMS-1 powders in a 4:1 weight ratio. The powders were mixed with different multiwall carbon nanotube concentrations of 0, 0.5, 1 and 2%MWCNTs in volume. The Fe–Cu-X%MWCNT nanocomposites obtained via high-energy ball milling. The ball mill (containing 20 hard stainless-steel balls) is of the same alloy as the planetary vial, which sustained an acceleration of 50 g with pressure reaching 5 GPa on a particle.

Three milling times 20, 60 and 120 min were taken. After processing in the ball mill, the nanocomposites were subjected to a 40% compression followed by annealing at 950 °C for 30 min in Argon flow. Then, the powder samples were subjected to cold rolling followed by annealing at 900 °C in Argon flow. The described cycle was repeated three times; thus, the samples were undergone to 80% thinning. Finally, the thickness of the obtained ribbons was near 0.3 mm.

2.2 *Characterization Equipment*

Different characterization techniques were used to investigate the changes and behaviors in both structural and thermal properties.

A NETZSCH 402C dilatometer (NETZSCH, Selb, Germany) with 3% accuracy was used in the study. The heating rate was near 10 °C/min. The thermal expansion coefficient was measured in the temperature range from 30 to 800 °C.

Differential scanning calorimetry (DSC) and thermogravimetric analysis (TGA) tests were conducted using a Jupiter STA 449 F3 calorimeter by NETZSCH (NETZSCH, Selb, Germany). The same heating rate as in the dilatometric measurements was used.

The X-ray diffraction patterns of the nanocomposite under study were obtained using the automated Philips X Pert Pro diffractometer with the filtered cobalt radiation $K\alpha_1 = 1.7909$ Å under the following scanning parameters: monitoring range [20–80°], with step scan 0.026°.

The Raman spectroscopy was carried out using a Raman Bruker Senterra spectrometer instrument, with a green laser (wavelength $\lambda = 532$ nm) from an argon ion.

3 Results and Discussion

3.1 Thermal Analysis

The Fe–Cu nanocomposite samples differ by milling time and the amount of multi-walled carbon nanotubes added to the metallic matrix. We proceed with the thermal analysis using dilatometric measurements, The regarded parameter is the coefficient of thermal expansion (CTE or α <T>).

Figure 1 shows the coefficient of thermal expansion α <T> of the Fe–Cu-X% MWCNTs nanocomposites ground for 1 and 2 h, respectively. As for the 1 h milling time the CTE variations for the samples of the 60-min milling time, shows that both 1 and 2% MWCNTs additions have the same changes with a slight increase from ambient temperature to around 450 °C followed by slight decrease in the vicinity of 650 °C, both values of CTE intersect at the end of the range near 800 °C. While the CTE for 0.5% sample is increasing steadily over the temperature range.

As for the 1% MWCNTs, it is lower than that of 2% MWCNTs. Perhaps due to the good dispersion of nanofillers in the 1% sample, leading to an amplification of the interatomic bonding between the host matrix and the incorporated CNTs. In the vicinity of 800 °C, the CTE of both 1 and 2% MWCNTs intersect. This situation is due to the strong thermal agitation causing a homogeneous dispersion of the MWCNTs.

The CTE behavior of the 120-min milling time, as for the Fe–Cu-1% MWCNTs nanocomposite is significantly lower than the other two over the entire temperature range. Certainly, this concentration is well-dispersed giving rise to a significant amount of bonding which translates into a force.

On the other hand, for the differential calorimetry accompanied with the weight variation (Thermogravimetry) analysis. Figure 2 shows both heat flow and weight change signals of the 1 h milling time.

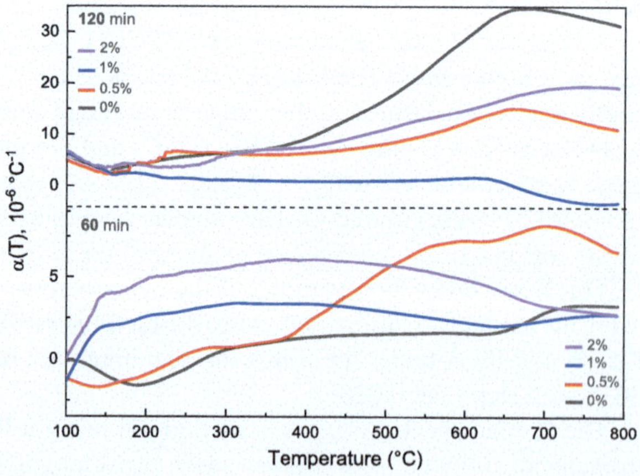

Fig. 1 α <T> (CTE) of the Fe–Cu-X% MWCNTs nanocomposites (1 and 2 h)

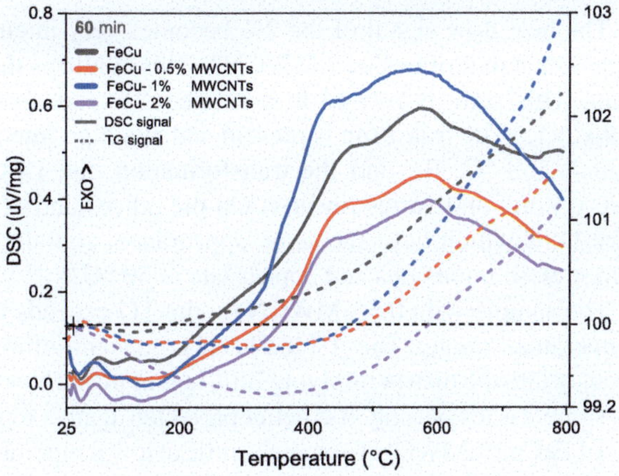

Fig. 2 Heat flow accompanied by thermogravimetry of the Fe–Cu nanocomposites (1 h time)

At low temperatures, the curves overlap. The signal of all samples has nearly the same shapes. An exothermic peak located around 600 °C and whose intensity is concentration dependent. FeCu1%MWCNT has a high intensity of capacity. At high temperatures, the DSC of the four samples tend to decrease and the differences between the intensities drop. The energy of thermal stirring is probably maximum, which intensely excites MWCNTs and promotes their distribution in all matrices. When the stirring time increases to 120 min, the roles are reversed.

The curves have the same shapes. The TG of FeCu1%MWCNT has the lowest weight loss, compared to that of the other samples, over the entire temperature range. Similarly, oxygen attacks this nanocomposite the least. The highest concentration sample is the less oxidized. Above 700 °C, the TG shows a very strong acceleration leading to intense oxidation. The sample containing the least number of nanotubes has the highest weight gain (oxidation) over almost the whole range. This might be due to the small amount of carbon to wrap the nanomaterial against oxygen (Fig. 3).

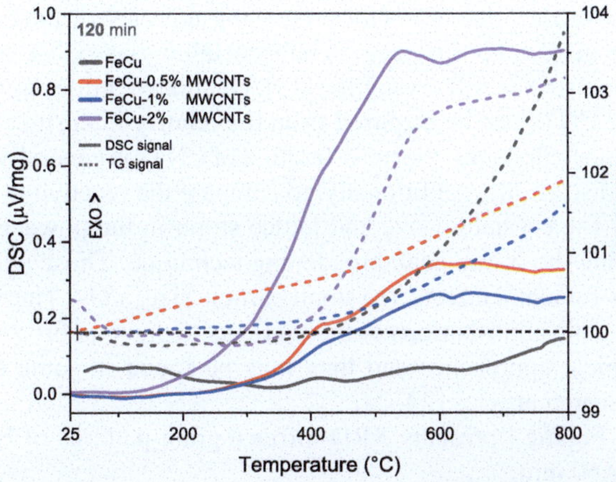

Fig. 3 Heat flow and thermogravimetry of the Fe–Cu-X% MWCNTs nanocomposites (2 h time)

The heat flow signal of the 2% becomes very intense. Its intensity is four times higher than that of FeCu1%MWCNT. This confirms that its calorific value is significant. This state is related to a refined structure and a reduced deterioration of MWCNTs. The transition is present and at a high intensity. This anomaly is located around 550 °C. Beyond the transformation, the DSC is approximately equal to 0.90 μV/mg and varies linearly. On the other hand, the two other nanocomposites have DSC which have the same appearance, contain very spread-out calorimetric peaks, close intensities and appear late at 50 °C.

The sample with 0.5%MWCNTs, the TG exceeds that of the 1% over the entire temperature range. The 1%MWCNTs concentration is likely diffused homogeneously into the matrix implying little loss of mass. As a result, the effect of oxygen on the sample is less strong. It is better protected against oxidation. On the other hand, the TG of FeCu2%MWCNTs changes little at low temperatures. From 400 °C, it presents an abrupt rise, leading to an abrupt change in mass. Above 550 °C, the thermal agitation is strong. This state leads to a homogeneous distribution of MWCNTs and thus a less accentuated change in the high temperature range. The comparative study of the TG of the three samples shows that 1%MWCNTs is the origin of the small TG. The effect of oxidation is the least strong compared to the other two nanomaterials. This shows that the effect of concentration plays a non-negligible role.

3.2 *Structural Analysis*

X-ray diffraction analysis

The position of the diffraction peaks gives access to the distance d_{hkl} between crystallographic planes, the structure and lattice parameters.

X-ray diffraction has been applied to study the reaction between metals and carbon components qualitatively. It provides vital information on MMN-CNT composites while also determining phase composition after processing. It delivers quick information on the influence of the processing approach on the changes in the phase existing in the matrix. For instance, in bulk-metallic-glass-CNT composites, it reveals whether the matrix is still crystalline [19]. Qualitative information regarding the alignment of the CNTs can be acquired from the relative intensities of the distinct peaks [20]. In nanocomposites, the mass fraction of CNTs is generally low, and only the (002) peak is discernible. Additionally, to compare the reactivity of different carbon species.

The crystallite size and lattice strain in the powder particles can be determined using the X-ray peak broadening technique. Those peaks are broadened due to (a) instrumental effects, (b) small particle size, and (c) lattice strain in the material. The specific contributions of these effects to the total broadening can be separated using typical approaches and they may be found in some of the textbooks [21] and/or research articles [22–24].

Figure 4 presents XRD diffractogram patterns of Fe–Cu nanocomposite milled for 60 min.

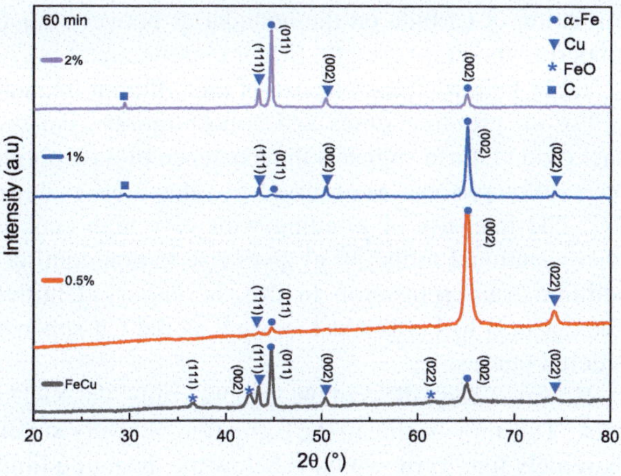

Fig. 4 XRD diagrams of Fe–Cu-X% MWCNTs nanocomposites (1 h time)

A presence of iron oxide only in the samples Fe–Cu without any MWCNTs, while the samples containing different concentrations of CNTs have the same planes corresponding to iron and copper, except for the 1% MWCNTs favoring the plane (200).

The presence of carbide in samples containing CNTs is immanent, but it was recorded only in the samples with 0.5% addition, also a trace of carbon in 1 and 2% and most notably in 0.5% MWCNTs. The average grain size is around 35 nm, the effect of concentrations appeared at a high concentration of 2% (Fig. 5).

A weak appearance for iron carbide presence between lower angles and 2θ = 35° for samples containing 0.5%, 1% and 2% MWCNTs respectively. As for grain size, it averages around 21.5 nm and it doesn't differ from the Fe–Cu without MWCNTs. This grain size is somehow stable, this might be due to, (i) the pining effect, the

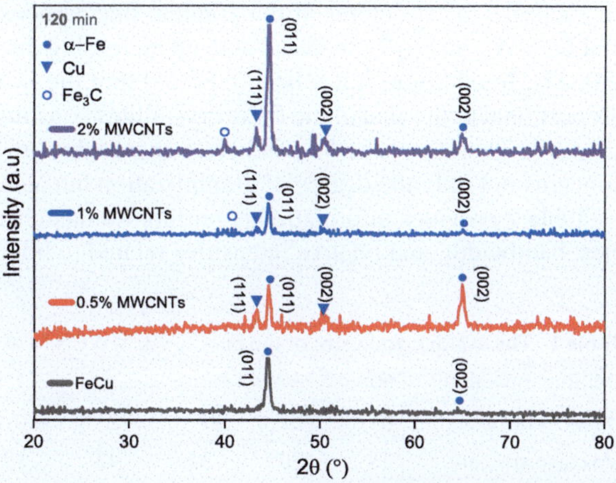

Fig. 5 XRD diagrams of Fe–Cu-X% MWCNTs nanocomposites (2 h time)

formation of carbide on the boundaries between the CNTs and the hosting matrix hindering the grain growth. (ii) Grain refinement.

Table 1 shows the grain size of the different treated samples.

For all grinding times and concentrations, the presence of more or less high background noise suggests the existence of an amorphous phase due to the sample preparation process, as observed by other authors during high-energy grinding [25–27]. The intensity of grinding with very high acceleration, twice as high as that of conventional mills (50 g) gives rise to non-equilibrium supersaturated phases in addition to amorphization. In [28], the authors identified several phases ranging from Fe(95)Cu(5) to Fe(83)Cu(17), such as the Cu supersaturated α(Fe, Cu) phase and ordered phases.

However, regardless of the grinding time, the XRD spectra do not show a carbide peak. This may be due to the formation of a very small amount of iron carbide in an extremely fine form, which is below the detection limit of the XRD apparatus used.

Raman spectroscopy

The influence of the dispersion process on CNTs structure was evaluated by Raman spectroscopy. Through this technique, it is possible to analyses the defect ratio, and thus measure the damage induced by the dispersion technique. The ratio of the intensity of the *D* band (I_D) to the intensity of the *G* band (I_G) is used to evaluate the quality of the CNTs [29]. An increase in this ratio indicates a growth in the number of CNTs defects, whereas a decrease indicates a change in the structure of the CNTs.

Each Raman band was analyzed by fitting with Breit–Wigner–Fano lines, and Lorentzian lines.

It is well-known in literature for graphite or composites containing graphitic additions behavior when goes under investigation using Raman spectroscopy, the observation of the presence of two characteristic peaks, D and G in the Fe–Cu nanocomposite with different concentrations of multiwall carbon nanotubes, the above.

The W_G bandwidth and the I_D/I_G intensity ratio have been frequently used to characterize the structures of the graphite like carbons [30–35]. It is widely known that both W_G and I_D/I_G increase with an increasing disorder in the graphite structure [30, 31]. In addition, it has been reported that the I_D/I_G intensity ratio is inversely proportional to the size of the finite crystallites constituting the graphite like carbons [31, 36]. Therefore, the W_G indicates the degree of graphitization, and the I_D/I_G concerns not only the degree of graphitization but also the crystallite size (Fig. 6).

Table 2 provides information regarding Raman characterization; peaks, frequencies, bandwidth, and relative intensities related to both D and G bands.

Table 1 The average grain size of the Fe–Cu-x% MWCNTs of different milling times

Milling time, min	60				120			
MWCNTs, %	0	0.5	1	2	0	0.5	1	2
Average size, nm	33	30	35	43	19	21	19	23
Strain, %	0.35	0.32	0.27	0.28	0.63	0.42	0.77	0.47

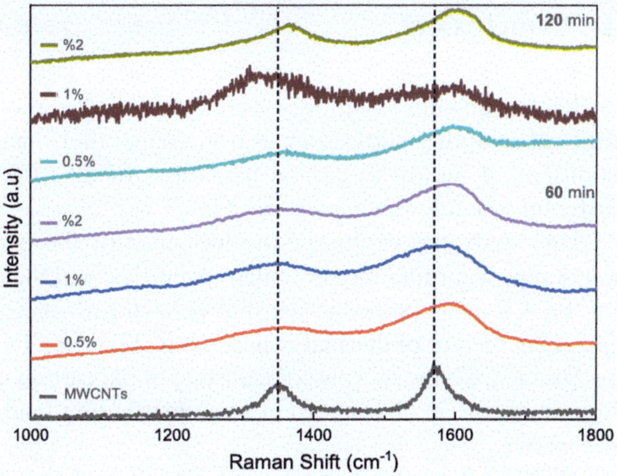

Fig. 6 Raman spectra of the Fe–Cu-X% MWCNTs nanocomposites (1 and 2 h time)

Table 2 Peak frequencies, bandwidths, and relative intensities of D And G bands in the Raman spectra of Fe–Cu-X% MWCNTs nanocomposite with different milling times and ν_D and ν_G are the peak frequencies of the D, And G bands, respectively

Milling time, min	MWCNTs, %	ν_D, cm^{-1}	W_D, cm^{-1}	ν_G, cm^{-1}	W_G, cm^{-1}	I_D/I_G
60	0.5	1338	1.07	1586	94	0.3
	1	1331	101	1580	119	0.38
	2	1337	113	1587	105	0.29
120	0.5	1346	73.1	1592	77.3	0.41
	1	1334	164.9	1581	153.4	1.11
	2	1364	52.4	1597	76.9	0.47

W_d, And W_g, are the widths [Full Widths at Half Maximums (FWHM)] of D And G bands, respectively. I_d/I_g Is the peak intensity ratio of the D band to the G band. Each band was Analyzed by fitting with a Lorentzian line

The obtained Raman spectra are in the range 1000–2000 cm^{-1}, the frequency values of the G band in between 1570 and 1599 cm^{-1} due to the A_{g1}, E_{1g} and E_{2g} vibrational modes, and the frequency of the D band in between 1334 and 1361 cm^{-1} arising from the disorder-induced A_{g1} mode. The width of the G band is related to the CNTs size distribution whereas the intensity of the D band decreases with the degree of graphitization of the tubes. In this work, Raman spectroscopy was mainly used to verify the multiwalled morphological state and the overall homogeneity of the sample. The obtained result can be merged with some techniques in control theory [37–43].

4 Conclusion

A classical powder metallurgy route can be used to produce a different aspect ratio of difficult solubility alloys such as iron-copper (4:1) nanocomposites with the volume additions of multiwall carbon nanotubes of 0.5, 1.0, and 2.0% respectively with different grinding times of 1 and 2 h.

This study has evaluated the influence of the dispersion of multiwall carbon nanotubes regarding the presented quantities, and the effect of milling time on both the Fe–Cu nanocomposite thermal behavior as of heat flow, weight variation and linear coefficient of thermal expansion (CTE or $\alpha <T>$), the phases presence as well as "grain size", the morphological state of the carbon nanotubes, which plays a very important role in altering grain size, distributing load and transfer heat throughout the matrix.

The heat flow and thermogravimetry show thermal stability and higher calorific capacity for the samples with longer milling time and higher concentrations of CNTs.

As for the coefficient of thermal expansion, improved CTE of the Fe–Cu-1% MWCN milled for two hours while the x-ray diffraction diagrams show a grain refinement for the two hours milling time, also a homogenous distribution of MWCNT (the absence of MWCNT clusters appearing as carbon-graphite). Raman spectroscopy, on the other hand, assessed the CNTs morphology integrity and damage state of the milling time, it revealed a higher defect density with longer milling time, with an exception for the 60 min it shows a lower defect density which indicates the healing and the recovery of CNTs.

However, the results of Raman spectroscopy show that damage to the structure of the MWCNT is higher with a longer milling time, thus compromising the beneficial effect of better dispersion.

The results obtained are of interest and are a certain contribution to the field of nanomaterials. They highlight the important role of multiwall carbon nanotubes in the Fe–Cu metallic matrix. Mechano-synthesis is shown to be a very efficient elaboration technique for the development of nanocomposites.

The use of a high-energy mechano-synthesis allows for the elaboration of Fe–Cu-based samples containing different concentrations of multiwall carbon nanotubes that are difficult to obtain by conventional elaboration methods.

The influence of the two parameters, grinding time and CNT concentrations is responsible for significant changes in the behavior of the samples. They are complementary and are responsible for the improvements of the thermal-structural characteristics of the Fe–Cu matrix containing different concentrations and ground for different times. They can be the origin of good performances or deterioration of the physical characteristics of the studied nanomaterials.

The multiple experiments used in this work are complementary and allows us to conclude that the thermo-structural properties of Fe–Cu-based nanocomposites have improved. These results are in agreement with other works. They showed that the mechanical properties of the same nanocomposites were significantly improved because of the high density of dislocations due to the mechanical treatments and whose mobility was blocked by the grain boundaries which became numerous. The dilatometry revealed that the coefficients of thermal expansion of all the samples noticeably decreased.

X-ray diffraction shows that the average crystallite dimensions deduced from the analysis of the Willamson-Hall diffraction line profiles vary between 19 and 43 nm. They change as a function of concentration and grinding time. Longer grinding leads to better refinement compared to the other two cases. As for the micro stresses, their range of variation is between 0.23 and 0.83. Grinding for one hour leads to low values in front of the other two cases. This suggests that the nanocomposites milled for one hour are less constrained and thus allow the MWCNTs to move in the Fe–Cu matrix. Overall, the different anneal after deformation seems to have mitigated the stresses at the crystallites.

References

1. Peigney A, Laurent C, Flahaut E, Bacsa RR, Rousset A (2001) Specific surface area of carbon nanotubes and bundles of carbon nanotubes. Carbon 39:507–514
2. Treacy M, Ebbesen T, Gibson J (1996) Exceptionally high Young's modulus observed for individual carbon nanotubes. Nature 381:678–680
3. Yu MF, Lourie O, Dyer MJ, Moloni K, Kelly TF, Ruoff RS (2000) Strength and breaking mechanism of multiwalled carbon nanotubes under tensile load. Science 287:637–640
4. Suh JY, Bae DH (2013) Mechanical properties of Fe-based composites reinforced with multi-walled carbon nanotubes. Mater Sci Eng A 582:321–325
5. Munir KS, Zheng Y, Zhang D, Lin J, Li Y, Wen C (2017) Microstructure and mechanical properties of carbon nanotubes reinforced titanium matrix composites fabricated via spark plasma sintering. Mater Sci Eng A 688:505–523
6. Housaer F, Beclin F, Touzin M, Tingaud D, Legris A, Addad A (2015) Interfacial characterization in carbon nanotube reinforced aluminum matrix composites. Mater Charact 110:94–101
7. Zhou W, Bang S, Kurita H, Miyazaki T, Fan Y, Kawasaki A (2016) Interface and interfacial reactions in multi-walled carbon nanotube-reinforced aluminum matrix composites. Carbon 96:919–928
8. Wang S, Liang R, Wang B, Zhang C (2008) Load-transfer in functionalized carbon nanotubes/polymer composites. Chem Phys Lett 457(4–6):371–375
9. Ying YC, Rainer S, Austin CY (1984) Thermodynamic analysis of the iron-copper system I: the stable and metastable phase equilibria. MTA A 15:1921–1930
10. Mazzone G, Antisari MV (1996) Structural and magnetic properties of metastable fcc Cu–Fe alloys. Phys Rev B 54:441–446
11. Suryanarayana C (2001) Mechanical alloying and milling. Progr Mater Sci 46:1–184
12. Hamzaoui R, Elkedim O, Gaffet E (2004) Milling condition effect on structure and magnetic properties for mechanically alloyed nanocrystalline Fe—10%Ni and Fe—20%Ni alloys. Mater Sci Eng A 381:363–371
13. Rochman NT, Kawamoto K, Sueyoshi H, Nakamura Y, Nishida T (1999) Effect of milling temperature and additive elements on an Fe–C system alloy prepared by mechanical alloying. J Mater Process Technol 89:367–372
14. Parswajinan C, Ramnath BV, Vetrivel M, Ranjith A, Rohit BK, Vasanth S, Vignesh S (2018) Determination of mechanical behaviour of Fe-CNT MMC. Mater Today Proc 5(1):1176–1185
15. Mani MK, Viola G, Reece MJ, Hall JP, Evans SL (2014) Fabrication of carbon nanotube reinforced iron based magnetic alloy composites by spark plasma sintering. J Alloy Compd 601:146–153

16. Wan H, An X, Kong Q, Wu X, Feng W, Wang H, Wu J, Lu C, Zha W, Sun H, Huang L (2021) Fabrication of ultrafine grained FeCrAl-0.6 wt.% ZrC alloys with enhanced mechanical properties by spark plasma sintering. Adv Powder Technol 32(5):1380–1389
17. Shaw LL (2000) Processing nanostructured materials: an overview. JOM 52(12):41–45
18. Revo SL, Sementsov YI, Lozovii FV, Ivanenko EA, Druga L (2008) Structure and resistance of Al-C nanocomposite material. Heat Treat Surf Eng 8:3
19. Bian Z, Pan MX, Zhang Y, Wang WH (2002) Carbon-nanotube-reinforced $Zr_{52.5}Cu_{17.9}Ni_{14.6}Al_{10}Ti_5$ bulk metallic glass composites. Appl Phys Lett 81:4739–4741
20. Cao A, Xu C, Liang J, Wu D, Wei B (2001) X-ray diffraction characterization on the alignment degree of carbon nanotubes. Chem Phys Lett 344:13–17
21. Norton MG, Suryanarayana C (1998) X-Ray diffraction: a practical approach. Plenum Press
22. Kneringer G, Roedhammer P, Wildner H (2001) Powder metallurgical high performance materials. In: Proceedings. Volume 2: P/M hard materials
23. Lönnberg B (1994) Characterization of milled Si_3N_4 powder using X-ray peak broadening and surface area analysis. J Mater Sci 29:3224–3230
24. Aymard L, Delahaye-Vidal A, Portemer F, Disma F (1996) Study of the formation reactions of silver-palladium alloys by grinding and post-milling isothermal annealing. J Alloys Compd 238:116–127
25. Boshko O, Dashevskyi M, Mykhaliuk O, Ivanenko K, Hamamda S, Revo S (2016) Structure and strength of iron-copper-carbon nanotube nanocomposites. Nanoscale Res Lett 11(1):1–8
26. Guérault H (2000) Propriétés structurales et magnétiques de poudres de fluorures nanostructurées MF3 (M = Fe, Ga) obtenues par broyage mécanique (Doctoral dissertation, Le Mans)
27. Kumar A, Banerjee U, Chowrasia MK, Shekhar H, Banerjee MK (2019) Effect of MWCNT Content on the structure and properties of spark plasma-sintered Iron-MWCNT composites synthesized by high-energy ball milling. J Mater Eng Perform 28(5):2983–3000
28. Boshko O, Nakonechna O, Dashevskyi M, Ivanenko K, Belyavina N, Revo S (2016) Effect of the carbon nanotubes on structure and magnetic properties of the Fe–Cu (4: 1) composites. Adv Powder Technol 27(4):1101–1108
29. Keszler AM, Nemes L, Ahmad SR, Fang X (2004) Characterisation of carbon nanotube materials by Raman spectroscopy and microscopy—a case study of multiwalled and single walled samples. J Optoelectron Adv Mater 6:1269–1274
30. Knight DS, White WB (1989) Characterization of diamond films by Raman spectroscopy. J Mater Res 4:385–393
31. Chieu TC, Dresselhaus MS, Endo M (1982) Raman studies of benzene-derived graphite fibers. Phys Rev B 26:5867–5877
32. Katagiri G (1996) Raman spectroscopy of graphite and carbon materials and its recent application. TANSO 1996:304–313
33. Lespade P, Marchand A, Couzi M, Cruege F (1984) Caracterisation de matériaux carbones par microspectrométrie Raman. Carbon 22:375–385
34. Katagiri G, Ishida H, Ishitani A (1988) Raman spectra of graphite edge planes. Carbon 26:565–571
35. Asari E, Kamioka I, Nakamura KG, Kawabe T, Lewis WA, Kitajima M (1994) Lattice disordering in graphite under rare-gas ion irradiation studied by Raman spectroscopy. Phys Rev B 49:1011–1015
36. Tuinstra F, Koenig JL (1970) Raman spectrum of graphite. J Chem Phys 53:1126–1130
37. Bounemeur A, Chemachema M, Essounbouli N (2018) Indirect adaptive fuzzy fault-tolerant tracking control for MIMO nonlinear systems with actuator and sensor failures. ISA Trans 1(79):45–61
38. Bounemeur A, Chemachema M (2021) Adaptive fuzzy fault-tolerant control using Nussbaum-type function with state-dependent actuator failures. Neural Comput Appl 33:191–208
39. Bounemeur A, Chemachema M (2023) General fuzzy adaptive fault-tolerant control based on Nussbaum-type function with additive and multiplicative sensor and state-dependent actuator faults. Fuzzy Sets Syst 7:108616

40. Abdelhamid B, Chemachema M, Bouzina S (2023) Fuzzy fault-tolerant control applied on two inverted pendulums with nonaffine nonlinear actuator failures. Int J Robot Control Syst 3(2)
41. Zahaf A, Bououden S, Chadli M, Chemachema M (2022) Robust fault tolerant optimal predictive control of hybrid actuators with time-varying delay for industrial robot arm. Asian J Control 24(1):1–5
42. Zahaf A, Bououden S, Chadli M, Boulkaibet I, Neji B, Khezami N (2023) Dynamic sensorless control approach for Markovian switching systems applied to PWM DC–DC converters with time-delay and partial input saturation. Sensors 23(15):6936
43. Abdelmalek Z, Abdelhamid B, Sofiane B, Boulkaibet I (2021) Fault diagnosis of uncertain hybrid actuators based model predictive control. In: Proceedings of the 4th international conference on electrical engineering and control applications: ICEECA 2019, 17–19 Dec 2019, Constantine, Algeria 2021. Springer, Singapore, pp 961–971

The Carbon Nanotubes, Graphene Nanoparticles Their Oxygen Modified Forms and Composites

Yuli Shi, Yuliia Hrebelna, Eugeniy Demianenko, Stanislav Makhno, Kateryna Ivanenko, Smail Hamamda, Mariya Terets, Mykola Kartel, and Yurii Sementsov

Abstract Multilayer graphene nanoparticles (GNPs) have unique properties and potential for application. They are more stable compared to single-layer graphene and are suitable for mass production by splitting precursors with a graphite-like structure. The problem is not in the production of GNPs, but in the regulation of its characteristics. The application of GNPs is hindered by their agglomeration due to van der Waals interactions. To avoid this, it is necessary to create a gap between GNPs. Carbon nanotubes (CNTs) are an ideal candidate for this. CNTs can reduce internal electrical resistance and improve overall electrical conductivity. Therefore, the production of GNPs@CNTs hybrids is of interest. In work, a simple and cheap method of GNP synthesis by anodic exfoliation of expanded graphite (EG) foil in a weak alkaline electrolyte is shown. A two-level method of regulating the structure of GNPs by adjusting the nanoscale cluster structure of EG by changing the parameters of its production process and modes of "secondary intercalation", i.e., anodic oxidation of the EG foil, is revealed. The method of synthesis of dispersions of GNPs@CNTs and film materials from them is described. The structural characteristics of GNPs, CNTs and their composites, determined by the methods of X-ray

Y. Shi · S. Makhno · K. Ivanenko · Y. Sementsov (✉)
Ningbo Sino-Ukrainian New Materials Industrial Technologies Institute, Kechuang Building, N777 Zhongguan Road, Ningbo 315211, China
e-mail: ysementsov@ukr.net

Y. Shi · S. Makhno · Y. Sementsov
Ningbo University of Technology, No 55-155 Cui Bai Road, Ningbo 315211, China

Y. Hrebelna · E. Demianenko · S. Makhno · M. Terets · M. Kartel · Y. Sementsov
Chuiko Institute of Surface Chemistry of NAS of Ukraine, 17 General Naumov Str., Kyiv 03164, Ukraine

K. Ivanenko
Institute of Macromolecular Chemistry, NAS of Ukraine, 48 Kharkiv Highway, Kyiv 02160, Ukraine

S. Hamamda
University of Mentouri Brothers Constantine 1, 325 Route Ain El BEY, 25017 Constantine, Algeria

S. Hamamda et al. (eds.), *Proceedings of the 2nd International Conference of Nanotechnology for Environmental Protection and Clean Energy Production*, Springer Proceedings in Materials 45, https://doi.org/10.1007/978-981-97-1916-7_3

diffraction, LCS, Raman spectroscopy, XPS, DTA, TG are presented. The energetic effects of the interaction of partially oxidized graphene-like planes with each other were elucidated by the methods of quantum chemistry. It was found that the reaction between hydroxyl and aldehyde groups of two interacting graphene-like planes is thermodynamically most probable.

Keywords Expanded graphite · Carbon nanotubes · Graphene nanoparticles · Anodic and chemical oxidation

1 Introduction

Graphene is the only graphitic plane in which sp^2-hybridized carbon atoms form a hexagonal lattice. It can be represented as a "building block" of graphite, nanotubes, graphite intercalated compounds (GIC), etc. Graphene, with its high specific surface area and unique optical, electronic and mechanical properties, has great potential for applications in many fields, from electronics to medicine. Unique properties are also possessed by multilayer graphene nanoparticles (GNPs), containing up to dozens of layers [1, 2]. They are more stable and suitable for mass production according to the "top-down" scheme that is, by splitting precursors with a graphite-like structure using a wide range of methods [3]. The problem is not in the very fact of obtaining GNPs, but in the possibility of managing their characteristics [3]. The practical application of graphene materials is limited by the agglomeration of graphene sheets due to their strong van der Waals interaction. One of the best approaches to realize the great potential of graphene is to add some space between the folded sheets of graphene. CNTs are an ideal candidate due to their high conductivity, specific surface area and, most importantly, their interaction with graphene sheets due to their similar carbon structure [4]. Therefore, the production of GNPs@CNTs hybrids is of great interest. Electrochemical (anodic) oxidation of graphite materials in electrolytes of various natures is one of the methods that makes it possible to control and regulate the properties of GNPs [5]. We also applied this method, but expanded graphite (EG) was chosen as a precursor. Why did you choose EG? Expanded (exfoliated) graphite [6] is a group of materials with a graphite-like structure, which are obtained by rapid heating at temperatures of 800–1100 °C GIC or their hydrolyzed forms. Decomposition and evaporation of the intercalant leads to a strong expansion along the axis perpendicular to the graphene planes. This effect is most pronounced for graphite salts with mineral acid anions, the molecules of which easily form gaseous products and make it possible to obtain EG powder with a specific surface area of 35–75 m^2/g (Fig. 1a, b). The positron annihilation method showed that EG samples contain nanopores with an average size of about 0.7 nm [6]. TEM images of EG scales [6] (Fig. 1c, d) showed that the structure contains fragments—folded formations, possibly cylindrical, conical or slotted, convolutions of the graphene plane. The sizes of nanostructured fragments that can be separated range from units to hundreds of nanometers.

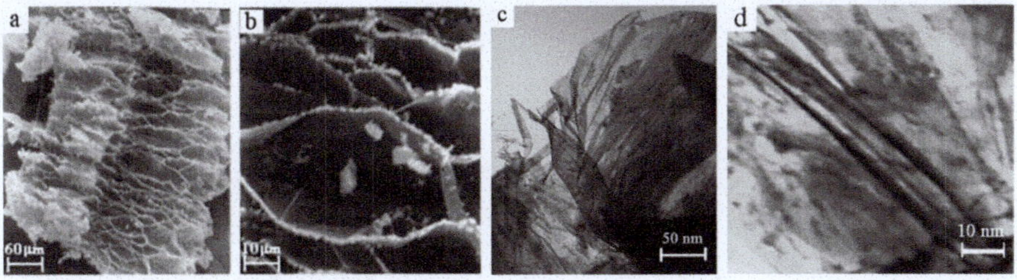

Fig. 1 SEM images of EG surface (**a**, **b**), TEM images (**c**, **d**)

The similarity of the structural elements of EG and SWCNT is confirmed by their Raman spectra [6]. The influence of technological synthesis parameters on the structural characteristics of EG was determined in [6]. The lattice parameter along the C-d_{002} axis can vary from 0.335 to 0.344 nm, the sizes of the regions of coherent X-ray scattering in the base plane L_a from 110 to 21 nm, along the C–L_c axis—40 to 22 nm. For EG obtained by different methods, the sizes and distribution of pores by size are different, and their values lie in the range from less than 1 to 100 nm.

This brief review shows that EG is a nanoscale cluster system that contains cylindrical, conical, slit-like defects with pore half-widths ranging from less than 1 to ~ 100 nm. The structural characteristics of EG are regulated by changing the technological parameters of the process of its production. Therefore, EG is the most attractive precursor of GNPs for industrial production. It is known that re-intercalation of EG in the classical sense does not occur. Diffusion of the electrolyte (intercalant) occurs along the existing defects. Secondary pseudo-intercalation of the dense EG material allows to "detect" nanoscale defects of the EG structure and to obtain a dispersion of GNPs in an aqueous electrolyte solution.

2 Materials and Methods

GNPs was obtained by secondary "intercalation" of electrodes from EG foil (according to TU U 26.8-30969031-002–2002, production of "TMSpetsmash" ltd. Kyiv city, Ukraine) in low-concentration alkaline electrolyte (KOH) with a current density of 6.0–60.0 mA/cm^2. The scheme of the device is shown in Fig. 2 [7, 8]. Figure 2b shows scheme of a vertical electrochemical reactor for oxidizing dispersed carbon materials. EG foil reinforced with glass fiber of different bulk density is used as a graphite tape. The foil is continuously passed at a speed of 0.5–2.0 cm/h through a solution of potassium hydroxide electrolyte, which moves at a controlled speed through the electrolytic bath, with electric field strength of 2–30 V/cm. The electrolytic bath is divided by a membrane into an anodic and a cathodic space.

The electrochemical reactor (scheme Fig. 2b) consists of a rectangular polypropylene housing 1, in which the cathode 2 and anode 5 are vertically placed, made of a sheet of pure iron (99.4%), the cathode is wrapped with a membrane-separating

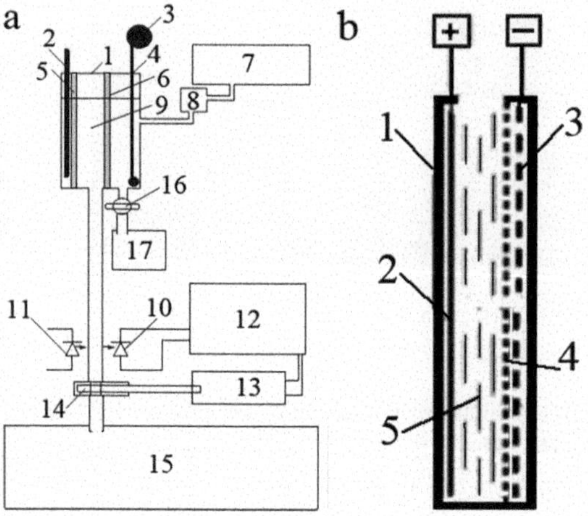

Fig. 2 **a** Block diagram of the installation for obtaining a suspension of graphene nanoparticles, where: 1—electrolytic bath, 2—cathode, 3—cassette device with a strip of fiberglass-reinforced graphite foil, 4—anode, 5—polypropylene membrane, 6—filter, 7—tank with electrolyte solution of potassium hydroxide, 8—float valve, 9—part of the electrolytic bath for suspension of graphene nanoparticles, 10—photodiode, 11—LED, 12—control unit, 13—control valve drive, 14—control valve, 15—tank for suspension of graphene nanoparticles, 16—tap for removing coarse suspension of EG, 17—container for coarse suspension of expanded graphite. **b** Scheme of a vertical electrochemical reactor: 1—housing, 2—anode, 3—cathode, 4—separator (membrane), 5—CNTs [7]

polypropylene fabric for filters. The space between the anode and the cathode is filled with a mixture of CNTs and sulfuric acid in a mass ratio of approximately 1:2. The gap between the electrodes is ~ 20 mm.

The nanoscale of GNPs was confirmed by studies of Raman spectra (Brucker RFS 100 s spectrometer, the source of radiation was an argon laser, $\lambda = 514.5$ nm), XRD patterns (powder diffractometer DRON-4 07 CuKα radiation in nickel according to the geometry of Bragg–Brentano surveys), thermogravimetric measurements (Derivatograph Q-1500 D (Hungary) in a static atmosphere, and SEM images (JSM-35, JEOL). The state of the CNTs and EG surface was monitored by the X-ray photoelectron spectroscopy (XPS) (spectrometer Thermo scientific K-Alpha). Aqueous dispersions of the obtained materials were analysed by particle size and mass using laser correlation spectroscopy (LCS) (spectrometer "ZetaSizer 3" Malvern Instrument (UK) with a correlator 7032 and helium–neon laser LG-111 with a power of 25 mW for wavelength $\lambda = 633$ nm). The autocorrelation function was processed using the program PCS-Size mode v 1.61.

Multiwall CNTs synthesized by the method of catalytic pyrolysis (CCVD), according to [7], conformed to the requirements of the standard TU U 24.1-03291669-009:2009. The average diameter of the CNTs was 10–20 nm, specific surface area determined by argon desorption was 200–400 m^2/g, and bulk density was within 20–40 g/dm^3. According to TEM, X-ray diffraction, and Raman spectroscopy data, the noticeable amount of amorphous carbon was not detected [7].

3 Experimental Results and Discussion

The dependence of the size and shape (the degree of deviation from sphere, reflecting the coefficients of polydispersity) on the parameters of the process of obtaining nanoparticles of graphene material are given in Table 1. There are two groups of particles in the solution that differ in size. Large particles (13 μm) were found in the suspensions, which eventually sedimented. The second group of particles sizing up to 450 nm was more stable (Fig. 3). Since the variance is unstable over time, the results should be considered qualitative. The average particle size is ~ 1350 nm. Two fractions: fraction I—size from 10 to 200 nm, quantity—99,9%, weight ~ 26%, the most probable size ~ 40 nm; fraction II—size from 200 to 3200 nm, quantity 0.1%, weight ~ 74%, the most probable size ~ 640 nm; the average polydispersity is ~ 0.6, i.e., the particles have a shape not very far from spherical. Thus, it is possible to adjust the size of graphene particles that are formed during anodic oxidation of EG. A suspension of GNPs diluted 20 times was studied by the SEM method. It was applied to the surface of substrates (films of gold or carbon on glass) and dried.

Two types of particles were detected (Fig. 3). The first type is a lamellar agglomerate with an ordered structure in the form of stacks or a pentagonal hollow prism (Fig. 3a), which may have a disclination origin. The second type of particles is formed into a laminar structure similar to a "flower" and consists of very thin plates (Fig. 3b). The minimum thickness of individual plates of the first type was 25–30 nm (Fig. 3a arrow). Particles of the second type were more often observed on gold substrates

Table 1 Particle sizes of GNPs obtained under different conditions of anodic oxidation of EG

No sample	Average size, nm	Fraction I size, nm			Fraction II size, nm			Poly-dispersity
		Min	Max	Prob	Min	Max	Prob	
1	2162	8	130	40	130	31,000	7954	1
2	623	12	300	54	300	25,000	4133	1
3	1529	11	200	35	200	68,000	16,500	1
4	3251	11	280	35	280	70,000	16,500	1
5	1264	12	300	55	300	43,000	10,048	1
6	1349	12	200	40	200	3200	642	0.6
7	7049	33	800	42	800	8000	1664	1

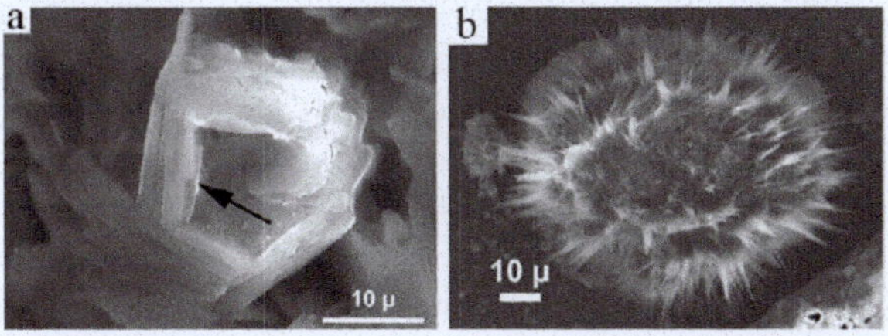

Fig. 3 SEM image of graphite structure on the carbon surface (**a**, **b**)

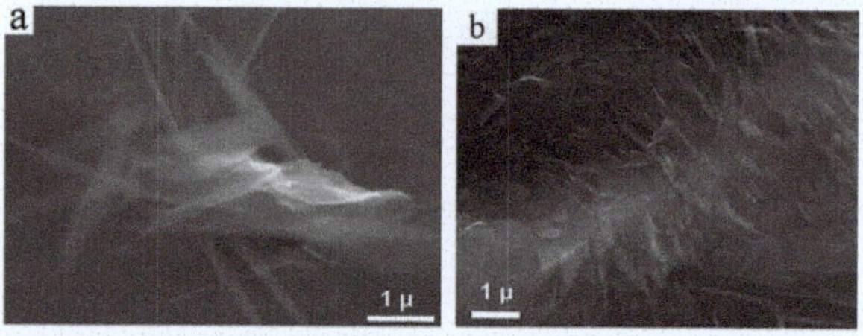

Fig. 4 SEM images of GNP samples continuous films; **a** "free" drying; **b** under vacuum ~ 1 atm in a limited volume

(Fig. 3b). The thickness of the plates was determined as 5–7 nm. And taking into account the characteristics of the microscope, we can assume that the thickness of the plates is at least half as much and is 3–4 nm, that is, about ten layers. Dispersions of high concentration form hard films of a complex structure (Fig. 4). In addition, depending on the drying conditions, the surface may be more embossed, with the appearance of individual GNPs and a smoother film.

The Raman scattering spectra of the GNPs synthesized by us are typical for graphene multilayers (Fig. 5). The symmetric, homogeneous 2D band exceeds the G band in intensity (the so-called "graphite" mode of symmetry E_{1g} in the G point of the Brillouin zone) at 1581 cm^{-1}. In the case of single-layer graphene, I_{2D}/I_g is ~ 4, and in our case it is 1.12. In addition, a weak band is registered at 2451 cm^{-1}, called D″ [9, 10] which consists of the sum of D and D^1 modes (D^1-sp^3 at 1060–1080 cm^{-1}), its appearance is caused by the process of inelastic scattering on the combination D, D″ phonons of the edges of the Brillouin zone. On the diffractogram of a GNPs film (Fig. 6a), dried from an aqueous solution, there is an intense peak characteristic of graphite (002) at an angle of 27°, which has a wide influx towards small angles and a less intense, obviously, second order (004), at angle 55.5°. These lines indicate the presence of polycrystalline graphite, and a significant expansion towards small angles corresponds to the presence of an X-ray amorphous phase, i.e. particles of a rather small size—GNPs. Figure 6b presents the dependences of weight loss (TG (1) and differential thermal analysis (DTA (2)) of GNPs films. Temperatures of 450, 620, 660, 690, 1020 K correspond to temperature anomalies in which the mass of the sample decreases during the thermal destruction of individual components: desorption of chemically bound water, destruction of amorphous carbon, melting of KOH, oxidation of polycrystalline graphite, respectively. In the temperature range of 450–620 K, there is a mass loss of about 18%, which indicates the disintegration of oxygen-containing functional groups and is accompanied by the release of water and carbon dioxide. In the range of 670–1020 K, about 50% of the original mass is lost, which is obviously the result of oxidation of graphene nanoparticles. After 1020 K, only polycrystalline graphite remains (loss of mass up to 17% by mass). Thus, gravimetric analysis (Fig. 6b) of annealed films (without volatiles), GNPs is about 75% wt., the rest is graphite particles.

Recently, we developed method of obtaining EG reinforced by CNTs [11] (Fig. 7a). CNTs "mounted" in the EG matrix are indicated by arrows. The use of a nanotube-reinforced EG foil as an anode in the scheme in Fig. 2a made it possible to obtain a suspension of GNPs@CNT. The size distribution of particles according

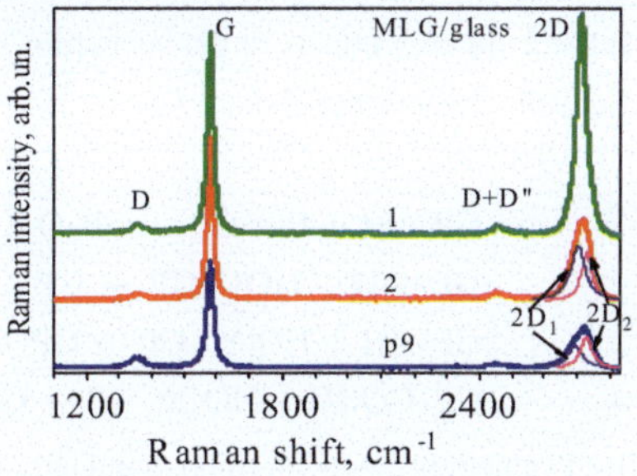

Fig. 5 GNPs Raman spectra on glass substrate measured over the different parts of the sample

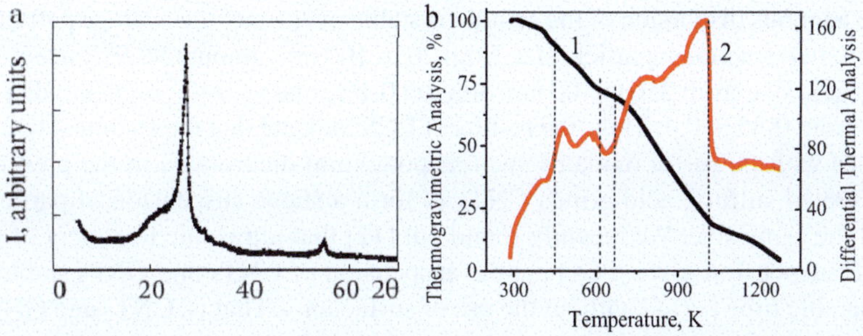

Fig. 6 GNPs film: **a** XRD pattern; **b** dependences of TG (1) and DTA (2) on temperature

to LCS data is shown in Fig. 7b. The dried suspension of GNPs@CNTs gives a continuous film (Fig. 7c).

During industrial production using the CVD method, CNTs are obtained in the of agglomerates with sizes of 20–500 μm. Strong acids and oxidizing agents are effective deagglomeration agents for CNT [7]. Multilayered CNTs were treated with a solution of potassium dichromate in sulfuric acid according to the well-known technology of oxidation of natural graphite [6]:

$$C_n + 25H_2SO_4 + K_2Cr_2O_7 \circledR 6C_{24}^{+}HSO_4{}^{-} \times 2,5H_2SO_4 \\ + Cr_2(SO_4)_3 + K_2SO_4 + 7H_2O + C_{n-144}$$

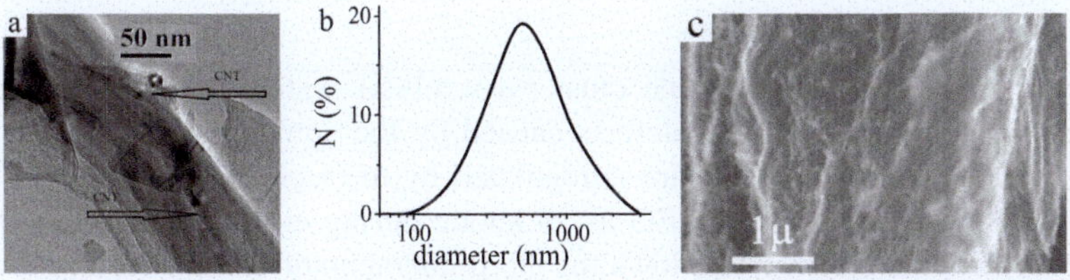

Fig. 7 **a** TEM image of a foil with EG-CNTs, **b** the size distribution of GNPs@CNT particles according to LCS data, **c** SEM images of GNPs@CNT foil

Table 2 Energy effects of the interaction of partially oxidized graphene-like planes

Scheme	Reactions schemes	ΔE reaction, (kJ/mol)		
		$C_{16}H_{10}$	$C_{24}H_{12}$	$C_{42}H_{16}$
1	GP–OH + HO–GP → GP–O–GP + H_2O	+ 22.2	+ 1.3	+ 0.1
2	**GP–OH + HOC–GP → GP–OC–GP + H_2O**	**– 17.1**	**+ 0.5**	**– 0.9**
3	GP–OH + HOOC–GP → GP–O–OC–GP + H_2O	+ 26.0	+ 21.9	+ 19.9
4	GP–COOH + HOOC–GP → GP–CO–O–OC–GP + H_2O	+ 74.5	+ 70.6	+ 69.3

Bold value represents thermodynamically the most probable reaction

The aqueous dispersion of the sample is stable over time: the average particle size ~ 50 nm; two fractions: particle size from 20 to 100 nm, amount 99.9%, mass ~ 10% and particle size from 250 to 500 nm, amount 0.1%, mass ~ 90%; high polydispersity ranges from 0.35 to 0.4. Thus, the results of LCS indicate that the chemical treatment of CNTs with powerful oxidants such as potassium dichromate in the presence of concentrated sulfuric acid grinds CNTs to form a stable suspension of particles of almost the same size. We recently found out [12] that ultrasonic treatment of CNTs in GNP suspension gives a compatible suspension of CNTs and GNPs stable for a considerable time period without the use of surfactants. That is, CNTs and GNPs are "surface-active" and stabilizing components in relation to each other, at least when the mass ratio of GNPs to CNTs is 0.1:1.0 and their content is 0.5–0.1% wt. in an aquatic environment. The dried suspension of GNPs@CNTs gives a continuous film in the same way as in Fig. 7c.

In Ref. [13], we used the methods of quantum chemistry to find out the energy effects of the interaction of partially oxidized graphene-like planes with each other and their dependence on the nature of the oxygen-containing functional groups present in them, as well as on the dimensions of the graphene-like planes themselves (Table 2).

The most thermodynamically probable reaction with the formation of a chemical bond is the interaction of two GPs that contain both hydroxyl and aldehyde groups (Table 2, scheme 2).

4 Conclusions

Expanded graphite is a nanoscale cluster-assembled system due to the presence of cylindrical, conical, slit-like defects caused by the convolution of one or more graphene layers. The EG structure is regulated by the technological conditions of its synthesis. The use of rolled EG as an anode during electrochemical oxidation in a weak alkaline electrolyte allows for the implementation of a two-level system of regulating the structure of graphene nanoparticles, which are obtained according to the "top-down" scheme using the secondary "intercalation" mechanism. That is, the first level of management of the structure and composition of EG, the second level—the parameters of the anodic oxidation process.

Oxidation of carbon nanotubes with strong oxidants allows for their deagglomeration and obtaining stable dispersions in aqueous media. Ultrasonic treatment of CNTs in a suspension of GNPs gives a compatible suspension of CNTs and GNPs, which is stable for a considerable time without the use of surfactants. That is, CNTs and GNPs are "surface-active" and stabilizing components in relation to each other. The dried suspension of GNPs as well as GNPs@CNTs gives a continuous (solid) film.

Quantum chemistry methods have shown that the reaction between hydroxyl and aldehyde groups of two interacting graphene-like planes is the most thermodynamically probable, regardless of their size.

References

1. Geim AK, Novoselov KS (2007) The rise of grapheme. Nat Mater 6:183–191
2. Jang B, Zhamu A (2008) Processing of nanographene platelets (NGPs) and NGP nanocomposites: a review. J Mater Sci 43(15):5092–5101
3. Graifer ED, Makotchenko VG, Nazarov AS, Kim SJ, Fedorov VE (2011) Graphene: chemical approaches to the synthesis and modification. Russ Chem Rev 80(8):751–770
4. Mani V, Chen SM, Lou BS (2013) Three dimensional graphene oxide-carbon nanotubes and graphene-carbon nanotubes hybrids. Int J Electrochem Sci 8(10):11641–11660
5. Tripathi P, Patel CRP, Shaz MA, Srivastava ON (2013) Synthesis of high-quality graphene through electrochemical exfoliation of graphite in alkaline electrolyte. Mat Sci. arXiv:1310.7371
6. Sementsov YI, Revo SL, Ivanenko KO, Hamamda S (2019) Expanded graphite and its composites. In: Kartel MT (ed) Akademperiodika, Kyiv
7. Sementsov YI (2019) Formation of structure and properties of sp^2-carbon nanomaterials and functional composites with their participation. Interservis, Kyiv
8. Sementsov YI, Makhno SM, Kartel MT, Hozhdzinskii SM, Ivanenko KO, Hrebelna YV, Ding A, Wang B, Wang D, Li T (2021) Patent application of Ukraine No a 2021 02264. Electrochemical method of obtaining graphene nanoparticles, 28 Apr 2021
9. Ferrari AC, Meyer JC, Scardaci V, Casiraghi C, Lazzeri M, Mauri F, Piscanec S, Jiang D, Novoselov KS, Roth S, Geim AK (2006) Raman spectrum of graphene and graphene layers. Phys Rev Lett 97(18):187401
10. Park JS, Reina A, Saito R, Kong J, Dresselhaus G, Dresselhausce MS (2009) G' band raman spectra of single, double and triple layer graphen. Carbon 47(5):1303–1310
11. Ding A, Kartel N, Wang D, Sementsov Y (2022) Patent CN No 110950628 B. Preparation method of carbon composite material, 15 Feb 2022
12. Makhno SM, Lisova OM, Gorbyk PP, Ivanenko KO, Sementsov YI, Kartel MT, Ding A, Wang B, Wang D, Li T (2023) Patent of Ukraine no. 152237. The method of obtaining a stable suspension of oxidized graphene nanoparticles and carbon nanotubes, 12 Jan 2023
13. Hrebelna Y, Demianenko E, Terets M, Sementsov Y, Lobanov V, Grebenyuk A, Kuts V, Zhuravskyi S, Khora O, Kartel M (2023) Quantum-chemical studies of the interaction of partially oxidized graphene-like planes with each other. Phys Chem Sol State 24(2):269–277. https://doi.org/10.15330/pcss.24.2.269-277

Estimation of Percolation Threshold and Its Influence on the Properties of Epoxy Resin-Based Polymer Composite Materials Filled Carbon Fibers and Carbon Nanotubes

Stanislav Makhno, Oksana Lisova, Petro Gorbyk, Yuli Shi, Kateryna Ivanenko, and Yurii Sementsov

Abstract Carbon fillers are widely used in the polymer composite materials to control electrophysical properties, which makes them promising for shielding/absorbing electromagnetic radiation in various bands, protection against radio interference and antistatic coatings, etc. The study of percolation transitions in such materials is a priority for such tasks. Epoxy resin-carbon fiber (ER-CF) and epoxy resin-carbon fiber-carbon nanotubes (ER-CF-CNT) systems with different filler contents were investigated. Electrophysical studies were carried out in the frequency range of 8–12 GHz by the non-contact method, and electrical conductivity at low frequencies of 0.1, 1 and 10 kHz was measured by the two-contact method. The bending strength was tested on a 2167 P-50 tensile tester. It was found that: the electrical conductivity at low frequencies and the complex permittivity at 9 GHz of the composites change significantly (percolation transition) in the range of 0.001–0.005 volume fraction of the combined filler; the values of the real and imaginary components of the complex dielectric constant of the composites are ~ 50, which indicates a significant level of interaction between the components of the system and a uniform distribution of the conductive component in the composite; the maximum values of the relative flexural strength are observed at the content of CF-CNT proportional to the percolation

S. Makhno · Y. Shi · K. Ivanenko (✉) · Y. Sementsov
Ningbo Sino-Ukrainian New Materials Industrial Technologies Institute, Kechuang Building, N777 Zhongguan Road, Ningbo 315211, China
e-mail: k_ivanenko@i.ua

S. Makhno · Y. Shi · Y. Sementsov
Ningbo University of Technology, No 55-155 Cui Bai Road, Ningbo 315211, China

S. Makhno · O. Lisova · P. Gorbyk · Y. Sementsov
Chuiko Institute of Surface Chemistry of NAS of Ukraine, 17 General Naumov Str., Kyiv 03164, Ukraine

K. Ivanenko
Institute of Macromolecular Chemistry, NAS of Ukraine, 48 Kharkiv Highway, Kyiv 02160, Ukraine

S. Hamamda et al. (eds.), *Proceedings of the 2nd International Conference of Nanotechnology for Environmental Protection and Clean Energy Production*, Springer Proceedings in Materials 45, https://doi.org/10.1007/978-981-97-1916-7_4

threshold in the systems. The composites have a high level of strength at a low content of fillers, which, makes them attractive for use as protective coatings for absorbing or shielding from microwave EMFs.

Keywords Epoxy resin · Carbon fibres and nanotubes · Electrophysical property

1 Introduction

Conductive fillers expand the range of applications of polymer composite materials due to their electrical and electromagnetic properties [1]. In particular, polymer composites with carbon fillers [2] are used for electromagnetic radiation shielding in various ranges or absorption [3], protection against radio interference and antistatic coatings [4]. For such systems, it is important that the electrically conductive filler forms a continuous network, that is, it is in a state of a continuous percolation cluster. The study of percolation processes in such composites is a priority for this task [5].

Polymer composites reinforced with carbon fibres (CF) are considered promising structural materials for automotive and aerospace applications due to their excellent mechanical properties, lightweight, and flexible design [6]. When used as a filler CF in the form of continuous threads, products are obtained that have exceptionally high strength characteristics compared to the original polymer. The improvement of the mechanical parameters of the composite depends on the aspect ratio (length/diameter) of the fibres, the ratio of the mechanical properties of the polymer and filler, and the efficiency of interaction at the polymer matrix–fibre interface [6]. Combined and hybrid fibrous fillers with different particles (aluminium oxide, graphene nanoparticles, titanium oxide and silicon carbide, metal nanoparticles, etc.) are often considered in the literature [1, 2, 7–11], synergism of properties is inherent in such fillers [4].

Furthermore, in the work [10], the mechanical properties (tensile and bending) of polyamide polymer composite samples reinforced with long carbon fibers were investigated, and the synergistic effect of incorporating both short and long carbon fibers (combined filler) into the polymer was studied. The tensile strength was higher for samples reinforced with fibers of different lengths compared to polymer composites with only continuous long fibers. In our opinion, in [10], the authors introduce new ideas and approaches to composite creation, as an extremely high fiber content of 20–35% by volume of the polymer is considered.

The results of [11] show that increasing the carbon fibre content in a polymer composite to 2% improves the strength by 20–24%, while further increasing the filler content reduces it.

For the purpose of creating polymer coatings with desired electrophysical and other operational indicators, research was conducted on the epoxy resin–carbon fiber (ER–CF) and epoxy resin–carbon fiber–carbon nanotubes (ER–CF–CNT) systems with varying filler content.

2 Materials and Methods

CNTs were used to the requirements of the standard TU U 24.1-03291669-009:2009. The method of their synthesis and the main characteristics are described in [7, 12].

CF was obtained by grinding carbon fabric UVM-24 (TU88-USSR 259-003-89). CF and CNT were dispersed in isopropyl alcohol for 2 min, and the suspension was added to ER LR 285 and mechanically stirred. Then, it was further mixed in a three-roll mixer until the mixture had a uniform color, and a curing agent was added. Final homogenization was performed, and samples of the required size were poured into moulds.

Composite materials with a volumetric content of CF and CNT ranging from 0.0005 to 0.01 volume fractions (vf) were used for the research.

Electrophysical studies were conducted in the frequency range of 8–12 GHz using a non-contact method, and electrical conductivity at low frequencies of 0.1, 1, and 10 kHz was measured using a two-contact method. The relative error of determining ε′, ε″, and σ did not exceed ± 5% [7]. The reflection coefficient R was measured when a 2 mm thick sample was placed on the surface of a metal plate.

Bending testing of layered composites was carried out by the three-point method in accordance with DSTU ISO 7438:2005 on a 2167 P-50 tensile machine with continuous automatic recording of "load-deformation" diagrams on a PC [7], with sample sizes of 50 × 20 × 2 mm.

3 Experimental Results and Discussion

In this work, the dependence of electrophysical characteristics and bending strength on the content of CF and combined filler (CF-CNT) in ES was investigated. Figure 1 shows the dependence of the logarithm of electrical conductivity, and Fig. 2 shows the dependence of the actual ε' and imaginary ε'' components of the complex dielectric constant of the studied systems on the content of the carbon component (CF and CF-CNT). At a filler content of 0.001–0.005 vf in the samples of all systems, a nonlinear dependence of electrical conductivity, ε′ and ε″, is observed, which is associated with the percolation transition. The obtained results were analyzed from the point of view of percolation theory [13] according to equations:

$$\sigma = \sigma_i(\phi - \phi_c)^t, \text{ when } \phi > \phi_c \quad (1)$$

$$\varepsilon' = \varepsilon_i'(\phi_c - \phi)^{-\upsilon}, \text{ when } 0 < \phi < \phi_c, \quad (2)$$

where σ_i and ε_i' are the electrical conductivity and permittivity of the composite; ϕ—volume content of the filler; ϕ_c is the volume content value corresponding to the percolation threshold; *t*, *v*—critical indices.

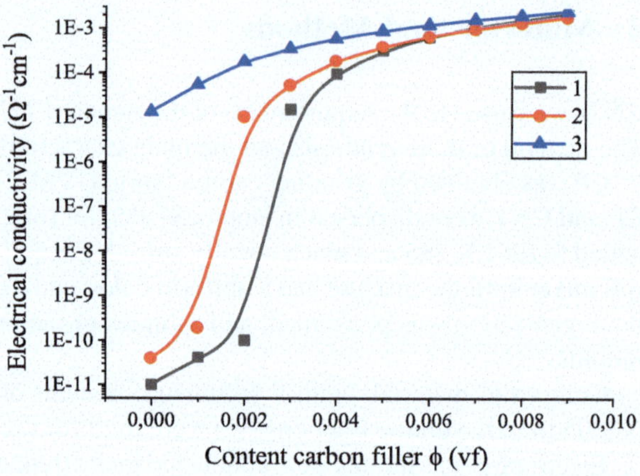

Fig. 1 Electrical conductivity (σ) on the content of CF in ER–CF system (1) and CNTs content in ER–0.001CF–CNT (2) and ER–0.003CF–CNT systems (3)

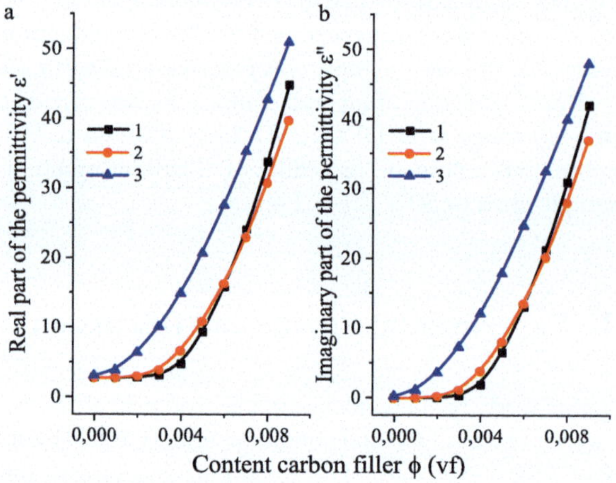

Fig. 2 The real ε' (**a**) and imaginary ε'' (**b**) parts of the permittivity on the content of CF in the ER–CF system (1) and on the content of CNTs in the ER–0.001CF–CNT (2) and ER–0.003CF–CNT systems (3)

The results of calculations according to (1), (2) are given in Tables 1 and 2. The dependences of electrical conductivity (Fig. 1) and permittivity (Fig. 2) in ER–CF systems on the content of CFs and ER–CF–CNTs with different content of CFs on the content of CNTs have a similar character. The values of the calculated percolation thresholds from the dependences of conductivity and dielectric constant have close values (Tables 1 and 2).

For the ER-CF system, the percolation threshold is ~ 0.003 volume fractions, for the ER-CF-CNT systems, the percolation threshold is lower, and the electrical conductivity around the percolation threshold is higher. For the ER-0.001CF-CNT system, the conductivity at a higher degree of filling than 0.006 vf is lower than that of the ER-CF system. The effect may be associated with a greater number of contacts and a higher overall contact resistance between CNTs agglomerates. If we take into account the total volume content of CFs and CNTs then the percolation threshold for

Table 1 Index of the percolation threshold of polymer composites ER–CF and ER–CF–CNT according to electrical conductivity data at a frequency of 1 kHz

Parameter	Sample		
	ER–CF	ER–0.001CF–CNT	ER–0.003CF–CNT
ϕ_c, percolation threshold	0.0031	0.002	0.00023
t, percolation index	1.65	1.81	1.68
σ_i, $(\Omega\ \text{cm})^{-1}$	10	14	7

Table 2 Index of the percolation threshold of polymer composites ER–CF and ER–CF–CNT according to the dielectric constant at a frequency of 9 GHz

Parameter	Sample		
	ER–CF	ER–0.001CF–CNT	ER–0.003CF–CNT
ϕ_c, percolation threshold	0.0026	0.0017	0.00016
v, percolation index	0.78	1.05	0.91
ε_i'	1.4	1.2	1.1

the ER-0.001CF-CNT and ER-0.003CF-CNT systems will be almost the same and equal to $\phi c = 0.003 \pm 0.0003$, and the values of the percolation indices $t = 1.8 \pm 0.2$, $\sigma i = 10 \pm 4$. Thus, their percolation parameters approach the parameters of the ER-CF system.

Deagglomeration and more uniform distribution of the conductive filler in the polymer matrix contributes to their wetting and increasing the area of the interfacial surface. This increases the degree of polarization of filler particles due to the formation of a double charged layer, as well as the number of polar particles per unit volume of the composite. An increase in the number of dipoles in the volume promotes effective interaction with an electromagnetic wave of the centimetre range. This is also manifested in an increase in the values of the complex permittivity even with a low content of filler in the composite. This approach to the creation of composites allows reducing the content of fillers, but provides high permittivity (~ 50). The parts of the complex permittivity have high values. This indicates a significant level of interaction between the components of the system and a uniform distribution of the conductive component in the composite.

The mechanical properties were investigated, namely the relative limit of bending strength of ER-CF composite systems depending on the content of CFs and ER-CF-CNT depending on the content of CNTs at the content of CFs of 0.001, 0.003 and 0.005 volume fractions (Fig. 3). Bending strength for all systems has maximum values in the range of 0.001–0.007 volume fraction of the filler. For the ER-CF systems, with the content of CF below the percolation threshold, the values of the bending strength limit increase, and with an increase in the CF content approximately above the percolation threshold, the values of the bending strength limit decrease (Fig. 3). Similar dependencies are observed for other systems. That is, there is a certain correlation between mechanical properties and electrical conductivity at low

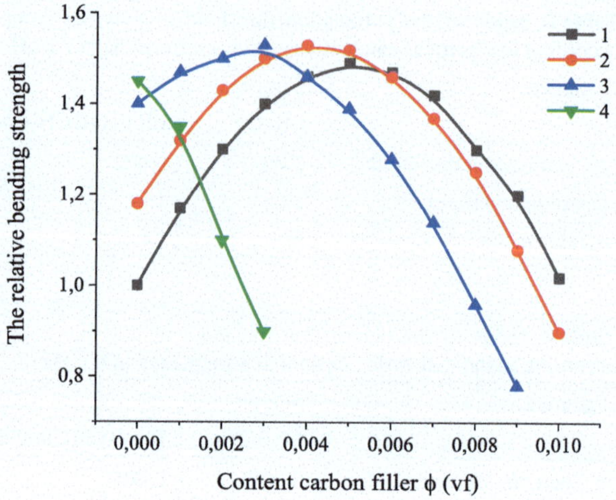

Fig. 3 The relative tensile bending on the CF content in ER in the ER–CF system (1) and the CNTs content in the ER–xCF–CNT systems where x(vf) = 0.001 (2); 0.003 (3); 0.005 (4)

frequencies in the region of the percolation transition. The maximum limit of the relative bending strength P_{max} is reached at a somewhat higher content of CFs and CNTs than the percolation threshold. Analyzing the results shown in Figs. 1 and 3, it can be assumed that there is a dependence between the content of CFs and CNTs at the maximum value of the limit of relative strength (P_{max}) for bending $\phi(P_{max})$ and the value of the percolation threshold ϕ_c [14]. This dependence can be expressed as:

$$\phi(P_{\max}) = \kappa_p \cdot \phi_c, \tag{3}$$

where κ_p is the proportionality coefficient, which depends on the aspect ratio (length to diameter ratio) of the filler and the uniformity of its distribution in the polymer volume.

By lowering the percolation threshold in polymer systems with a filler having a high aspect ratio, it is possible to achieve a high level of electrical conductivity and dielectric losses at microwave frequencies at a lower filler content, as well as to obtain higher strength characteristics. One of the methods of reducing the percolation threshold is to reduce the content of CNTs in ER when they are introduced [7, 14], dispersed and stretched fibres and tubes [15] in the polymer volume.

The percolation thresholds determined from the dependences of the logarithm of the electrical conductivity on the volume fraction of the filler are located near the lower bend of the curves (Fig. 1), and the value of ϕ(Pmax) (Fig. 3) is located near the upper bend (Fig. 1). The coefficient κ_p for the considered systems, in the general case, should be a function of the angle of inclination of the curve section between two fractures on the dependences (Fig. 1). With a low content of CFs in the ER-CF system, the addition of CNTs increases the mechanical properties of composite to the content of CNTs $\phi(P_{max})$, and then decreases with an increase in the content of CNTs. If the content of CFs exceeds the value of $\phi(P_{max})$ (Fig. 3), curve 4 (0.005 CF), the introduction of CNTs reduces the mechanical properties of composites.

According to [16], the relationship between the dimensions of the filler (CF or CNT), length l, diameter d and the percolation threshold ϕ_c can be written as:

$$\phi_c \approx d/l. \quad (4)$$

Accordingly, Eq. (3) will have the form:

$$\phi(P_{\max}) \approx \kappa_p \cdot d/l. \quad (5)$$

Equations (4) and (5) will be fulfilled the better the more uniform will be the distribution of stretched (deagglomerated) filler particles (CFs or CNTs). From calculations according to (4), it was established that with a fibre thickness of approximately 6 μm and a percolation threshold of 0.0031 vf (for the ER-CFs system), the length of the fibres should be about 2 mm, which is close to the actual length of the CFs that were introduced into the ER. This indicates a sufficiently uniform distribution of CFs in the binder, and also indirectly indicates the aligned state of the fibres.

Figure 4 shows the dependences of the coefficients of absorption (A) (Fig. 4a) and reflection (R) (Fig. 4b) on the frequency of electromagnetic radiation (EMR) for the ER–CF and ER–CF–0.001CNT systems with a CF vf: 0.003 (1, 1′), 0.005 (2, 2′), 0.008 (3, 3′) (Fig. 4).

Composites with a CFs content of 0.005 vf in ER have the maximum values of the absorption coefficient A (Fig. 4). When a larger amount of CF is added into the polymer, the value of ε' increases, correspondingly, the reflection coefficient R also increases, that is, a significant part of the EMR energy will not be absorbed by the sample. According [8]:

$$A = 1 - R - T, \quad (6)$$

where T is the transmission coefficient.

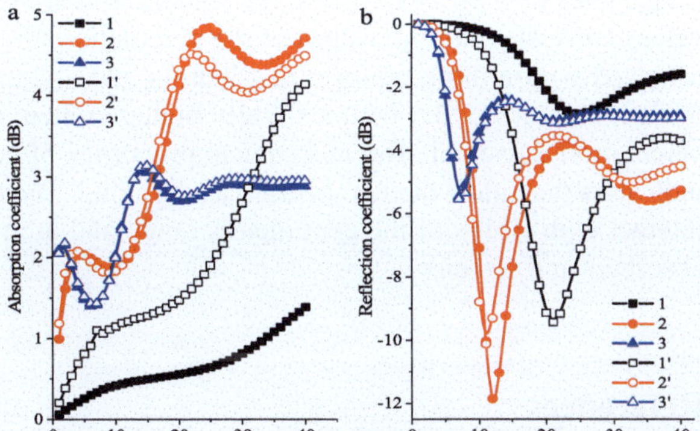

Fig. 4 The absorption A (**a**) and reflection R (**b**) coefficients on the metal surface on the EMF frequency at the specified experimentally determined values ε', ε'' shown in Fig. 1 for a thickness of 2 mm in the systems ER–xCF and ER–xCF–0.001CNT where x(vf) = 0.003 (1, 1′); 0.005 (2, 2′); 0.008 (3, 3′)

The EMR attenuation coefficient in composite materials will grow according to an exponential law, which for the electric field strength vector E will have the form [17]:

$$E = E_0 \cdot \exp(-k''h), \tag{7}$$

where: E_0 is the electric field intensity vector of the incident electromagnetic wave; k'' is the EMR attenuation coefficient, which depends on the value of ε'', h is the thickness of the sample.

Figure 4b shows an increase in the reflectance with an increase in the content of CFs in the composite, since ε' also increases. At some frequencies for which the sample thickness satisfies the condition [8]

$$h = \frac{\lambda}{4\sqrt{\varepsilon'}}, \tag{8}$$

where λ is the EMV wavelength, which affects a significant decrease in the reflection coefficient, which is associated with the interference attenuation of the wave reflected from the front and back surfaces of the sample.

4 Conclusions

The dependence of the electrophysical and mechanical characteristics of the ER-CF and ER-CF-CNT systems on the content of CF and CF-CNT was studied. It was established that the reduction of the percolation threshold to 0.003 volume fraction of fillers in these systems could be achieved by reducing the concentrations of CF and CNT with a more uniform distribution of them in the matrix. This leads to higher electrical conductivity at low frequencies and dielectric loss at microwave frequencies with lower CF and CNT content. It is shown that the maximum strength values correlate with the values of the percolation threshold in the systems and are achieved with a more uniform distribution and alignment of CNTs in the volume of the polymer matrix, with a slightly higher content of CF and CF-CNT than the percolation threshold. The mechanical properties of composites have performance indicators that allow them to be used as a coating. The resulting composites actively interact with EMFs of the centimetres range and are promising for the creation of absorbers and screens.

References

1. Hong X, Shi W, Zheng H, Liang D (2019) Effective carbon nanotubes/graphene hybrid films for electron field emission application. Vacuum 169:108917–108923

2. Ye J, Hu B, Jin Y, Wang Z, Xi Y, Fang L, Pan Q (2020) Interface engineering integrates fractal-tree structured nitrogen-doped graphene/carbon nanotubes for supercapacitors. Electrochim Acta 349:136372
3. Dróżdż PA, Xenidis N, Campion J, Smirnov S, Przewłoka A, Krajewska A, Haras M, Nasibulin A, Oberhammer J, Lioubtchenko D (2022) Highly efficient absorption of THz radiation using waveguide-integrated carbon nanotube/cellulose aerogels. Appl Mater Today 29:101684
4. Sementsov Y, Makhno S, Kartel M, Wang B, Dovbeshko G, Styopkin V, Nedilko S (2017) Graphene nanoparticles and graphene nanoparticles—polyamide 12/12 composites. Int J Innov Sci Eng Tech 4(8):71–78
5. Bhagat AB, Ghosh AK (2022) Estimation of rheological percolation threshold and influence of fibre length on properties of polypropylene/sisal fibre composites having near critical fibre length. Polymer 258:125304
6. Zhang N, Huang D, Chen X, Ye C, Li B, Zhu S, Fan Z, Liu H, Liu J (2023) Improving the interlaminar bonding and thermal conductivity of polymer composites by using split-radial mesophase pitch-based carbon fiber as reinforcement. Compos B Eng 252:110509
7. Makhno SM, Lisova OM, Mazurenko RV, Gorbyk PP, Ivanenko KO, Kartel MT, Sementsov YuI (2023) Electrophysical and strength characteristics of polychlorotrifluoroethylene filled with carbon nanotubes dispersed in graphene suspensions. Appl Nanosci. https://doi.org/10.1007/s13204-023-02902-6
8. Makhno SM, Lisova OM, Gunya GM, Gorbyk PP, Kartel MT (2023) Synthesis and electrophysical properties of nanostructured composites NiCo/$BaTiO_3$ and NiCo/TiO_2. Himia, Fizika ta Tehnologia Poverhni 14(2):173–181
9. Cherniuk OA, Demianenko EM, Terets MI, Zhuravskyi SV, Makhno SM, Lobanov VV, Kartel MT, Sementsov YI (2020) Study of the mechanism of influence of carbon nanotubes surface chemistry on the mechanical properties of fiberglass. Appl Nanosci 10(12):4797–4807
10. Saeed K, McIlhagger A, Harkin-Jones E, McGarrigle C, Dixon D, Shar MA, McMillan A, Archera E (2022) Characterization of continuous carbon fibre reinforced 3D printed polymer composites with varying fibre volume fractions. Compos Struct 282:115033
11. Jotiram GA, Palai BK, Bhattacharya S, Aravinth S, Gnanakumar G, Subbiah R, Chandrakasu M (2022) Investigating mechanical strength of a natural fibre polymer composite using SiO_2 nano-filler. Mater Today Proc 56(3):1522–1526
12. Melezhik AV, Sementsov YuI, Yanchenko VV (2005) Synthesis of fine carbon nanotubes on coprecipitated metal oxide catalysts. Russ J Appl Chem 78(6):917–923
13. Kirkpatrick S (1973) Percolation and conduction. Rev Mod Phys 45:574–588
14. Makhno SM, Lisova OM, Ivanenko KO, Sementsov YI, Horbyk PP, Kartel MT, Shi Y, Wang B, Wang D, Li T (2023) Patent application of Ukraine No a202300157. Method of determining the content of carbon filler for obtaining the maximum strength characteristics of polymer composites by measuring their electrical conductivity, 17 Jan 2023
15. Elenchezhian M, Vadlamudi V, Kumar Banerjee P, Dave C, Mahmood A, Raihan R, Reifsnider K (2017) Quality assessment of adhesive bond based on dielectric properties. In: SAMPE 2017 conference proceedings. Seattle, WA
16. Balberg I, Anderson CH, Alexander S, Wagner N (1984) Excluded volume and its relation to the onset of percolation. Phys Rev B 30(7):3933
17. Nikolsky VV, Nikolskaya TI (1989) Electrodynamics and propagation of radio waves, 3rd edn. Nauka, Moscow

Preparation, Structure and Luminescence Properties of “Microcrystalline Cellulose—$K_3Tb(PO_4)_2$” Composites

Vitalii Chornii, Serhii G. Nedilko, Kateryna Terebilenko, Valeria Zozulia, Volodymyr Boyko, Yaroslav Zhydachevskyy, and Andrzej Suchocki

Abstract The paper deals with preparation and optical properties of the oxide-polymer composite consisting of the luminescent $K_3Tb(PO_4)_2$ polycrystals embedded into microcrystalline cellulose (MCC) matrix. The band gap of the materials under study has been estimated to be 3.76 eV for MCC and 3.15 eV for $K_3Tb(PO_4)_2$@MCC, respectively. The photoluminescence spectra of the composite depend on excitation wavelengths: cellulose-related emission dominates under excitation at 405 nm while simultaneous emission from cellulose matrix and Tb^{3+} ions observed for case of $\lambda_{ex} = 300$ nm. At the same time, the composite sample reveals quite intensive green photoluminescence regardless on λ_{ex}. The main band of luminescence excitation undergoes red-shift when moving from MCC to $K_3Tb(PO_4)_2$@MCC. These results point out an energy transfer from cellulose to oxide particles. In general, observed peculiarities of the diffuse reflectance and photoluminescence properties reveal interaction between matrix and filler in the composites under study. The specific emission spectra of the composite suggest its possible application at document protection branch.

Keywords Composite · Microcrystalline cellulose · Photoluminescence

V. Chornii (✉) · S. G. Nedilko · K. Terebilenko · V. Zozulia
Taras Shevchenko National University of Kyiv, Kyiv 01601, Ukraine
e-mail: vchornii@nubip.edu.ua

V. Chornii · V. Boyko
National University of Life and Environmental Sciences of Ukraine, Kyiv 03041, Ukraine

S. G. Nedilko · Y. Zhydachevskyy · A. Suchocki
Institute of Physics, Polish Academy of Sciences, 02-668 Warsaw, Poland

S. Hamamda et al. (eds.), *Proceedings of the 2nd International Conference of Nanotechnology for Environmental Protection and Clean Energy Production*, Springer Proceedings in Materials 45, https://doi.org/10.1007/978-981-97-1916-7_5

1 Introduction

Composites have become an essential issue for a range of modern industrial, clinical, and scientific applications. Mainly, composite materials consist of two or more constituent parts with wide discrepancies in their mechanical, chemical and physical properties. To satisfy the wide range of modern demands the main body of the material is commonly based on polymer with functional particles dispersed inside. In case of nanoparticles used for filling nanocomposites are formed [1]

Nowadays, a huge number of polymer composites are known: some of them are well-known and other ones are just being developed and are promising for application [1]. Polymer composites can be quite flexible, which makes it possible to develop flexible electronic devices based on them: displays, solar batteries, etc. [2, 3]. Unfortunately, most of the polymer composites are products of petroleum processing, and therefore they are harmful to people and the environment. At the contrary, polymer materials based on cellulose are products of plant processing, or the result of the activity of microorganisms (bacteria), and therefore, such polymer materials are environmentally safe. That is why significant attention has been paid to the development of composite materials based on cellulose and its derivatives [4].

In this paper, composites are prepared on the basis of microcrystalline cellulose (MCC), which is widely used in various technological and industrial fields, in particular, in food, pharmaceutical, cosmetic, etc. [5]. Despite the significant area of application, interest in new composite materials containing MCC has been continuously grown up [6]. However, it should be noted there have been developments of composites where MCC fragments are used as reinforcing fillers [7], while developments of composites where MCC is a matrix are few [8]. However, the production of solid composites based on MCC is much simpler than the production of solid composites, for example, based on nanocellulose (NC) [9].

Importantly, the choice of composite filler is determined by considerations regarding the expected properties of the composite, because the purpose of developing of the latter one is to design a material with properties different from those of the matrix. In particular, fillers in the form of particles of oxide compounds are used to modify the optical properties of cellulosic polymers. Namely, oxides containing luminescent ions of rare earth elements (REEs) are of particular interest [10].

Regarding oxide fillers, the cellulose composite acquires new luminescent properties, and therefore can be used for transformation and visualization of light flows, concentration of solar radiation, sensorics, marking, etc. [11]. The use of trivalent terbium ions in oxide compounds—fillers is due to its specific properties, such as wide spectral composition of luminescence (from blue to red light), high radiation intensity and long-lasting afterglow—up to seconds. Micro/nano-sized crystalline powders of $K_3Tb(PO_4)_2$ are used in our work as a filler containing luminescent triply charged terbium ions (Tb^{3+}). The optical luminescent properties of phosphate compounds containing terbium ions were studied earlier [12].

Comprehensive studies of the luminescent properties of composite materials, where the matrix is microcrystalline cellulose, and the filler is luminescent micro/nanosized particles of $K_3Tb(PO_4)_2$ oxide, have not been conducted so far.

Therefore, the main tasks of this work are to find out the characteristics of optical absorption (scattering), luminescence and excitation of luminescence of composites "MCC—$K_3Tb(PO_4)_2$" and to identify manifestations of interaction between the matrix of MCC and $K_3Tb(PO_4)_2$ filler according to the specified characteristics.

2 Experimental

2.1 *Samples Preparation*

As initial pure microcrystalline cellulose (MCC) in a form of commercial MCC tablets (manufactured by FARMAKOM, Ukraine) have been used in this study. The polycrystalline $K_3Tb(PO_4)_2$ has been synthesized by solid state method from K_2CO_3, K_2HPO_4 and Tb_2O_7 in platinum crucibles at 1300 K. The samples of $K_3Tb(PO_4)_2$@MCC have been prepared from MCC and oxide powders those have been simultaneously grinded in a low-speed mill during 7 days. The mass of oxide used is equal to 10% of the total mass of $K_3Tb(PO_4)_2$@MCC composite. The UV–Vis spectroscopy has been performed for the samples in a powdered form while PL properties are measured for pressed samples.

2.2 *Equipment*

The UV–Vis diffuse reflectance spectra as well as photoluminescence properties (emission and its excitation spectra, kinetics, quantum yield) of the studied samples have been obtained using Spectrofluorimeter Fluorolog 3.11, UV/Vis/NIR (JobinYvon-Horiba). The PL kinetics were studied with using of sub-nanosecond ($\Delta\tau < 0.01$ ns) pulsed light emitting diodes. All the measurements have been performed at room temperature.

3 Results and Discussion

3.1 *Band Gap Determination*

The diffuse reflectance spectra have been measured in the 200–800 nm spectral range for powdered MCC, "$K_3Tb(PO_4)_2$@MCC", and $K_3Tb(PO_4)_2$ samples. It has been found that pure MCC demonstrate high diffuse reflection (above 85%) over

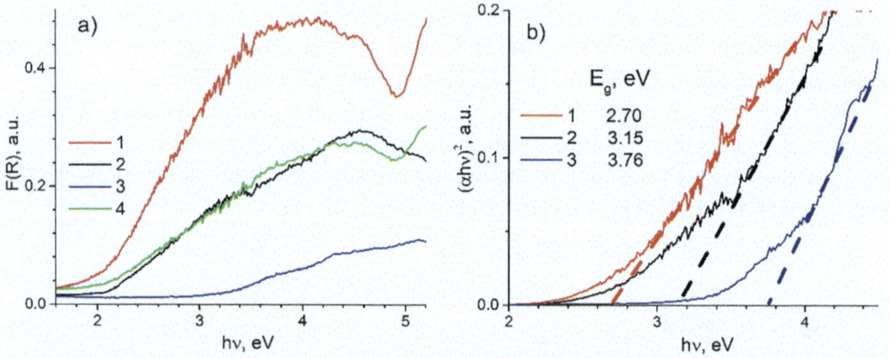

Fig. 1 **a** The Kubelka–Munk functions and **b** Tauc plots for $K_3Tb(PO_4)_2$ (1), $K_3Tb(PO_4)_2$@MCC (2), MCC (3) powders, and linear combination of curves 1 and 3 (4)

all the visible spectral region. The reflectivity of $K_3Tb(PO_4)_2$ phosphor is near 70% in this region, while the composite sample $K_3Tb(PO_4)_2$@MCC shows intermediate case of refection intensity between MCC and oxide. The diffuse reflectance spectra have been further analyzed from viewpoint of materials band gap, E_g. The Tauc plot method is well-known technique for estimation of E_g for polycrystalline materials [13]. The Kubelka–Munk theory [14] was applied to diffuse reflectance spectra of the samples in order to obtain absorption coefficient:

$$F(R) = \frac{(1 - R)^2}{2R} = \frac{K}{S} \tag{1}$$

where, R is diffuse reflectance of a sample normalized to non-absorbing standard, K is proportional to absorption coefficient of the sample, S is an effective scattering coefficient of the sample. The calculated spectra $F(R)$ and Tauc plots are shown in Fig. 1a, b, respectively.

According to literature data [15], the electronic transitions in the PO_4^{3-} molecular anions take place at photon energies above 7 eV. Thus, the band at 2–5 eV region of $K_3Tb(PO_4)_2$ absorption spectrum can be associated with some defects in the crystalline compound. An effective light absorption due to defect-related centers in used oxide takes place for hν > 2.7 eV (corresponds to wavelength λ ≈ 459 nm). In case of MCC a light absorption is related with organic chromophores, those are presented in cellulose-based materials: carbonyl, carboxyl, methylol groups, etc. [16]. These chromophores are "regular" moieties of any MCC also, therefore the E_g obtained value of 3.76 eV can be regarded as band gap of the studied cellulose matrix.

As it follows from Fig. 1, the case of $K_3Tb(PO_4)_2$@MCC has "intermediate" absorption spectrum between those corresponding to pure MCC and oxide filler. In order to analyze effect of interaction between cellulose and oxide, a curve $F_4(R)$ in Fig. 1a was fitted as a linear combination of curves for MCC $F_1(R)$ and oxide $F_3(R)$:

$$F_4(R) = 0.38297 * F_1(R) + 1.1073 * F_3(R) \quad (2)$$

It can be seen that absorption of composite is not just linear combination of absorption of $K_3Tb(PO_4)_2$ and MCC powders. So, the $K_3Tb(PO_4)_2$@MCC composite is not described by "rule of mixture". The differences between curves 2 and 4 on Fig. 1a can be assumed as evidence of an interaction between cellulose matrix and oxide filler. The calculated band gap value for $K_3Tb(PO_4)_2$@MCC sample ($E_g \approx$ 3.15 eV) is determined by both cellulose-related chromophores and defects of the $K_3Tb(PO_4)_2$ crystals. Taking into account the results of diffuse reflection, the further PL studies have been performed at excitation wavelengths $\lambda = 300$ nm (photon energy $\approx$ 4.13 eV), that is above E_g for MCC and 405 nm ($h\nu \approx$ 3.06 eV) that is close to E_g for composite.

3.2 *Photoluminescence Spectroscopy*

The room temperature PL spectra of the studied samples are shown in Fig. 2. The photoluminescence of pure MCC is more effectively excited at 405 nm than at 300 nm (Fig. 2a). Such result can be explained by presence of different functional chromophore groups in the MCC, and some of those groups are active in light absorption and they are luminescent groups (fluorophores), too [16]. It is obviously, they reveal very weak emission at 300 nm excitation. This fact indicates a significant role of non-radiation processes which influenced noted luminescence. The PL spectra of MCC under excitation at 300 nm can be fitted by two Gauss components with maxima near 3.09 and 2.55 eV, respectively. The later component has quite large full width at half maximum (FWHM)—0.72 eV that likely due to overlapping of few PL components. Decomposition of PL spectra obtained under excitation at 405 nm has shown two components at 2.50 eV (FWHM = 0.48 eV) and 2.13 eV (FWHM = 0.57 eV). So, the pure MCC has at least three types of luminescence centers.

Only weak wideband luminescence at 520–640 nm range was registered for the PL excitation at 405 nm in the case of the $K_3Tb(PO_4)_2$ powders (Fig. 2c). Similar luminescence of various phosphates compounds was ascribed earlier to defect-related centers of phosphate matrix like ones created at oxygen vacancies sites [17]. The narrow bands of Tb^{3+} emission was observed when excitation took place at 300 nm (region of $^7F_6 \rightarrow ^5H_J$ electronic transitions in terbium ions [12]).

The PL spectra of $K_3Tb(PO_4)_2$@MCC composite consist of narrow bands of Tb^{3+} ions emission and wide bands of cellulose luminescence. The interactions between MCC matrix and oxide filler are manifested in the PL spectra. Particularly, under excitation at 405 nm narrow band of Tb^{3+} ions clearly seen at 544 nm (Fig. 2b). In case of PL excitation at 300 nm a redistribution of intensity between Tb^{3+}—related bands can be noted when moving from "free" oxide to composite. These results can be ascribed to energy transfer from cellulose to filler. It is worth noting, that specific luminescence spectra of Tb^{3+} ions in composite comparing with "free" oxide when

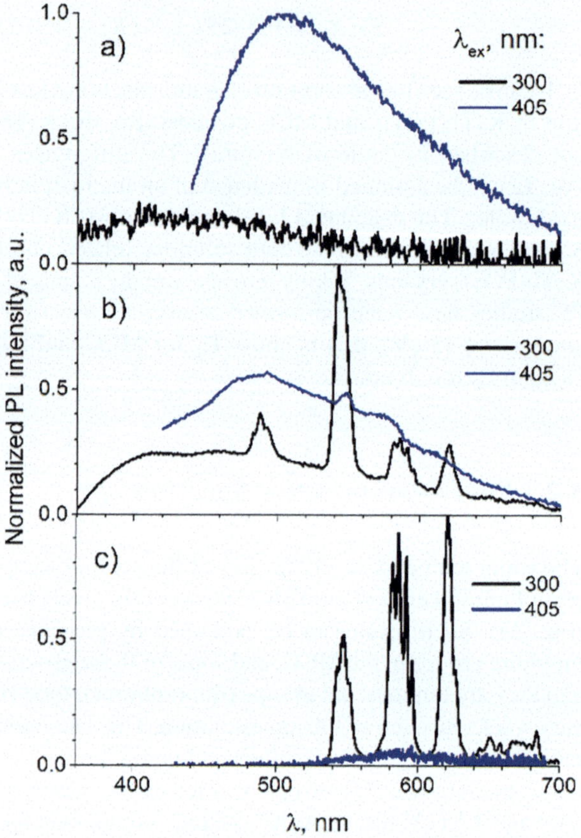

Fig. 2 PL spectra of the pure MCC (**a**), $K_3Tb(PO_4)_2$@MCC (**b**), and $K_3Tb(PO_4)_2$ samples (**c**)

excitation takes place at 405 nm suggest a possible application of such material in document security.

The PL excitation spectra for pure MCC and $K_3Tb(PO_4)_2$@MCC composite samples are shown in Fig. 3a. A blue luminescence of pure MCC is the most effectively excited at 337 nm (spectrum 1 in Fig. 3a). In case of a composite, a maximum of the blue PL excitation band is shifted to 370 nm. This shift can be explained by assumption that an energy of exciting photons near $\lambda \sim 300$ nm is absorbed by both cellulose and $K_3Tb(PO_4)_2$ filler in the case of composite samples. Consequently, excitation of blue cellulose emission became less effective.

An excitation of red luminescence of cellulose ($\lambda_{em} = 610$ nm) in MCC and $K_3Tb(PO_4)_2$@MCC samples can be done at the region of wide complex band with general maxima near 400 nm. In case of composite this band is modified by light absorption of oxide as intensity drops near 480 nm corresponds to ${}^7F_6 \rightarrow {}^5D_4$ electronic transitions in Tb^{3+} ions. It is clear from comparison of $K_3Tb(PO_4)_2$@MCC luminescence and its excitation spectra that emission from cellulose matrix can be re-absorbed by Tb^{3+} ions of the filler.

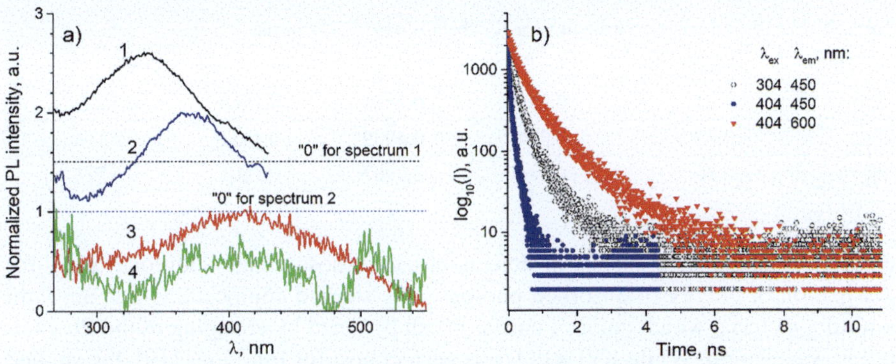

Fig. 3 **a** PL excitation spectra of MCC (1, 3) and $K_3Tb(PO_4)_2$@MCC (2, 4) samples registered at $\lambda_{em} = 450$ (1, 2) and 610 nm (3, 4); **b** PL decay curves for $K_3Tb(PO_4)_2$@MCC samples.

The $K_3Tb(PO_4)_2$@MCC sample demonstrate fast luminescence. The decay curves of blue and red luminescence obtained for excitations by nanosecond pulses at ultraviolet and violet regions are shown in Fig. 3b. All the PL decay curves was successfully fitted by double exponential decay functions:

$$I = A_1 e^{-\frac{t}{\tau_1}} + A_2 e^{-\frac{t}{\tau_2}} \tag{3}$$

where A_1 and A_2—amplitudes, τ_1 and τ_2—PL time constants. The parameters of fitting are collected in Table 1. It is known [18], that the PL time constants τ for Tb^{3+} emission in $K_3Tb(PO_4)_2$ are of ~ 2 ms as corresponding to forbidden *f-f* radiation transitions. Thus, observed fast luminescence is related either with cellulose-related centers or with defect-related centers in oxide filler.

The photoluminescence quantum yield (PLQY) of composite was obtained for PL emission in the 450–700 nm spectral range. The chromaticity coordinates for composite sample were calculated for CIE 1931 (x, y) and CIE 1976 (u′, v′) color space. The results are collected in Table 2.

All the excitation wavelengths in Table 2 correspond to cellulose matrix absorption ($h\nu < E_g$ for phosphate) but narrow bands of Tb^{3+} ions emission was observed in the PL spectra during PLQY measurements. As it follows from analysis of the PL properties, the energy transfer from cellulose matrix to oxide filler takes place in composite at various excitation wavelengths. Such low quantum yields for $K_3Tb(PO_4)_2$@MCC

Table 1 Parameters of PL decay curves fitting for $K_3Tb(PO_4)_2$@MCC sample

λ_{ex}, nm	λ_{em}, nm	A_1, r.u	τ_1, ns	A_2 r.u	τ_2, ns
304	450	2100 ± 5	0.12 ± 0.01	572 ± 5	0.50 ± 0.01
404	450	1343 ± 6	0.04 ± 0.01	456 ± 5	0.15 ± 0.01
404	600	1627 ± 6	0.24 ± 0.01	936 ± 6	0.89 ± 0.01

Table 2 PLQY and color coordinates for $K_3Tb(PO_4)_2$@MCC sample

λ_{ex}, nm	PLQY, %	x	y	u′	v′
260	1.04	0.3520	0.5581	0.1566	0.5585
282	0.23	0.2941	0.4939	0.1411	0.5331
370	0.16	0.3867	0.5907	0.1661	0.5707

can be explained by presence of defects in oxide those create additional ways for dissipation of energy of absorbed photons. The studied composite is a green light emitting material which color of emission can be tuned by changing of the PL excitation wavelength. While $K_3Tb(PO_4)_2$ is perfect host for Eu^{3+} ions [19], the studied composite can be also the base for elaboration of multicolor luminescence covers.

4 Conclusions

The samples of the $K_3Tb(PO_4)_2$@MCC composite have been obtained by simple method of simultaneous low speed grinding of its component with subsequent pressing. The analysis of the optical properties has showed that studied $K_3Tb(PO_4)_2$@MCC samples are not a simple mechanical mixture of its constituents. The interaction between cellulose matrix and oxide filler manifested in diffuse reflectance spectra as well as in photoluminescence properties. Particularly, the energy transfer from cellulose to oxide-related emission centers take place in composite under study. The specific emission spectra of the composite suggest its possible application at document protection branch.

Acknowledgements The study was financed by National Research Foundation of Ukraine (project № 2022.01/0168) and supported by Institute of Physics Polish Academy of Sciences

References

1. Shivalkar S, Ranjan S, Sahoo AK (2023) Polymeric nanocomposites: synthesis, characterization, and recent applications. In: Singh DK, Singh S, Singh P (eds) Nanomaterials. Springer, Singapore, pp 267–295
2. Song JK, Do K, Koo JH, Son D, Kim DH (2019) Nanomaterials-based flexible and stretchable bioelectronics. MRS Bull 44(8):643–656
3. Khelil S, Barama N, Naoui Y, Bouleklab MC, Dorbani T, Nedilko S, Revo S, Hamamda S (2023) Influence of the MWCNTs on the properties of the HDPE+ X% MWCNTs nanocomposites. Appl Nanosci: 1–10
4. Sharma A, Thakur M, Bhattacharya M, Mandal T, Goswami S (2019) Commercial application of cellulose nano-composites—a review. Biotechnol Rep 21:e00316

5. Trache D, Hussin MH, Chuin CTH, Sabar S, Fazita MN, Taiwo OF, Hassan TM, Haafiz MM (2016) Microcrystalline cellulose: isolation, characterization and bio-composites application—a review. Int J Biol Macromol 93:789–804
6. Miyashiro D, Hamano R, Umemura K (2020) A review of applications using mixed materials of cellulose, nanocellulose and carbon nanotubes. Nanomaterials 10(2):186
7. Rathinavelu R, Paramathma BS, Divkaran D, Siengchin S (2023) Physicochemical, thermal, and morphological properties of microcrystalline cellulose extracted from *Calotropis gigantea* leaf. Biomass Conversion Biorefinery: 1–18
8. Spoljaric S, Genovese A, Shanks RA (2009) Polypropylene–microcrystalline cellulose composites with enhanced compatibility and properties. Compos A Appl Sci Manuf 40(6–7):791–799
9. Kargarzadeh H, Mariano M, Huang J, Lin N, Ahmad I, Dufresne A, Thomas S (2017) Recent developments on nanocellulose reinforced polymer nanocomposites: a review. Polymer 132:368–393
10. Carlos LD, Ferreira RA, Bermudez VDZ, Ribeiro SJ (2009) Lanthanide-containing light-emitting organic–inorganic hybrids: a bet on the future. Adv Mater 21(5):509–534
11. Erdman A, Kulpinski P, Olejnik K (2016) Application of nanocomposite cellulose fibers with luminescent properties to paper functionalization. Cellulose 23:2087–2097
12. Drabik J, Kowalski R, Marciniak L (2020) Enhancement of the sensitivity of single band ratiometric luminescent nanothermometers based on Tb^{3+} ions through activation of the cross relaxation process. Sci Rep 10(1):11190
13. Sharma S, Kumar M, Laref A, Siqueiros JM, Herrera OR (2023) Recent advances in interfacial engineering and band-gap tuning of perovskite multiferroic heterostructures for high performance photovoltaic applications. Mater Lett 331:133490
14. Landi S Jr, Segundo IR, Freitas E, Vasilevskiy M, Carneiro J, Tavares CJ (2022) Use and misuse of the Kubelka-Munk function to obtain the band gap energy from diffuse reflectance measurements. Solid State Commun 341:114573
15. Saito S, Wada KI, Onaka R (1974) Vacuum ultraviolet reflection spectra of KDP and ADP. J Phys Soc Jpn 37(3):711–715
16. Rosenau T, Potthast A, Krainz K, Hettegger H, Henniges U, Yoneda Y, Rohrer C, French AD (2014) Chromophores in cellulosics, XI: isolation and identification of residual chromophores from bacterial cellulose. Cellulose 21:2271–2283
17. Gieszczyk W, Bilski P, Mrozik A, Kłosowski M, Marczewska B, Sas-Bieniarz A, Perzanowski M, Zorenko T, Zorenko Y (2020) Intrinsic and dopant-related luminescence of undoped and Tb plus Tm double-doped lithium magnesium phosphate ($LiMgPO_4$, LMP) crystals. Materials 13(9):2032
18. Finke B, Schwarz L (1995) Luminescence decay of rare earth ions in an orthophosphate matrix. Radiat Eff Defects Solids 135(1–4):407–411
19. Mikhailik VB, Kraus H, Dorenbos P (2009) Efficient VUV sensitization of Eu^{3+} emission by Tb^{3+} in potassium rare-earth double phosphate. Physica Status Solidi (RRL)–Rapid Res Lett 3(1):13–15

Formulation and Optimization of Sustained Release Microparticles of Ketoprofen

Fatima Zohra Badaoui, Souha Meziani, Lina Benmegoura, and Souha Djoumana Hedna

Abstract Microparticles are one of the drug delivery systems used to sustain the release of drugs. The purpose of this work was to forumlate sustained-release microparticles of ketoprofen (KTP) for ocular delivery. The microparticles were prepared by the method of quasi emulsion-solvent diffusion using the design of experiments, the MINITAB 16 was used to analyze the results. EC:KTP mass ratio (%), PVA concentration (m/v %) and stirring speed (rpm) were selected as factors. Encapsulation efficiency rate (EE%) and in vitro release after 6h (REL6%) were selected as responses. The central composite design (CCD) was used for the formulation optimization. The results indicated that the optimal microparticles formulated has an EE = 78.89%, REL6% = 39.98% for a EC:KTP mass ratio = 80.59%, PVA concentration = 0.52% and stirring speed = 477.11 rpm. The optical microscope was used to determine the morphology of the optimized microparticles, they were of spherical shape. The in vitro dissolution at 35 °C in a medium with pH = 7.4 gave a sustained release in which 36.98% of ketoprofen was released after 6 h. According to the results obtained, the microparticles of ketoprofen can be used for the ocular sustained release.

Keywords Microparticles · Ketoprofen · Sustained release · Design of experiment · Ocular delivery

---

F. Z. Badaoui (✉) · S. Meziani · L. Benmegoura · S. D. Hedna
Department of Pharmaceutical Engineering, Faculty of Processes Engineering, Salah Boubnider University-Constantine 3, 25000 Constantine, Algeria
e-mail: fatimazohra.badaoui@univ-constantine3.dz

F. Z. Badaoui
Process Engineering Laboratory for Sustainable Development and Health Products, Constantine, Algeria

S. Hamamda et al. (eds.), *Proceedings of the 2nd International Conference of Nanotechnology for Environmental Protection and Clean Energy Production*, Springer Proceedings in Materials 45, https://doi.org/10.1007/978-981-97-1916-7_6

1 Introduction

Microencapsulation is a technique for preserving the quality of sensitive substances by enclosing liquids or solids in a membrane. It is a method to produce materials with interesting new properties, consisting of a coating material containing an active substance. The product obtained by this process is called "microparticles", which are differentiated by their morphology and structure [1].

Microparticles (MP) are known for offering prolonged or controlled release of the entrapped active ingredient through the use of the polymeric matrix. In addition, the active ingredient is also protected from direct contact with various physiological fluids and/or enzymes, thus preventing its immediate degradation. MP are generally prepared using biodegradable or erodible polymers that modify the release mode of the drug [2].

Ketoprofen (KTP) is a non-steroidal anti-inflammatory derived from aryl carboxylic acid, a member of the propionic group. It has several properties as peripheral and central analgesic, antipyretic, anti-inflammatory and enables short-term inhibition of platelet functions, it has a short half-life, therfore it requires several administrations in the same day [3].

In order to enhance efficacy and avoid adverse effects of ketoprofen, it has been formulated in polymeric microparticles for prolonged ophthalmic release.

2 Materials and Methods

2.1 Materials

Ketoprofen was a gift from LNCPP-Constantine. Dialysis bags (cut-off 12 kDa), Polyvinylalcohol (87–90% hydrolyzed, average mol wt. 30.000–70.000) and Ethylcellulose (viscosity 22cP, 48% ethoxyl) were purchased from Sigma Aldrich USA. All other ingredients and solvents were of analytical grade.

2.2 Methods

Preparation of Microparticles

The quasi-emulsion solvent diffusion (ESD) was the method used for the preparation of KTP microparticles [4]. To 5 mL of a mixture of ethanol and dichloromethane (v:v; 1:1), ethylcellulose (EC) and KTP were added under vigorous magnetic stirring. To obtain an emulsion O/W, the solution obtained was emulsified with a mechanical propeller stirrer in 50 mL of the aqueous solution of PVA. Under mechanical stirring at room temperature, the organic solvents were left to evaporate for 2 h in order to obtain the microparticles suspension. The microparticles were separated by vacuum filtration, washed and dried. The filtrate was used for non encapsulated KTP analysis.

Characterization of Microparticles

Entrapment Efficiency (EE)

The indirect method was used to determine the KTP entrapment efficiency and was calculated as mentioned below. The amount of non-encapsulated KTP present in the filtrate was analyzed by UV–visible spectroscopy (Shimadzu, UV 1800, Japan) at 255 nm.

$$\text{EE} = \frac{\text{Initial weight of feeding drug} - \text{Weight of not encapsulated drug in filtrate}}{\text{Initial weight of feeding drug}} \times 100 \quad (1)$$

In Vitro Dissolution

The in vitro dissolution test was realized in conditions representing the ocular medium (50 rpm, 35 °C ± 0.5). The dry powder of the microparticles formulated which contain a predetermined quantity of KTP was dissolved with Phosphate Buffered Saline (PBS, pH = 7.4) and placed into cellulose dialysis bags. The dialysis bags were submerged in 100 ml of the dissolution medium PBS (pH = 7.4). At each time interval, 3 mL of the dissolution medium were withdrawn, filtered, diluted and tested by UV–visible spectrophotometry to determine the quantity of the KTP released.

Morphology

An optical microscope was used to observe the morphology of the optimized microparticles. To observe the morphology of MP, the dry powder of the microparticles was placed on a glass slide and overlaid by a lamella.

Experimental Design

A Central Composite Design was used for the KTP-loaded microparticles preparation. Minitab® was the software used for data analysis.

The mass ratio of EC:KTP (X_1: $\frac{mEC}{mKTP+mEC} \times 100$), the PVA concentration (X_2: w/v %) and the speed of agitation (X_3: rpm) were the factors evaluated. Table 1 describe the levels used for each factor studied. Two responses were chosen: were the entrapment efficiency (Y1) and the release of KTP after 6 h (Y2).

Table 2 summarizes the results of CCD with factors levels. The model generated by the design was quadratic non-linear and presented as follow:

$$\begin{aligned} Y = {} & A_0 + A_1X_1 + A_2X_2 + A_3X_3 + A_{12}X_1X_2 \\ & + A_{13}X_1X_3 + A_{23}X_2X_3 + A_{11}X_1^2 + A_{22}X_2^2 + A_{33}X_3^2 \end{aligned} \quad (2)$$

where

Table 1 Dependent and independent variables in CCD

Factors: independent variables	Coded and real values		
	Low (− 1)	Intermediate (0)	High (+ 1)
X_1: mass ratio of EC:KTP (%)	71.42	75.71	80
X_2: PVA concentration (%)	0.4	0.5	0.6
X_3: speed of agitation	400	600	800
Responses: dependent variables	*Constraints*		
Y1: entrapment efficiency (EE %)	Maximize (30–90)		
Y2: release of KTP after 6 h (REL6%)	Minimize (40–60)		

Table 2 Data and factors levels of CCD

EXP	X1 (%)	X2 (%)	X3 (rpm)	Y1 (%)	Y2 (%)
1	71.42	0.4	400	39.157	48.448
2	80	0.4	400	61.741	41.207
3	71.42	0.6	400	69.540	61.169
4	80	0.6	400	79.038	54.968
5	71.42	0.4	800	50.320	76.938
6	80	0.4	800	55.445	68.634
7	71.42	0.6	800	51.978	77.257
8	80	0.6	800	67.253	74.724
9	68.495	0.5	600	33.874	44.896
10	82.924	0.5	600	84.340	45.654
11	75.71	0.305	600	46.087	49.076
12	7571	0.66	600	70.447	50.868
13	75.71	0.5	263.64	65.532	34.465
14	75.71	0.5	939.35	69.256	75.910
15	75.71	0.5	600	61.373	40.505
16	75.71	0.5	600	72.080	23.472
17	75.71	0.5	600	78.387	31.609
18	75.71	0.5	600	63.981	49.530
19	75.71	0.5	600	60.367	53.938
20	75.71	0.5	600	81.456	35.288

- Y: the obtained response
- A_0: an intercept
- A_1, A_2, A_3: the coefficients of linear regression
- A_{12}, A_{13}, A_{23}: the coefficients of interactive regression
- A_{11}, A_{22}, A_{33}: the coefficients of quadratic regression

- X_1, X_2, X_3: the studied factors
- X_1^2, X_2^2, X_3^2: the quadratic effects
- X_1X_2, X_1X_3, X_2X_3: the interaction between the variables [5].

To select the desirable optimum region, the method named desirability function was used. Table 1 shows the constraints used in order to choose an optimum formulation.

First, the individual desirability function d1 and d2 were calculated by transformation of responses Y1 and Y2 as follow:

$$\text{d1, d2} = \left(\frac{Y1 - L}{T - L}\right) \tag{3}$$

U, L are the upper and lower acceptable values of response, T is the target value desired. The individual desirability functions vary over the range $0 \leq (\text{d1, d2}) \leq 1$.

The following equation was used for calculation of the overall desirability value (D):

$$D = \sqrt{d1 \times d2} \tag{4}$$

3 Results and Discussion

3.1 Experimental Design

The R^2, adjusted R^2, and p-values of the model were the quality indicators for the obtained model. The highest R^2 values, insignificant lack of fit p-value ($p > 0.05$) and significant model *p*-value ($p < 0.05$) for Y1 and Y2 were observed with the quadratic model without interactions as mentioned in Table 3.

The regression equations of the fitted models that describe the experimental responses are presented as follow:

$$\begin{aligned} \text{Y} = {} & 69.66 + 10.05X1 + 7.47X2 - 1.33X3 - 4.09\text{X1}^2 \\ & - 4.38\text{X2}^2 - 1.16\ \text{X3}^2 - 4.27\text{X2X3} \end{aligned} \tag{5}$$

Table 3 Results of the regression analysis

	Adjusted R^2	R^2	*p*-value	
			Lack of fit	Model
Y1	0.77	0.64	0.689	0.004
Y2	0.74	0.62	0.760	0.003

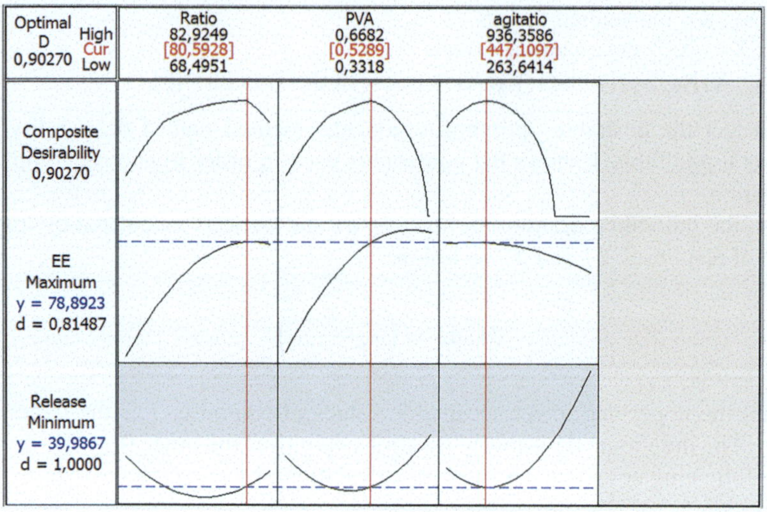

Fig. 1 Factors and responses of optimized KTP microparticles

$$Y2 = 38.66 - 1.68X1 + 2.62X2 + 11.82X3 + 4.75X1^2 + 6.41X2^2 + 8.25\,X3^2 \quad (6)$$

The optimum MP formulation shows an EE (Y1) of 78.89% and a release of KTP after 6 h (Y2) of 39.98%, at levels of factors: EC:KTP ratio X1 = 80.52%, PVA concentration X2 = 0.52% and stirring speed X3 = 447.10 rpm. According to Fig. 1, the overall desirability value D = 0.90 with d1 = 0.81 and d2 = 1.

3.2 *Morphology*

The microscopic observation of the optimized KTP microparticles shows a spherical shape (Fig. 2).

3.3 *In Vitro Release Test*

Over 6 h, only 36.98% of the KTP was released from the optimized microparticles (Fig. 3). The release of KTP from the optimized microparticles demonstrate a biphasic behavior, where 31.04% of KTP were released after one hour, and then the appereance of a plateau. This behavior can be explained by the release of the encapsulated KTP

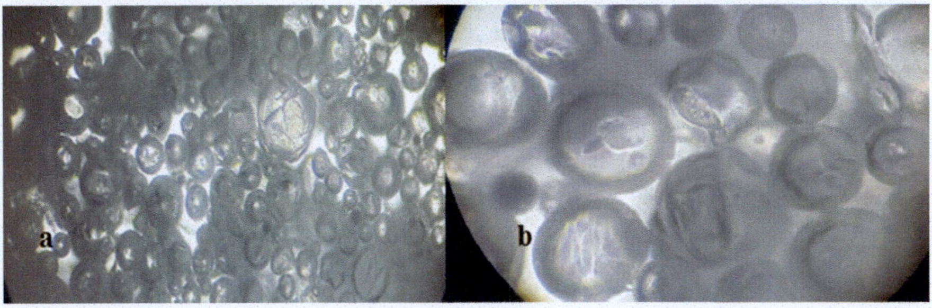

Fig. 2 Microscopic observation of the optimized Ketoprofen microparticles **a** × 10 (b) × 40

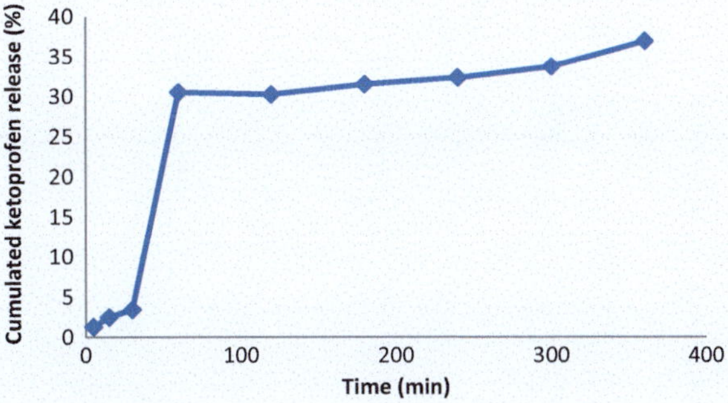

Fig. 3 Profile of in vitro release of Ketoprofen microparticles

inside the microparticles after the discharge of the adsorbed KTP on the MP surface [6, 7].

4 Conclusion

The optimized microparticles obtained have an EE of 78. 89% and a sustained release over 6 h REL6% of 39.98% with a spherical shape. Ketoprofen loaded Ethylcellulose microparticles showed an interesting ocular anti-inflammatory sustained effect.

References

1. Richard J, Benoît JP (2000) Microencapsulation, Techniques de l'Ingénieur, Traité génie des procédés J 2:210
2. Soni G, Yadav KS, Gupta MK (2019) Design of experiments (DoE) approach to optimize the sustained release microparticles of Gefitinib. Curr Drug Del 16:364–374
3. Das SK, Khanam J, Nanda A (2016) Optimization of preparation method for ketoprofen-loaded microspheres consisting polymeric blends using simplex lattice mixture design. Mater Sci Eng C 69:598–608

4. Ratnaparakhi DU, Patil SS, Patil SV (2015) Formulation and evaluation of microspongeof benzyol peroxide by quasi emulsion solvent diffusion method. Int J Pharmaceut Res 7(2):38–43
5. Vera Candioti L, De Zan MM, Cámara MS, Goicoechea HC (2014) Experimental design and multiple response optimization using the desirability function in analytical methods development. Talanta 124:123–138
6. Deshmukh K, Poddar SS (2012) Tyrosinase inhibitor-loaded microsponge drug delivery system: new approach for hyper pigmentation disorders. J Microencapsul 29(6):559–568
7. Gavini V, Murthy MS, Kumar PK (2014) Formulation and in vitro evaluation of nanoparticles drug delivery system loaded with 5-fluorouracil. Res J Pharmaceut Dosage Forms Technol 6(4):243–248

Effect of Fe_3O_4 Nanoparticle on Electronic Cooling: Numerical Investigation of Nanofluid Flow Through a Heat Sink

Touba Bouacida and Rachid Bessaih

Abstract Nanotechnology is a branch of science and engineering concerned with the design, production and use of structures, and systems through manipulating of atoms and molecules on the nanoscale, which has become necessary in electronics to follow rapid development in this sector, where precision and high performance are required. To learn about how adding nanoparticles affects heat transfer, flow of water and nanofluid inside heat sink filled with copper foam at a constant porosity is compared numerically in this paper. Fe_3O_4–H_2O nanofluid is selected as a coolant with constant concentration and diameter of nanoparticles. To enhance simulation accuracy, Eulerian two-phase modeling is used to predict the behavior of nanofluid. The simulation results were validated by comparing them with experimental data found in the literature, and good agreement was reached. Results clearly confirmed that nanofluid improved heat transfer by 50% and surface temperature reduced by 14 °C compared with water as a coolant. Opposite position of copper foam inside heat sink leads to an enhancement of Nusselt number, thus improving cooling efficiency.

Keywords Electronic cooling · Heat sink · Nanoparticles

1 Introduction

Adding nanoparticles to base liquids is a technology that is attracting the interest of scientists due to its effective results in numerous studies, particularly in heat transfer domain [1] and renewable energies [2]. On the other hand, electronics is one of the fields that have followed this technology to solve the problem of heat dissipation of electronic components, which is one of several causes of damage and short lifespan. Researchers have confirmed that increasing the temperature of these components by 10 °C reduces their reliability and efficiency by 50% [3]. In order to remove above

T. Bouacida (✉) · R. Bessaih
LEAP Laboratory, Department of Mechanical Engineering, University of Mentouri-Constantine 1, Constantine, Algeria
e-mail: touba.bouacida@doc.umc.edu.dz

S. Hamamda et al. (eds.), *Proceedings of the 2nd International Conference of Nanotechnology for Environmental Protection and Clean Energy Production*, Springer Proceedings in Materials 45, https://doi.org/10.1007/978-981-97-1916-7_7

problems and improve thermal management capacity, Faraji et al. [4] proposed a heat sink in the presence of a variable-phase material using 4 types of single and hybrid nanoparticles. They found that hybrid nanoparticles boosted the melting period of variable-phase material and reduced the temperature to 11 °C. For the same purpose, Zhao et al. [5] experimentally studied the effect of TiO_2 nanoparticle concentration ratio (0–0.5%) and Re number (472–1198) on CPU cooling. They demonstrated that nanoparticle concentration between 0.3 and 0.4% produced best results in terms of cooling efficiency. Nanoparticles were not limited to electronics, but were also applied to storage [6] and solar heat exchangers [7, 8]. According to a research study conducted by Khan et al. [9], biosynthetic nanoparticles can significantly boost solar panel efficiency. It can also be used in various medical and biological fields.

Based on the aforementioned research studies, its content is focused on how to improve performance and efficiency of devices using nanoparticle technology. Similarly, this study aims to enhance cooling efficiency of electronic components by employing nanoparticles and porous media. The flow of water and nanofluid inside copper foam-filled heat sink was numerically compared in this paper.

2 Modeling and Mathematical Equations

This study aims to simulate the 3D flow of water and Fe_3O_4–H_2O nanofluid in a heat sink used to cool electronic components as a CPU. The dimensions of the area study is H × L × W (12mm × 45mm × 36mm) respectively, heat sink is equipped with copper foam at a constant porosity ($\varepsilon = 0.8$) whose dimensions are h × l × w (6mm × 22.5mm × 2.5mm). Fe_3O_4-H_2O nf was used as a coolant with a constant concentration and diameter of nanoparticles ($\phi = 0.3\%$ and dn = 20 nm, respectively), its characteristics are illustrated in Table 1. Laminar flow ($400 \leq Re \leq 1100$) of water and nf are applied to the heat sink to dissipate heat from the bottom surface of CPU subjected to a heat flux of 220 kW/m^2. Eulerian two-phase modeling is used to predict the behavior of nf in the studied domain.

To simplify the study, the following assumptions were adopted: nanofluid flow is 3D, laminar, incompressible and stationary, the effect of gravity is negligible, and thermo-physical properties of solid are considered constant. The continuity equation, the equations of motion, and the energy equation are [11]:

$$\frac{\partial u}{\partial x} + \frac{\partial v}{\partial y} + \frac{\partial w}{\partial z} = 0 \tag{1}$$

Table 1 Physical properties of Fe_3O_4–H_2O nanofluid [10]

Physical properties	ρ (kg/m^3)	Cp (J/kg K)	k (W/m K)	μ (Pa s)
Water	998.1	4179	0.613	0.0013004
Fe_3O_4 nanoparticle	5180	670	80	–

$$\frac{1}{\varepsilon^2}\left(u\frac{\partial u}{\partial x}+v\frac{\partial u}{\partial y}+w\frac{\partial u}{\partial z}\right)=-\frac{1}{\rho}\frac{\partial P}{\partial x}+\frac{\upsilon}{\varepsilon}\left(\frac{\partial^2 u}{\partial x^2}+\frac{\partial^2 u}{\partial y^2}+\frac{\partial^2 u}{\partial z^2}\right)-\left(\frac{\upsilon}{K}+\frac{C}{\sqrt{K}}\sqrt{u^2+v^2+w^2}\right)u \quad (2)$$

$$\frac{1}{\varepsilon^2}\left(u\frac{\partial v}{\partial x}+v\frac{\partial v}{\partial y}+w\frac{\partial v}{\partial z}\right)=-\frac{1}{\rho}\frac{\partial P}{\partial y}+\frac{\upsilon}{\varepsilon}\left(\frac{\partial^2 v}{\partial x^2}+\frac{\partial^2 v}{\partial y^2}+\frac{\partial^2 v}{\partial z^2}\right)-\left(\frac{\upsilon}{K}+\frac{C}{\sqrt{K}}\sqrt{u^2+v^2+w^2}\right)v \quad (3)$$

$$\frac{1}{\varepsilon^2}\left(u\frac{\partial w}{\partial x}+v\frac{\partial w}{\partial y}+w\frac{\partial w}{\partial z}\right)=-\frac{1}{\rho}\frac{\partial P}{\partial z}+\frac{\upsilon}{\varepsilon}\left(\frac{\partial^2 w}{\partial x^2}+\frac{\partial^2 w}{\partial y^2}+\frac{\partial^2 w}{\partial z^2}\right)-\left(\frac{\upsilon}{K}+\frac{C}{\sqrt{K}}\sqrt{u^2+v^2+w^2}\right)w \quad (4)$$

C is the coefficient of inertia which is expressed as [11]: $C=\frac{1.75}{\sqrt{150\varepsilon^3}}$.

$$u\frac{\partial T}{\partial x}+v\frac{\partial T}{\partial y}+w\frac{\partial T}{\partial z}=\frac{K_{eff}}{\rho Cp}\left(\frac{\partial^2 T}{\partial x^2}+\frac{\partial^2 T}{\partial y^2}+\frac{\partial^2 T}{\partial z^2}\right) \quad (5)$$

$$k_{eff}=\varepsilon k_{nf}+(1-\varepsilon)k_s \quad (6)$$

Nusselt number can be computed by:

$$Nu=\frac{k_{nf}}{k_{bf}}\cdot\left(\frac{\partial T}{\partial x}\cdot\frac{D_h}{T-T_{in}}\right) \quad (7)$$

Average Nusselt number:

$$\overline{Nu}=\frac{1}{A}\int Nu\ dA \quad (8)$$

ANSYS 14.0 [12] software has been used for this numerical study. Coupled algorithm based on finite volume method was used to discretize the partial differential Eqs. (1–5). Constant velocity inlet, constant heat flux, and atmospheric pressure

Fig. 1 Mesh domain

outlet boundary conditions were applied to the computational domain. A mesh of 432,000 elements was adopted in this simulation (see Fig. 1).

3 Validation

To verify the reliability of this domain, results were compared with experimental results of Chen et al. [13], for copper foam of porosity 0.9 and laminar flow of TiO_2–H_2O nanofluid (Re: 414–1179) at constant nanoparticle concentration ($\phi = 0.3\%$) as illustrated in Fig. 2. Numerical results are in good agreement with the experimental ones, showing a maximum relative error of less than 4%.

Fig. 2 Validation of surface temperature by experimental results of Chen et al. [13]

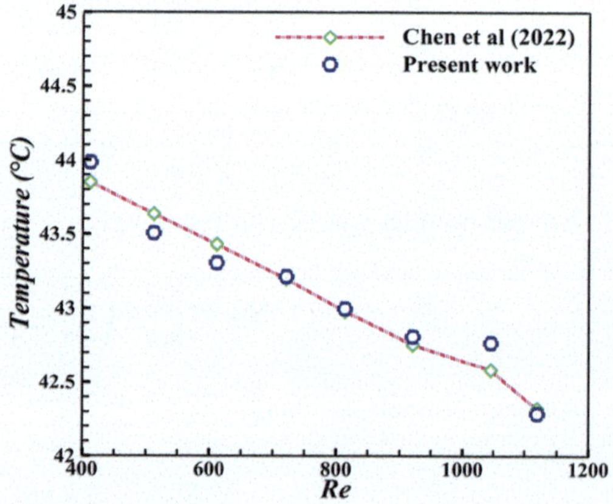

4 Results and Discussion

Laminar forced convection of Fe_3O_4–H_2O is simulated for Reynolds number range $400 \leq Re \leq 1100$, nanoparticle concentration $\phi = 0.3\%$ and diameter of nanoparticle dn = 20 nm. Results are presented in terms of Nusselt number, temperature, temperature and velocity counters detailed in this section.

Figure 3 shows a comparison of surface temperature exposed to heat flux cooled by water and nanofluid. Temperature decreases as flow accelerates, but cooling is efficient with nanofluid as surface temperature drops to 304.8 °C, versus using water where temperature is 318 °C at Re = 1100. That is due to penetration of nanofluid into pores of copper foam loaded with oxidized iron nanoparticles, which transmit heat rapidly on account of their high conductivity compared to pure water molecules.

The variation of average Nussult number in term of Re is presented in Fig. 4. In accordance with the majority of previous research [10], Nu_{avg} increases with flow velocity. This is due largely to a rise in heat exchange coefficient h as a result of forced convection of nf into copper foam, which creates a large surface area for heat exchange, leading to a lower convection/conduction ratio. The use of nanofluid enhances heat transfer by 50% compared to water at Re = 600.

Understanding the thermal cooling of heat sink, Fig. 5 plots temperature contours at Re = 900 for both cooling fluids (a) nanofluid, (b) water). Isotherms are denser around heated surfaces, explaining heat transfer between a heat sink and CPU. In porous regions, temperatures are more steady, indicating that copper foam has been able to remove excess energy internally from the CPU through greater conductive heat transfer. Isotherm length in Fig. 5a is less than in Fig. b where water flows into heat sink. When nanoparticles are added, the thickness of thermal boundary layer on bottom surface of heat sink decreases due to collisions between particles and porous medium. All this improves the efficiency of heat exchange, so temperature drops significantly.

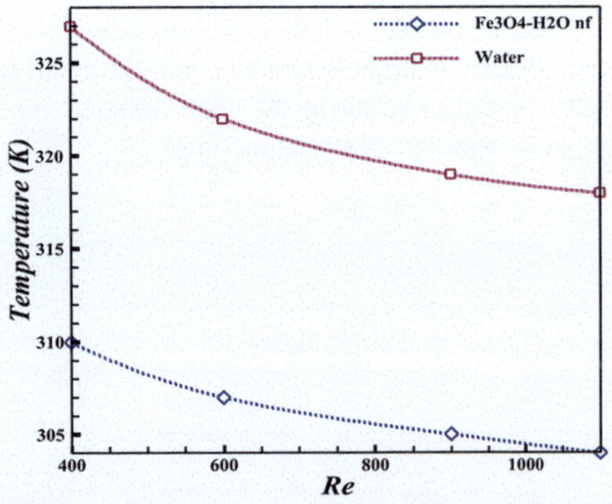

Fig. 3 Heated surface temperature versus Re number

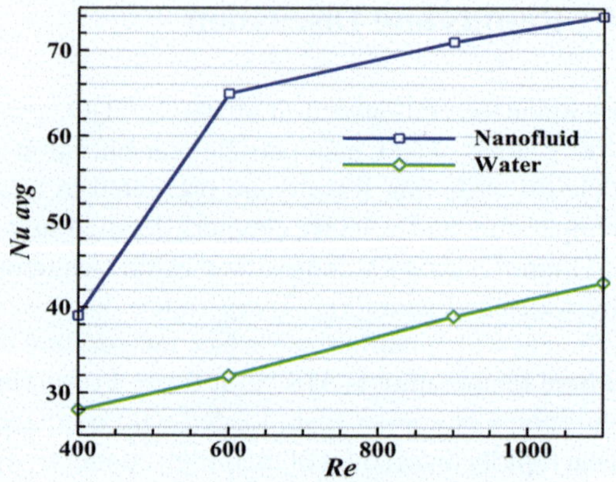

Fig. 4 Average Nusselt number as a function of Re number

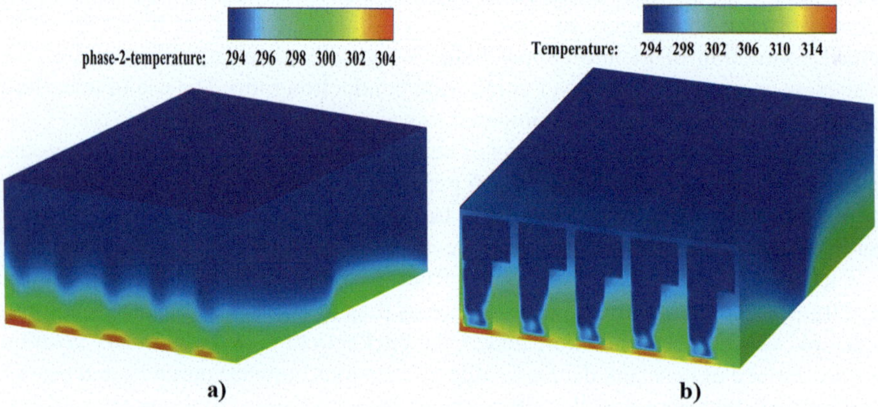

Fig. 5 Heat sink temperature contours at Re = 900

Figure 6 present velocity contour of water inside heat sink at Re = 900. It is clear that velocity is highest in areas without metal foam. This can be explained by foam pores acting as a barrier to water passage, in addition to the collisions occurring between water and porous medium.

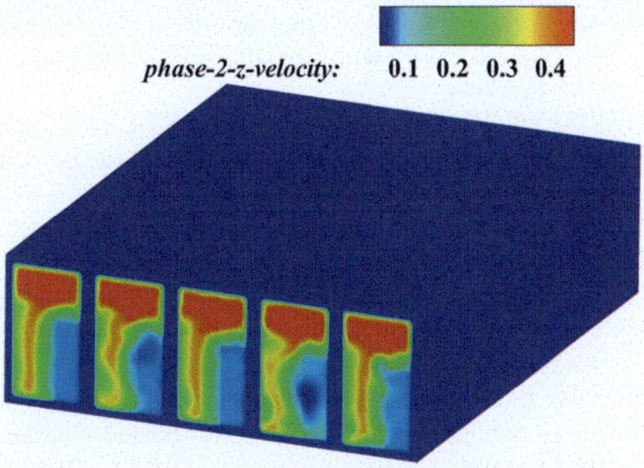

Fig. 6 Velocity contour in the outlet of heat sink at Re = 900

5 Conclusion

Addition of nanoparticles to base fluids is an attractive technique that has been used in many fields and proven effective in a numerous investigations. This numerical study aims to improve cooling efficiency of electronic components using nanoparticles and porous media. The flow of water and nanofluid ($\phi = 0.3\%$) inside a heat sink filled with copper foam ($\varepsilon = 0.8$) was compared. Main results obtained can be summarized as follows:

- The addition of Fe_3O_4 nanoparticles to the base fluid at constant concentration improves heat transfer by up to 50% compared with water.
- Using copper foam leads to an enhancement of Nusselt number at constant porosity.
- The presence of nanoparticles reduced heat sink temperature to 14 °C, thus improving cooling efficiency.

References

1. Nageswara Rao B, Sreedhar T (2023) Heat transfer analysis of nanofluids for fixed volume fraction of pipe in pipe heat exchanger using CFD. Mater Today Proc
2. Mahian O et al (2021) Recent advances in using nanofluids in renewable energy systems and the environmental implications of their uptake. Nano Energy 86:106069
3. Sharma CS, Zimmermann S, Tiwari MK, Michel B, Poulikakos D (2012) Optimal thermal operation of liquid-cooled electronic chips. Int J Heat Mass Transf 55(7–8):1957–1969
4. Faraji H, Yıldız Ç, Arshad A, Arıcı M, Choukairy K, El Alami M (2023) Passive thermal management strategy for cooling multiple portable electronic components: hybrid nanoparticles enhanced phase change materials as an innovative solution. J Energy Storage 70
5. Zhao N, Guo L, Qi C, Chen T, Cui X (2019) Experimental study on thermo-hydraulic performance of nanofluids in CPU heat sink with rectangular grooves and cylindrical bugles based on exergy efficiency. Energy Convers Manag 181:235–246

6. Khan SA et al (2023) Computational analysis of natural convection with water based nanofluid in a square cavity with partially active side walls: applications to thermal storage. J Mol Liq 382:122003
7. Dao QD, Tran NA, Doan TH (2022) Liquid crystal semiconductor C6TBTAPH2 for hole transport materials in pervoskite solar cells: fabrication, characterization, and simulation. Opt Mater 132:112820
8. Shakoor A et al (2023) Fabrication and characterization of TiO_2: ZnO thin films as electron transport material in perovskite solar cell (PSC). Phys B Condens Matter 654:414690
9. Khan F, Karimi MN, Khan O (2023) Green technologies and sustainability exploring the scalability and commercial viability of biosynthesized nanoparticles for cooling panels with the help of artificial intelligence and solar energy systems. Green Technol Sustain 1(3):100036
10. Qi C, Li K, Li C, Shang B, Yan Y (2020) Experimental study on thermal efficiency improvement using nanofluids in heat sink with heated circular cylinder. Int Commun Heat Mass Transf 114:104589
11. Mukeshkumar PC, Kumar A (2023) Numerical study on the performance of Al_2O_3/water nanofluids as a coolant in the fin channel heat sink for an electronic device cooling. Mater Today Proc
12. ANSYS fluent theory guide, ANSYS, Inc., 2600 ANSYS Drive Canonsburg, PA 15317, Jan 2014
13. Chen T, Wang Y, Qi C, Chen L, Tang Z (2022) Effects of metal foam filling heights and tilt angles on the cooling performance of a heat sink filled with nanofluids. Int Commun Heat Mass Transf 138:106326

Enhancing Photoluminescence and Crystallinity Through Aluminum Doping in Sol–Gel Zinc Oxide Thin Films

Ibrahim Yaacoub Bouderbala and Imed-Eddine Bouras

Abstract This study investigates to see how aluminum (Al) doping affects the optical and photoluminescent properties of sol–gel zinc oxide (ZnO) thin films. The sol–gel method was used to produce Al-doped ZnO films, which were further investigated using X-ray diffraction, UV–VIS spectrophotometry, and photoluminescence studies. According to the findings, increased Al doping concentrations resulted in reduced crystallinity and a blue shift in the UV emission peak, which was ascribed to the Moss-Burstein effect. Al doping prohibited the visible emission band, but higher annealing temperatures enhanced film quality and the UV-to-visible emission ratio. This research looks at the link between Al doping, crystallinity, and photoluminescence in Al-doped ZnO thin films.

Keywords Aluminum doping · Zinc oxide thin films · Photoluminescence

1 Introduction

Zinc oxide (ZnO) has attracted considerable attention for its multifunctional utility in a wide range of applications, including antireflection coatings, transparent solar cell electrodes [1, 2], thin film gas sensors [3, 4], spintronic devices [5, 6], and light-emitting diodes [7, 8]. This attractiveness comes from its large band gap, exceptional chemical and thermal stability, and exceptional electrical and optoelectronic properties as an II-VI semiconductor with a high exciton binding energy. ZnO shows

I. Y. Bouderbala (✉)
Applied Optics Laboratory, Institute of Optics and Precision Mechanics, Ferhat Abbas University, Setif 1, Setif, Algeria
e-mail: ibrahimyaacoub.bouderbala@univ-setif.dz

I.-E. Bouras
National School of Biotechnology "Toufik Khaznadar" (ENSB) Constantine, University of Constantine 3, Ali Mendjeli, Algeria

S. Hamamda et al. (eds.), *Proceedings of the 2nd International Conference of Nanotechnology for Environmental Protection and Clean Energy Production*, Springer Proceedings in Materials 45, https://doi.org/10.1007/978-981-97-1916-7_8

tetrahedral coordination of Zn atoms with O atoms in its usual hexagonal configuration with values a = 3.25 and c = 5.12, resulting in hybridization of Zn d-electrons with O p-electrons [9].

Chemical doping has a significant impact on the electrical and optical properties of ZnO, making the control of defects essential to practical ZnO applications. Doped ZnO thin films have a wide range of applications, including transparent conducting electrodes [10], insulating or dielectric layers [11], and spintronic devices [4]. Notably, Al-doped ZnO (AZO) films have drawn interest as transparent conducting electrodes due to their excellent conductivity and optical transparency [10]. Intrinsic defects present in undoped ZnO, such as Zn vacancies, interstitial Zn, O vacancies, interstitial O, and antisite O (O_{Zn}) [12], contribute to acceptor or donor levels inside the band gap, altering ZnO's luminescence properties considerably [13]. When external dopants, such as Al, are introduced, the defect structure changes because Al atoms either replace for Zn atoms or occupy interstitial positions. As a result, understanding the effect of Al doping on the characteristics of AZO is essential.

The sol–gel method is used in this work because of its precise compositional control, molecular-level uniformity, and lower crystallization temperatures. Nacib et al. [14] and Sharmin et al. [15] have recently investigated the electrical and optical properties of c-axis-oriented sol–gel AZO films. We expand the investigation in this paper to investigate the optical and photoluminescent behavior of sol–gel ZnO films, taking into consideration the impacts of aluminum doping and subsequent annealing.

2 Experimental Details

2.1 Deposition of Thin Films by Sol–Gel Process

The basic starting material for the current investigation was zinc acetate 2-hydrate (99.5% purity, $Zn(CH_3COO)_2H_2O$). The solvent was absolute ethanol, and the sol stabilizer was diethanolamine ($HN(CH_2CH_2OH)_2$, DEA). Aluminum was used as a dopant, using $Al(NO_3)_39H_2O$ with 0.0025 and 0.0050 M. In a 100 mL conical flask, dissolve 0.5 M of zinc acetate 2-hydrate in 0.1 L of 100% ethanol. DEA was steadily added drop by drop during 30 min of vigorous stirring, keeping a constant DEA/Zn^{2+} molar ratio of 1. The solution became clear after about 20 min of stirring. Before being used for thin film coating, it was constantly swirled overnight.

The dip coating process was used to create thin AZO films around 300 nm thick on supporting substrates. Prior to deposition, rigorous cleaning was performed using distilled water, ethanol, and acetone, followed by distilled water washing, and finished with drying using a nitrogen gas gun. After immersing the cleaned substrates in the sol, they were removed at a rate of 1.12 mm/s, yielding homogeneous films. After every dip coating, the films were direct dried at 200 °C for 10 min. This method was done six times followed by an hour of annealing at 400 and 600 °C in an air environment.

2.2 *Characterization of Thin Films*

X-ray diffraction (XRD) was used to identify crystal phases. A Philips PW3710 diffractometer with CuK1 radiation at 40 kV and 45 mA was used to obtain XRD patterns. The optical properties of films were examined using a UV–VIS spectrophotometer (HP 8453) and normal incident transmittance measurements. A SPEX 750 M PL spectrometer was used to perform photoluminescence (PL) measurements. A 325 nm He-Cd laser with a power output of 4 mW at room temperature was used to excite the samples.

3 Results and Discussion

Figure 1 depicts the X-ray diffraction (XRD) patterns of AZO thin films after 1 h of annealing at 400 °C in an air environment. All samples are polycrystalline, having a hexagonal structure (JCPDS 36-1451), and distinct peaks observed within the reported range. The XRD investigation revealed no further crystalline phases. Figure 1 shows that when doping concentrations increase, the intensity of the diffraction peaks decreases but the Full Width at Half Maximum (FWHM) of these peaks increases. This pattern indicates a decrease in film crystallinity. Because of the fast diffusion of Zn interstitials, grain formation in ZnO is often enhanced at temperatures over 327 °C [16]. However, in the case of trivalent cation doping, the concentration of zinc interstitials is reduced for charge balance, resulting in slowed ZnO grain development and a concomitant drop in crystallinity [17]. This reduction in crystallinity indicates that Al has been integrated into the ZnO structure.

Figure 2 shows the optical transmission spectra of AZO films after 1 h of annealing at 400 °C in an air environment. Within the visible range, all films exhibit an average optical transparency of more than 80%. The oscillations seen in the spectra are caused mostly by interference between the thin film layers. These interferences fringes patterns are seen in the spectra of all samples, highlighting the film's excellence and homogeneity. They are caused by the flat surface of the films and the interaction with the FTO substrate. This implies that the film surfaces reflect with little scattering or absorption inside the film [18, 19]. Multiple reflections between the lower substrate-contacting surface and the layer's free surface cause interference fringes to form across all wavelengths. As a result, the transmission spectrum peaks and minima vary as a function of wavelength. The envelope approach makes it easier to determine the optical thickness and refractive index of a semiconductor by utilizing its characteristic transmission spectrum profile. The refractive index and thickness of substrate must be determined for this approach. A refractometer was used to measure the glass's

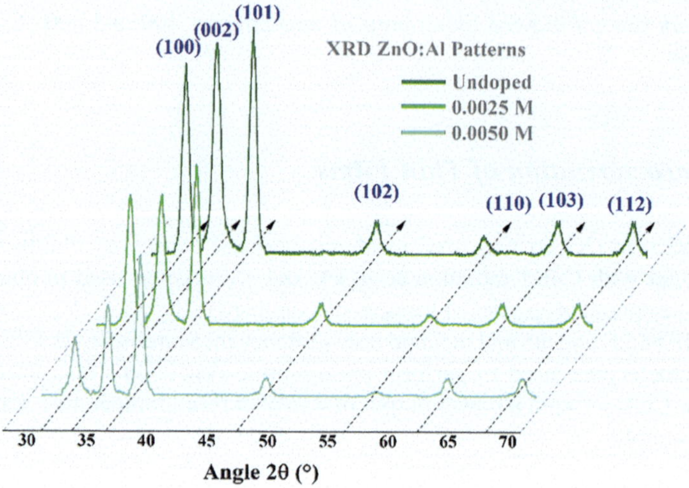

Fig. 1 XRD patterns of AZO films annealed at 400 °C with diffrent doping concentrations

refractive index, which was 1.52. With these characteristics determined, the indices and thicknesses of the ZnO nanostructures may be calculated.

The envelope approach is to trace envelop curves around the transmission curve, which connects the maxima (T_M) and minima (T_m) as continuous functions of wavelength. This method, based on the Swanepoel method, makes use of T_M and T_m arising from the interference fringes. The approach is particularly successful in calculating the refractive index and thickness in the low-absorption and transparent portion of

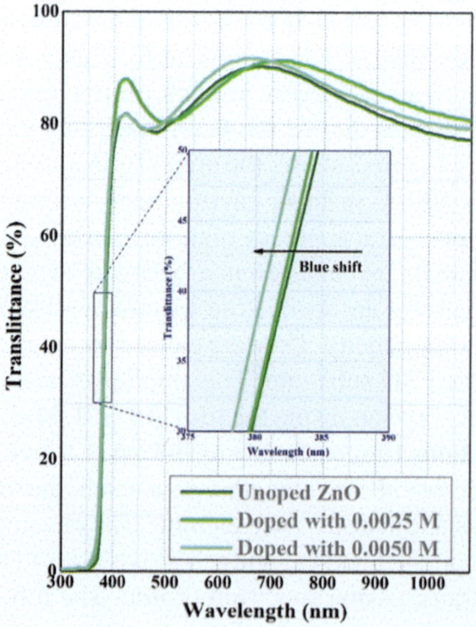

Fig. 2 Transmittance spectra of AZO with different Al concentration

the transmission spectrum. The thickness is determined from the following equation [20]:

$$d = \frac{\lambda_1 \lambda_2}{2(\lambda_1 n_2 - \lambda_2 n_1)} \tag{1}$$

n_1 and n_2 are the refractive indices of the layer for the wavelengths λ_1, λ_2 and are taken from the relationship:

$$n = \left[N + \left(N^2 - n_S\right)^{\frac{1}{2}}\right]^{\frac{1}{2}} \tag{2}$$

In addition, N can be calculated by the relation:

$$N = \frac{2n_S(T_M - T_m)}{T_M T_m} + \frac{n_S^2 + 1}{2} \tag{3}$$

With n_S is the index of the substrate and T_M and T_m are the transmission values for the upper and lower envelope respectively for a chosen wavelength. The basic equation for interference fringes is given by [20]:

$$2d = m\lambda \tag{4}$$

where, the order numbers *m* is integer for maxima and half integer for minima.

The results shown in Table 1 show that there is a notable tendency of increased thickness with increasing Al atom concentrations. This increase in the thickness correlates to a decrease in transmission, as shown in Fig. 2. The increase of Al-doped ZnO thin films can be attributed to the formation of atoms' clusters as the aluminum is incorporated in the ZnO lattice during the sol–gel formation [21–23].

From UV spectra, there is a visible blue shift in the UV absorption edge as the Al doping concentration increases, as shown in the inset of Fig. 2. This alteration indicates that the optical band gap (Eg) is increasing. ZnO has a direct band gap in its wurtzite structure, and the absorption edge associated with a direct interband transition is described by the Tauc's formula [24].

Figure 3 shows the $(h\nu)^2$ plotted versus photon energy ($\alpha h\nu$) for all samples. The linear relationship found at higher photon energy implies that the AZO films are fundamentally direct-transition-type semiconductors. The optical band gap is obtained by extrapolating the linear component of the curve to zero. Figure 3 shows

Table 1 Refraction index and thickness of the ZnO estimated by the envelope method

	Refraction index	Thickness (nm)
Undoped	1.852	275.14
Doped with 0.0025 M	1.873	281.38
Doped with 0.0050 M	1.906	289.47

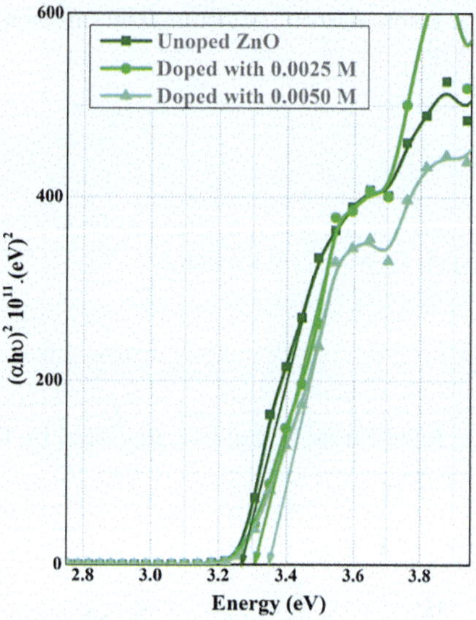

Fig. 3 Tauc's plot of AZO with different Al concentration

that E_g is 3.27, 3.31, and 3.34 eV for AZO thin films with Al doping concentrations of 0, 0.0025, and 0.0050 M.

The Moss-Burstein shift [25] refers to the increasing of the band gap. According to the Moss-Burstein hypothesis, donor electrons occupy states near the bottom of the conduction band in strongly doped zinc oxide films. Valence electrons require greater energy to move to higher energy levels inside the conduction band due to the Pauli exclusion principle and the vertical character of optical transitions. As a result, the optical band gap of doped zinc oxide films is larger than that of undoped zinc oxide films.

Figure 4 displays the Photoluminescence (PL) spectra of AZO films annealed at temperatures ranging of 400 and 600 °C. Two emission bands exist: a strong and narrow UV luminescence about 382 nm and a wide visible emission ranging from 450 to 640 nm, resulting in the green-yellow region [26]. Temperature and Al doping concentration have an effect on these emission bands. Notably, when Al doping increases, the UV emission almost blue-shifts and deteriorates, whereas the visible emission is inhibited. Elevating the annealing temperature improves UV emission and raises the UV-to-visible emission ratio, resulting in improved film quality.

Free exciton emission is responsible for the UV emission peak at 382 nm. The blue shift in the UV peak with increased Al doping corresponds to the trends found in the transmittance spectra, which are caused by the Moss-Burstein shift, which increases the band gap. The prior XRD data indicate that the attenuation of UV emission is potentially connected to poor film crystallinity produced by the addition of Al. The reduction in structural flaws is responsible for the increase in the UV-to-visible emission ratio.

This result clearly suggests that the integration of Al as dopant with different concentration caused a compromise in the crystalline structure of the all the films,

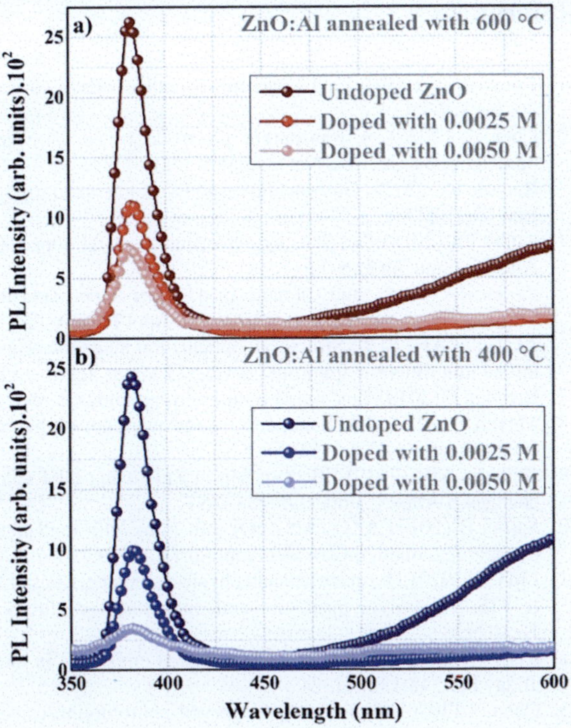

Fig. 4 PL spectra of AZO with different Al concentration **a** annealing at 400 °C and **b** annealing at 600 °C

which is consistent with the observed XRD. As a result, our results present a complete view of the effect of Al doping on the photoluminescent behavior of Al-doped ZnO thin films, revealing a link between variations in crystallinity, PL intensity, and Al concentration [27].

4 Conclusion

Finally, our study gives information on the complex relationship of aluminum doping, crystallinity, and photoluminescence characteristics in sol–gel zinc oxide thin films. The results of the study highlight the impact of Al doping concentrations on film quality, with greater doping resulting in impaired crystallinity and changed emission characteristics. The connection between doping and optoelectronic behavior is demonstrated by the blue shift in the UV emission peak, which is connected to the Moss-Burstein shift owing to increased band gap. Furthermore, increasing UV emission and the UV-to-visible emission ratio through higher annealing temperatures reveals an approach to better film quality. These findings add to our understanding of doped semiconductor materials and their prospective uses in optoelectronic devices.

References

1. Consonni V (2019) ZnO nanowires for solar cells: a comprehensive review. Nanotechnology 30(36):362001
2. Chen Z (2022) A transparent electrode based on solution-processed ZnO for organic optoelectronic devices. Nat Commun 13(1):4387
3. Ramola RC (2022) Gas sensing response of ion beam irradiated Ga-doped ZnO thin films. Sci Rep 12(1):22351
4. Kolhe PS (2018) Gas sensing performance of Al doped ZnO thin film for H_2S detection. J Alloy Compd 74(8):6–11
5. Zargar RA (2018) Study of nanosized copper-doped ZnO dilute magnetic semiconductor thick films for spintronic device applications. Appl Phys A 124:1–9
6. AL Naim AF (2020) Structural, optical, and magnetic properties of Co-doped ZnO nanocrystalline thin films for spintronic devices. J Mater Sci Mater Electron 31:3613–3621
7. Rahman F (2019) Zinc oxide light-emitting diodes: a review. Opt Eng 58(1):010901
8. Moyen E (2020) ZnO nanoparticles for quantum-dot-based light-emitting diodes. ACS Appl Nano Mater 3(6):5203–5211
9. Gallegos MV (2019) Effect of Mn in ZnO using DFT calculations: magnetic and electronic changes. J Alloy Compd 795:254–260
10. Sarma B (2019) AZO (Al: ZnO) thin films with high figure of merit as stable indium free transparent conducting oxide. Appl Surf Sci 479:786–795
11. Lin CF (2020) The electrical and physical characteristics of Mg-doped ZnO sensing membrane in EIS (electrolyte–insulator–semiconductor) for glucose sensing applications. Res Phys 16:102976
12. Rai H (2022) A review on defect related emissions in undoped ZnO nanostructures. Mater Today Proc 48:1320–1324
13. Chen X (2020) Significantly improved photoluminescence properties of ZnO thin films by lithium doping. Ceram Int 46(2):2309–2316
14. Necib K (2018) Investigation of the effects of thickness on physical properties of AZO sol-gel films for photonic device applications. J Alloy Compd 735:2236–2246
15. Sharmin A (2019) Depositions and characterization of sol–gel processed Al-doped ZnO (AZO) as transparent conducting oxide (TCO) for solar cell application. J Theor Appl Phys 13:123–132
16. Guan W (2017) Theoretical and experimental investigations of the thermoelectric properties of Al-, Bi-and Sn-doped ZnO. Mater Sci Semicond Process 66:247–252
17. Fujihara S (2003) Ga-doping effects on electrical and luminescent properties of ZnO:(La, Eu) OF red phosphor thin films. J Appl Phys 94(4):2411–2416
18. Rahman F (2023) Natural light-trapping nanostructures on thermally-grown cupric oxide. Mater Res Bull 159:112094
19. Tecaru A (2010) Zinc oxide thin films prepared by spray pyrolysis. J Optoelectron Adv Mater 12:1889–1893
20. Bouderbala IY (2018) Optical properties of Cu_2O electrodeposited on FTO substrates: effects of Cl concentration. J Electron Mater 47:2000–2008
21. Oubaha M (2012) Optical properties of high refractive index thin films processed at low-temperature. Opt Mater 34(8):1366–1370
22. Goktas A (2017) Tuning of structural, optical and dielectric constants by various transition metal doping in ZnO: TM (TM= Mn Co, Fe) nanostructured thin films: a comparative study. Ceram Int 43(1):704–713
23. Mohamed SH (2010) Photocatalytic, optical and electrical properties of copper-doped zinc sulfide thin films. J Phys D Appl Phys 43(3):035406
24. Bouderbala IY (2018) Optical, structural, and photoelectrochemical properties of nanostructured Cl-doped Cu_2O via electrochemical deposition. Solid State Sci 83:161–170
25. Mayekar J (2022) A review on the synthesis of doped and undoped ZnO thin films and understanding effect of doping on the band gap of doped zinc oxide thin films on the basis of Burstein-moss effect. Int J Res Analyt Rev 9(4):971–974

26. Bouderbala IY, Guessoum A, Rabhi S, Bouhlassa O, Bouras IE (2024) Optical band-diagram Urbach energy tails associated with photoluminescence emission in defected ZnO thin films deposited by sol–gel process dip-coating: effect of precursor concentration. Applied Physics A 130(3). https://doi.org/10.1007/s00339-024-07366-1
27. Nian Q (2014) Ultraviolet laser crystallized ZnO: Al films on sapphire with high hall mobility for simultaneous enhancement of conductivity and transparency. Appl Phys Lett 104(20):201907

Structural Characterization of Chorine Doped Cuprous Oxide Thin Films by X-Ray Diffraction and X-Ray Fluorescence

Imed-Eddine Bouras and Ibrahim Yaacoub Bouderbala

Abstract This study investigates the structural changes in Cl-doped Cu_2O thin films through X-ray diffraction (XRD). The incorporation of Cl^- ions into the Cu_2O lattice is analyzed at varying concentrations. XRD patterns reveal shifts in the (111) peak due to Cl^- ion interaction, indicative of interstitial and substitutional incorporation. The lattice parameter and interatomic distance are compared between theoretical and experimental values, reflecting volume changes caused by Cl^- ion presence. At lower concentrations, interstitial incorporation contracts the lattice, while higher concentrations lead to lattice expansion through substitutional incorporation. These findings provide insights into the structural modifications induced by Cl^- doping, crucial for understanding the conductive properties of the thin films.

Keywords Cu_2O thin films · Cl^- doping · Structural modification

1 Introduction

Extensive study into the creation and manipulation of nanostructures with special electronic properties has been stimulated by the quest for effective and adaptable semiconductor materials [1, 2]. Cuprous oxide (Cu_2O) nanostructures, one of these, have drawn a lot of interest because of their fascinating properties and prospective uses in optoelectronics, catalysis, and energy storage devices [3, 4]. A possible way to modify Cu_2O nanostructures' characteristics for particular purposes is by controlled electrodeposition [5]. In order to understand the role of doping and their impact on

I.-E. Bouras (✉)
National School of Biotechnology "Toufik Khaznadar" (ENSB) Constantine, University of Constantine 3, Ali Mendjeli, Algeria
e-mail: bouras-imededdine@ensbiotech.edu.dz

I. Y. Bouderbala
Applied Optics Laboratory, Institute of Optics and Precision Mechanics, Ferhat Abbas University, Setif 1, Setif, Algeria

S. Hamamda et al. (eds.), *Proceedings of the 2nd International Conference of Nanotechnology for Environmental Protection and Clean Energy Production*, Springer Proceedings in Materials 45, https://doi.org/10.1007/978-981-97-1916-7_9

the final properties, this research explores the complex interplay of electrochemical processes and crystal lattice alterations during the electrodeposition of Cu_2O films.

Due to their abundance, nontoxicity, and low cost, cuprous oxide (Cu_2O) thin films with high oriented crystalline structure along (111) plane are extensively researched for practical applications [6, 7]. Because of the work of Kennard et al., [8] Cu_2O is the first known material to exhibit p-type semiconductor behavior since 1917. Cu_2O is one of the most promising metal oxides for the production of optoelectronic devices like solar cells because of its straight band gap of 2 eV [9]. Sol–gel approach [10], chemical vapor deposition (CVD) [11, 12], oxidation of metallic copper [13], chemical bath deposition [14], and electrochemical deposition [15–19] are a few techniques that have been utilized to synthesize Cu_2O nanostructures.

The incorporation of doping agents, specifically chloride ions (Cl^-), into the Cu_2O lattice, is a significant aspect of this work. Dopants can significantly change a material's electronic structure, resulting in changes to the conductivity type and structural characteristics [20]. This study attempts to elucidate the mechanisms underlying the change from p-type to n-type conductivity in Cu_2O thin films by closely examining the impact of Cl^- ions on the crystal lattice [20].

To customize semiconductor structures' characteristics for particular applications, dopant ions are crucially important. Cu_2O has drawn attention for doping investigations because of its potential for optoelectronic and energy applications. In this situation, X-ray diffraction (XRD) appears as an effective method to explain the structural alterations brought on by dopants. Particularly, the influence of Cl^- ions on conductivity and optoelectronic behavior has drawn attention to the inclusion of Cu_2O thin films with Cl^- ions. By examining the crystalline lattice and locating shifts in diffraction peaks, XRD enables us to learn more about the locations and interactions of the dopant ions.

This study uses XRD analysis to look at the microstructural development of Cl^- doped Cu_2O thin films. Cl^- ion concentration-dependent variations in the (111) peak provide information about the incorporation processes, including whether substitutional or interstitial sites are involved. To measure the structural changes brought on by Cl^- doping, theoretical values for the lattice parameter and interatomic distances will be compared with experimental data. Optimizing the characteristics of Cu_2O thin films for varied purposes requires a thorough understanding of these structural alterations.

2 Experimental Details

Cu_2O films were prepared using electrochemical cell with three electrodes. A platinum electrode and a saturated calomel electrode (SCE) were used as counter electrode and reference electrode respectively. A fluorine tin oxide (FTO)-coated conducting glass substrate was used as working electrode. The substrates were cleaned ultrasonically in acetone and methanol for 20 min, then rinsed in distilled water and dried naturally. Different concentrations of sodium chloride (NaCl) varying from 0.002 to 0.010 M were dissolved in distilled water with 0.1 M of copper nitrate ($CuNO_3$), after that, 1 M of lactic acid ($C_3H_6O_3$) was added to all electrolytes as a complexing agent to prevent the precipitation of copper when adding NaOH to the solution to adjust the pH to 9.5. The temperature and time deposition were maintained at 70 °C and 25 min respectively.

Using a Rigaku Ultima IV and JEOL JDX 3530 diffractometer (40 mA, 45 kV) with CuK1 radiation ($\lambda = 0.15406$ nm) in a Bragg–Brentano geometry (2θ) between 27° and 75°, the structures of Cu_2O thin films were examined. It has become imperative to investigate how Cl^- ions are incorporated into the Cu_2O crystal lattice. X-ray fluorescence (XRF) investigations were carried out for this purpose utilizing a JEOL—JSX-3201 M spectrometer (150 kV, 1 mm diameter collimator).

3 Results and Discussion

A voltametric investigation is required to determine the potentials of the current oxidation–reduction reactions in the electrolyte and to select the proper potential to deposit Cu_2O by chronoamperometry in order to explore the Cu_2O nanostructures electrodeposition process. By supplying a constant voltage between the FTO and the counter electrode referred to a SCE, Cu_2O nanostructures are electrodeposited.

Several publications [21–23] have anticipated electrochemical reaction to produce CuCl in the presence of Cl^- ions in the deposition bath. Assuming that Cl^- doped Cu_2O was formed by the insertion of Cl^- into the Cu_2O lattice, the low solubility of CuCl suggests that even small concentrations of Cu^+ and Cl^- in the solution will react and precipitate. Since chlorine atoms have a radius greater than the oxygen, chlorine atoms positioned in an interstitial position act as donors. In the case that chlorine is positioned in substitutional site, they will act as acceptors. It is well known that the copper and oxygen sites in cause doping, which results in p and n doping, respectively [24].

We used photocurrent measurements to identify the different semiconductors' conductivity types. Only a system made up of a light source, a lighting switch, a three-electrode cell containing 0.8 M Na_2SO_4, and a potentiostat may be used to perform this approach. We see cathodic current generated by some films, one of which is undoped and the others two doped with 0.002 and 0.004 M of chloride ions as shown on Fig. 1a, indicating that the depletion zone is negatively charged. As a result, in the presence of light, the holes flow in the direction of the semiconductor layer and the electrons in the direction of the electrolyte. This demonstrates clearly that these semi-conductors exhibit p-type conductivities. The depletion zone of the other films, on the other hand, produces an anodic current as we can see on Fig. 1b, which is positively charged. In this case, n-type conduction is clearly obtained with increasing the Cl^- concentration because holes go toward the electrolyte while electrons move toward the layer.

Figure 2a demonstrates how the Cu/O atomic ratio changes depending on the amount of dopant present as well as the proportion of Cl^- ions included in the

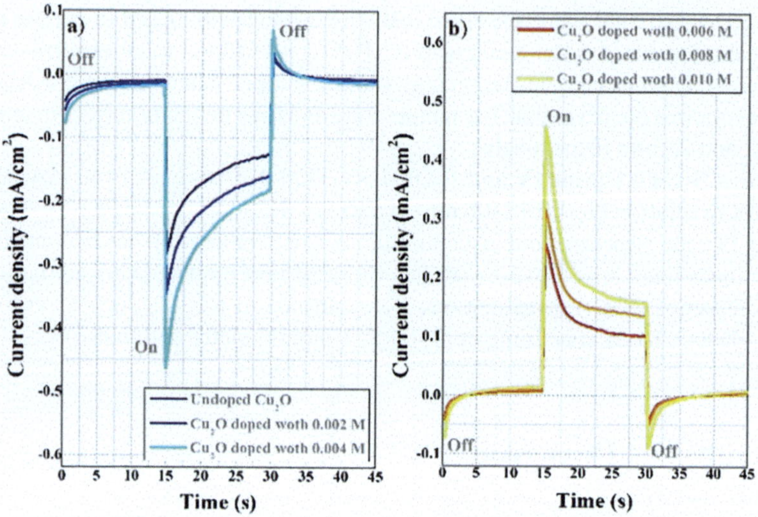

Fig. 1 Photocurrent measurements for undoped and doped Cu_2O nanostructures with different Cl^- concentration **a** Cu_2O-p, **b** Cu_2O-n

crystalline lattice. These findings show that copper and oxygen atoms are abundant in all thin layers of Cu_2O and that the concentration of Cl^- ions in the deposition bath affects how quickly Cl^- ions are incorporated into the crystal structure.

Additionally, according to the JCPDS Sheet No. 00-005-0667, XRF investigation showed that the Cu/O atomic ratio of the two type-p thin layers is less than the theoretical ratio of Cu_2O, which is equal to 2. This suggests that there are copper vacancies (VCu) in the crystal lattice, which explains why the conduction is p-type. However, compared to the predicted ratio, the atomic ratio of the other thin layers

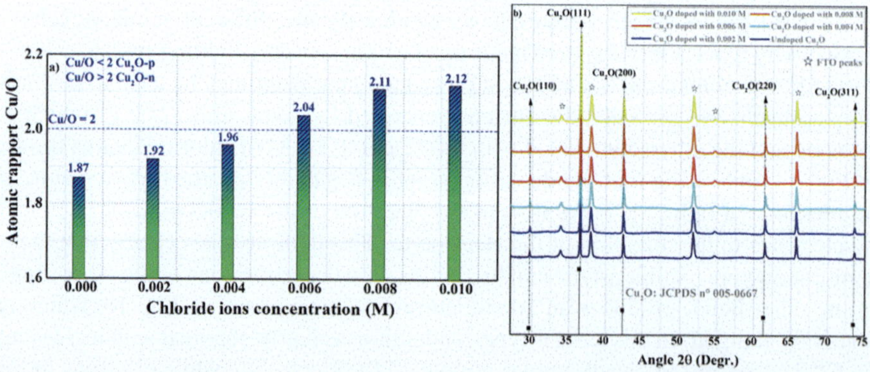

Fig. 2 **a** Copper/oxygen atomic ratio and rate of incorporated chloride ions in the crystal lattice of Cu_2O, **b** XRD diffraction patterns of thin Cu_2O layers prepared at different chloride ions concentrations

is higher. This explains why there is a change in conduction when Cl^- ions are incorporated into substitutional sites (O^{2-}).

X-ray diffraction (XRD) investigations were carried out to identify existing phases and analyze the microstructural development of the thin Cu_2O layers. Figure 2b displays the X-ray diffraction patterns of doped and undoped Cu_2O thin layers produced by electrodeposition at E = – 0.3 V/ECS. The tetragonal phase of the FTO substrate is indicated by the peaks denoted by stars.

The (111) orientation crystallizing in the cubic phase is indicated by the existence of a Cu_2O peak at 36.5° (JCPDS No. 00-005-0667), according to XRD studies. Cu_2O thin layers grow perpendicular to the (111) plane since all the other peaks show a small intensity comparing to the (111), and no other peaks representing other phases or impurities were found, which mean that our film are pure Cu_2O. Han et al. [25] discovered the same outcome when they placed thin layers of Cl-doped Cu_2O on a gold (Au) substrate; they only got one Cu_2O peak that corresponded to the (111) orientation. The crystalline planes' surface energies, which control crystal development along these planes, are the cause for that.

Calculating the texture coefficient $T_C(hkl)$ given by the following calculation [16] is necessary to confirm the Cu_2O thin films' preferred orientation:

$$T_C(hkl) = n\frac{I_m(hkl)/I_0(hkl)}{\sum_1^n I_m(hkl)/I_0(hkl)} \tag{1}$$

where n is the total number of diffraction peaks taken into account in the evaluation (5 in our case), I_m is the measured relative intensity of the peak corresponding to the hkl diffraction, I_0 is the relative intensity from the identical diffraction in the standard powder sample. Figure 3 illustrates how the concentration of chloride ions changes the textural coefficient of Cu_2O thin films along different crystallographic planes. While the texture coefficient along the other planes nearly stabilizes, it increases with an increase in Cl^- concentration for the major peak orientation of (111), which is higher than that for the other planes. The (111) peak intensifies as Cl^- increases, which is related to an increase in the number of grains along that plane and suggests improved crystallinity.

A decrease in full width at half maximum (FWHM) is seen when comparing the deposited samples, which causes an increase in crystallite size and the creation of microstructural changes in Cu_2O nanostructures. By applying the traditional Scherrer's relation [20, 26] to the (111) peaks, the average crystallite size was determined:

$$D = \frac{0.9\lambda}{\beta\cos\theta} \tag{2}$$

where θ is the diffraction angle, β is the full width at half maximum (FWHM) given in radians, and λ is the X-ray beam's wavelength (λ = 1.5406). For the undoped Cu_2O films and doped with 0.002, 0.004, 0.006, 0.008, and 0.01 M of Cl^- concentrations, the crystallite size (D) was determined to be 70.14, 73.42, 77.89, 84.57, 89.15, and

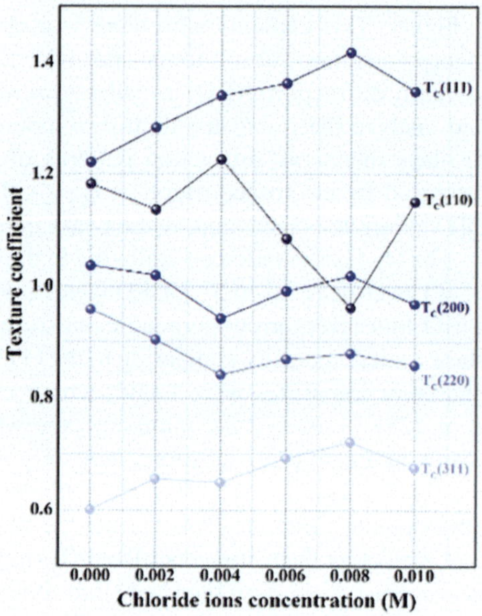

Fig. 3 Texture coefficient of all diffranciton peak for all Cu_2O thin film undoped and doped with different concentration of chloride ions

90.11 nm, respectively. The crystallite size increase with the chloride ions indication the successful of incorporation of chloride ions in the structure of cuprous oxide by changing the structural properties.

The relative surface energies for a cubic structure like Cu_2O are 111, 100, and 110. Bravais' law [27] and the Gibbs-Wulff growth theory indicate that faces with greater surface energies will crystallize more quickly, reducing their surface area. Lower surface energy faces will develop more slowly, increasing their surface area. Cu_2O thin layers will grow perpendicular to the (111) face in our situation because it has a lower surface energy due to the intensity of the (111) peak compared to other orientations. The (110) and (100) faces will disappear progressively due of the quickest crystal development, while the (111) will gradually appear because of the slower growth rate.

Depending on the deposition conditions, thin layers of Cu_2O with the (111) orientation will be easier to create than layers with other orientations (or crystalline planes). The (111) peak becomes more intense for a concentration of 0.008 M compared to other depositions, showing good crystallinity. As the concentration of Cl^- ions rises, it can be seen from the diffraction patterns that the FWHM of the (111) peaks decreases, suggesting an increase in the average size of the crystallites and creating microstructural changes in the crystalline lattice. Dislocations and/or stresses, which are present in these alterations, are reduced as the average size of the crystallites rises.

Cu^+ ions will combine with Cl^- ions to create CuCl during the deposition of thin layers of Cu_2O at varied Cl^- ion concentrations, shifting the (111) peak for the various Cl^- doped Cu_2O layers. The non-dimensionalized curves of the (111) peak

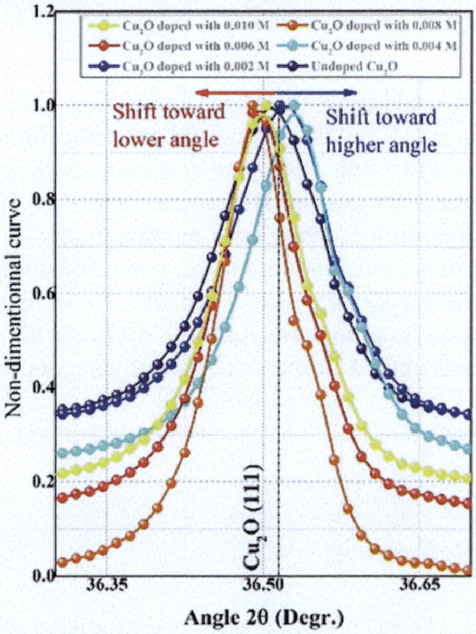

Fig. 4 Non-dimentionnal curve of Cu_2O (111) peak of all films

are shown in Fig. 4. Because most Cl^- ions are in interstitial sites, a level of acceptor is produced above the valence band, as shown by the peak shift of the Cu_2O doped with 0.002 and 0.004 M towards higher angles. This explains p-type conductivity [28] quite effectively.

The peak shift is seen when the Cl ions in the deposition bath is increased; this time, the peak shift is towards lower angles, suggesting that the majority of Cl^- are incorporated in substitutional oxygen sites, resulting in a switch from p-type to n-type conductivity. As a result, a donor level below the conduction band is created [29].

When comparing the theoretical values of the lattice parameter (a = 4.269 Å) and the interatomic distance (d_{hkl}) with the experimental values of Cu_2O ($a_{undoped}$ = 4.2574 Å, $a_{0.002}$ = 4.2562 Å, $a_{0.004}$ = 4.2557 Å, $a_{0.006}$ = 4.2596 Å, $a_{0.008}$ = 4.2599 Å, $a_{0.010}$ = 4.261 Å), we notice a slight difference, which is attributed to the constraints associated with the deposition method. The incorporation of Cl^- ions into the Cu_2O lattice doped with 0.002 and 0.004 M chlorine ions led to a decrease in the lattice parameter, resulting in a contraction of the volume of the Cu_2O lattice. This strongly indicates the insertion of Cl^- ions into the interstitial sites of the Cu_2O lattice. As the concentration of Cl^- ions in the deposition bath increases, we observe an increase in the lattice parameter, leading to lattice expansion and an increase in the volume of the Cu_2O lattice. This confirms the incorporation of Cl^- ions into the substitutional oxygen sites.

4 Conclusion

In conclusion, XRD has provided valuable insights into the structural changes induced by Cl^- doping in Cu_2O thin films. The shifts in the (111) peak position as a function of Cl^- ion concentration clearly demonstrate the different incorporation mechanisms. The lattice parameter variations and interatomic distances further confirm these structural modifications. At low concentrations, interstitial incorporation contracts the lattice, while substitutional incorporation at higher concentrations leads to lattice expansion. These findings highlight the role of Cl^- in altering the crystalline structure of Cu_2O thin films, which in turn impacts their electrical and optical properties. This study contributes to the fundamental understanding of dopant-induced structural changes and their implications for Cu_2O's potential applications in optoelectronics and energy devices.

References

1. Smith J (2018) Advances in semiconductor nanostructures. Annu Rev Mater Sci 45(2):87–101
2. Anderson K (2019) Semiconductors in modern electronics. Semicond Sci Technol 34(8):83001
3. Li Y (2019) Emerging applications of copper oxide nanostructures. Nano Rev 150:456–467
4. Brown A (2017) Catalytic properties of Cu_2O nanostructures. Catal Today 42(6):2165–2172
5. Johnson R (2022) Electrodeposition techniques for nanostructure synthesis. J Electroanal Chem 123(4):245–259
6. Sun S (2018) Cuprous oxide (Cu_2O) crystals with tailored architectures: a comprehensive review on synthesis, fundamental properties, functional modifications and applications. Prog Mater Sci 96:111–173
7. McShane CM (2012) Junction studies on electrochemically fabricated p–n Cu_2O homojunction solar cells for efficiency enhancement. Phys Chem Chem Phys 14(17):6112–6118
8. Kennard EH (1917) An effect of light upon the contact potential of selenium and cuprous oxide. Phys Rev 9(1):58–63
9. Elfadill NG (2016) Preparation of p-type Na-doped Cu_2O by electrodeposition for a p-n homojunction thin film solar cell. Semicond Sci Technol 31(6):065001
10. Kumar N (2021) Structural and optical properties of sol–gel derived CuO and Cu_2O nanoparticles. Mater Today Proc 41:237–241
11. Gupta N (2013) Deposition and characterization of nanostructured Cu_2O thin-film for potential photovoltaic applications. J Mater Res 28:1740–1746
12. Ahirrao PB (2011) Nanocrystalline p-type-cuprous oxide thin films by room temperature chemical bath deposition method. J Alloy Compd 509(18):5551–5554
13. Zhao W (2011) Electrodeposition of Cu_2O films and their photoelectrochemical properties. CrystEngComm 13(8):2871–2877
14. Bijani S (2009) Low-temperature electrodeposition of Cu_2O thin films: modulation of micro-nanostructure by modifying the applied potential and electrolytic bath pH. J Phys Chem C 113:19482–19487
15. Herbadji A (2019) The concentration effect of complexing agent on the morphology and optoelectronic properties of electrochemically deposited n-type Cu_2O thin films. J Electron Mater 48:4830–4839
16. Herbadji A (2019) Effect of copper sulfate concentration on the electrochemical nucleation process, growth and properties of n-type Cu_2O thin films. Russ J Electrochem 55:1336–1349

17. Laidoudi S (2013) Growth and characterization of electrodeposited Cu_2O thin films. Semicond Sci Technol 28(11):115005
18. Jrajri K (2022) Electrodeposition of Cu_2O thin film onto copper substrate by linear sweep voltammetry at low duration: effect of bath pH. Biointerface Res Appl Chem 12(6):7715–7724
19. Budi S (2022) Structural analysis of nanocrystalline Cu_2O prepared by electrodeposition at different temperatures. Int J Electrochem Sci 17(3):22033
20. Bouderbala IY (2018) Optical properties of Cu_2O electrodeposited on FTO substrates: effects of Cl concentration. J Electron Mater 47:2000–2008
21. Zhan C (2020) Critical roles of doping Cl on Cu_2O nanocrystals for direct epoxidation of propylene by molecular oxygen. J Am Chem Soc 142(33):14134–14141
22. Yu L (2019) Visible-light driven CO_2 reduction coupled with water oxidation on Cl-doped Cu_2O nanorods. Nano Energy 60:576–582
23. Nie J (2022) Energy band reconstruction mechanism of Cl-doped Cu_2O and photocatalytic degradation pathway for levofloxacin. J Clean Prod 363:132593
24. Bouderbala IY (2018) Optical, structural, and photoelectrochemical properties of nanostructured Cl-doped Cu_2O via electrochemical deposition. Solid State Sci 83:161–170
25. Han X (2009) N-Type Cu_2O by electrochemical doping with Cl. Electrochem Solid-State Lett 12(4):H89
26. Bouderbala IY, Guessoum A, Rabhi S, Bouhlassa O, Bouras IE (2024) Optical band-diagram Urbach energy tails associated with photoluminescence emission in defected ZnO thin films deposited by sol–gel process dip-coating: effect of precursor concentration. Appl Phys A 130(3). https://doi.org/10.1007/s00339-024-07366-1
27. Xue J (2012) Crystallization behavior and formation mechanism of dendrite Cu_2O crystals. CrystEngComm 14(23):8017
28. Musa A (1998) Production of cuprous oxide, a solar cell material, by thermal oxidation and a study of its physical and electrical properties. Sol Energy Mater Sol Cells 51(3–4):305–316
29. Biccari F (2010) Chlorine doping of Cu_2O. Sol Energy Mater Sol Cells 94:1947–1952

Experimental Study on the Effect of Drying on the Kinetics of Cubic Beetroot Slices Using an Industrial Electric Food Dehydrator

Soufounizia Boultif, Noureddine Belghar, and Foued Chabane

Abstract Fresh beetroots need appropriate storage methods due to their high moisture content, since failure to do so would result in their decay and subsequent disposal. Drying is often regarded as one of the most effective methods of preservation. This study investigates the impact of using an industrial electric food dehydrator on beetroot slices of varying thicknesses, namely 3, 4, and 5 mm. When subjected to an airflow at a temperature of 50 °C for 225 min, the initial weights of the objects were 30 grammes, 31 grammes, and 45 grammes, respectively. The final results indicated a significant reduction in the importance of the beets, with measurements of 0.5 g, 0.1 g, and 0.6 g, respectively, which suggests that the water removal rate was fast. Furthermore, the beetroot transformed its physical structure, adopting a chip-like shape while retaining its original colour. The electric food dehydrator is efficacious in addressing all three aspects mentioned.

Keywords Drying · Beetroot · Kinetics

1 Introduction

Red beetroots are essential vegetables due to their nutritional and physiological benefits. Its high magnesium and potassium content and low sodium concentrations improve human ionic equilibrium. Betacyanins reduce oxidative stress and free

S. Boultif (✉) · F. Chabane
Laboratoire de Génie Mécanique (LGM), Faculty of Science and Technology, University of Biskra, 07000 Biskra, Algeria
e-mail: Soufounizia.boultif@univ-biskra.dz

F. Chabane
e-mail: f.chabane@univ-biskra.dz

N. Belghar
Laboratoire Génie énergetique et matériaux LGEM, Faculty of Science and Technology, University of Biskra, 07000 Biskra, Algeria
e-mail: n.belghar@univ-biskra.dz

S. Hamamda et al. (eds.), *Proceedings of the 2nd International Conference of Nanotechnology for Environmental Protection and Clean Energy Production*, Springer Proceedings in Materials 45, https://doi.org/10.1007/978-981-97-1916-7_10

radical damage. They combat viruses and germs, prevent cancer, and prevent heart disease [1].

Besides its many active components, beetroot is a superfood with many health advantages. Antioxidants in beetroots inhibit tumour growth and kill malignant cells. Evidence suggests it is most beneficial for prostate, colon, and breast cancer, but it also promotes pancreatic cancer recovery [2].

Due to its high fibre and saponin content, beetroot cleanses the intestines and blood and aids in natural detoxification. Optimising nutrition absorption requires improved digestion and regular, full waste evacuation. Beets help reduce deficiencies-related illnesses. It optimises liver function and treats cirrhosis, hepatitis, and other liver diseases. Beetroot reduces digestive tract inflammation and relieves discomfort [3]. Drying reduces water activity to preserve this superfood, preserve its nutritious worth, and prevent bacterial and fungal contamination: Drying involves evaporating water to transfer bulk. In recent years, it has been used to preserve food by eliminating water. As health-conscious people learn more about natural products, beets (Beta vulgaris) have gained popularity as a functional food. Numerous research have examined this topic, including Shekhar Pandharipande and Bhushan Bele's artificial neural network study on sodium chloride content, time, and temperature's effects on beetroot slice drying and weight loss. The findings show that salt solution osmotic dehydration can dewater beetroot slices [4].

In their paper, Kulwinder Kaur and A. K. Singh explain their beetroot drying and quality trials with air temperature and microwave power. Osmotically prepared materials were dehydrated using a convection Tray. High air temperature (75 °C) and microwave power (1080 W) increased drying rates and lowered drying time [5].

Whereas Adam Figiel et al. made 18-mm-diameter beetroot slices with variable thicknesses in 40Bx chokeberry juice at 50 °C for two hours before vacuum-microwave (VM) drying at various microwave power levels. This increased its bioactivity. Increased sample specific surface area led to shorter drying periods, lower material temperatures, and better dried product quality [6].

Bhupinder Singh and Bahadur Singh Hathan's work aimed to enhance the beetroot juice spray drying procedure. Using the response surface method, many parameters were checked. The models were shown to be correct when the experimental data matched the expected values [7].

Recently, some new methods have been looked into. For example, Vallespir et al. studied how freezing before drying and ultrasonic waves affected beetroot microstructure, drying curves, and bioactive compounds. Results showed that both methods improved drying mass transfer [8].

A cutting-edge hybrid dryer was employed for drying trials. However, Dominik Mierzwa et al. used high-power airborne ultrasound, microwaves, and hot air to study how drying red beets in short bursts affected process kinetics and quality indicators. Hybrid drying periodically improves drying speed and output quality while lowering drying duration and power usage, the result being a more porous, pleasant-colored, and crispy vegetable product [9]. In another work by Atul Dhiman et al., Besides freeze-drying and spray-drying, they discuss various beetroot treatments. It showed beetroot can form stable microcapsules. By making beetroot more tasty, consumers can get more nutrition from these products [10]. SantanuMalakar et al. also tested an evacuated tube sun dryer (ETSD) with and without beetroot slices. ETSD and sun drying were used to compare beetroot slice drying kinetics, mass transfer phenomena, colour changes, and bioactive component preservation. ETSD dried many food commodities quickly and retained quality attributes [11]. The goal of the study by Yan Liu et al. was to evaluate the effects of hot air drying beetroot slices at 50–100 °C on their physicochemical and drying properties. Raising temperatures may cut drying time from 12.5 to 4.5 h. Beetroot slices dried at 60 °C showed the last colour variation [12].

To determine the features of beetroot dried in a cabinet Tray dryer at 600 °C with pretreatment, Vipul Chaudhary and Vivek Kumar conducted an experimental investigation. The findings showed that the pretreatments T3 (bleaching) and T4 (steam blanching) were determined to be of higher quality than other pretreatments. It also demonstrated that, compared to other powders, these two pretreatments offer the highest functional qualities for beetroot powder [13].

Nikola Tomic et al. aimed to compare the sensory properties and acceptance of dried, ready-to-eat beetroot snacks due to various drying methods: CO_2-drying, freeze-drying, and frying. Direct scCO_2-drying without a pre-cooking step emerged from the sensory evaluation as a promising alternative drying process in making dried beetroot snacks. The main drawbacks of freeze-drying were look and flavour [14]. Finally, Manhal Gobara Hamid and Abdel Azim Ahmed Mohamed Nour's papers examine how drying methods affect beetroot slice quality. The chemical arrangement, total energy, mineral content, and nitrate content of dried slices were substantially higher than fresh slices [15].

2 Materials and Methods

2.1 *Raw Materials*

The experiment employed a local variety of fresh beetroot from the Biskra neighbourhood market. To remove soil and other accumulated material, they were carefully cleaned with water, then peeled and weighed. Finally divided into 4×3 by 3, 4, and 5 mm-thick cubic slices.

2.2 *Method*

A sealed stainless steel Nabertherm drying oven was used for the studies. Ventilated ovens improve room thermal homogeneity. The ventilation may be adjusted to suit the tank. Horizontal air circulation and a continuous back wall exhaust with a front adjustment control maintain temperature homogeneity within 5 °C. Self-diagnostic

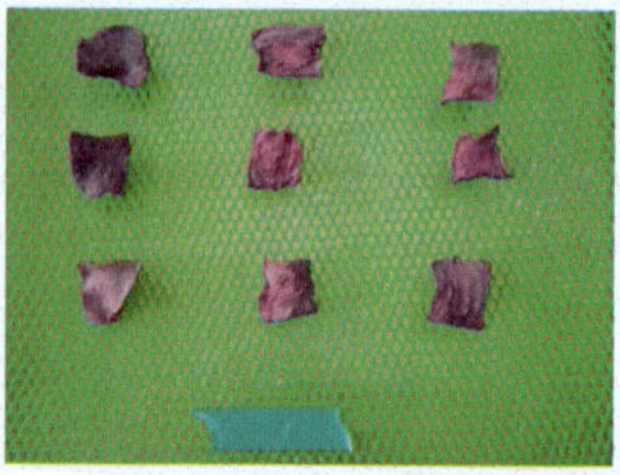
Fig. 1 3 mm cubic slices

PID microprocessors control drying parameters. Max operational temperature is 3000 °C, range is + 30 °C above ambient to 3000 °C. Multi-level loading. After setting the drier to the right temperature, the samples were weighed and placed on three trays on the two tiers of the 50 °C electric food dehydrator. Slices of 3, 4, and 5 mm thickness weighed 30, 31, and 45 g in dry-ing trials. Each drying trial measured the product's mass by removing it from the drying chamber for 40 s until a steady weight was attained. These steps occurred every 15 min. Figures 1, 2 and 3 show the finished result after 225 min.

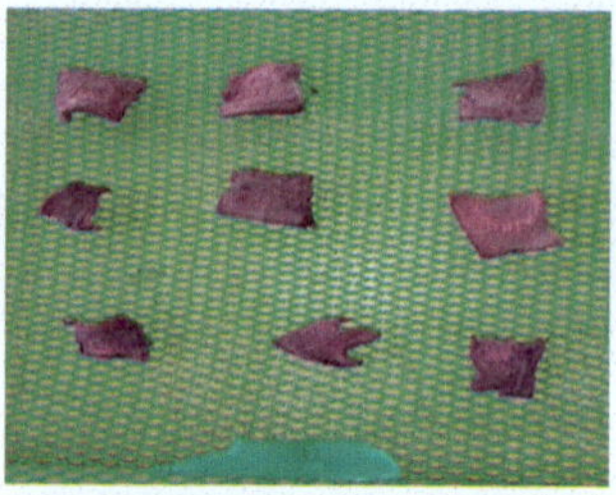
Fig. 2 4 mm cubic slices

Fig. 3 5 mm cubic slices

3 Nutritional Properties Before Drying

3.1 *Before Drying*

As shown in various studies, betta vulgaris, or beets, have several health-enhancing properties. They are rich in iron, vitamin C, and numerous vitamins and minerals. They also have the most incredible sugar content of any vegetable. They notably contain magnesium, potassium, betaine, and boron.

3.2 *After Drying*

According to Manthal Gobara Hamid's study, the chemical composition, total energy, minerals, and nitrate of the Tray-dried beet slices can be increased more than those of fresh beet slices [15]. Tray drying ensures a high content of it all due to the moderate drying temperature of 50 °C, which guarantees that all nutrients are conserved after drying.

4 Results and Discussion

The mass of water vapour is depicted in Fig. 4. As a time function, and we can see that m_{v1} begins at 7 g before changing until it reaches 0 g at 170 min. m_{v2} begins at 8.8 g and fluctuates until it reaches 0 g at 175 min. Finally, m_{v3} begins at 9 g before varying till it reaches 0 g at 230 min (Fig. 5).

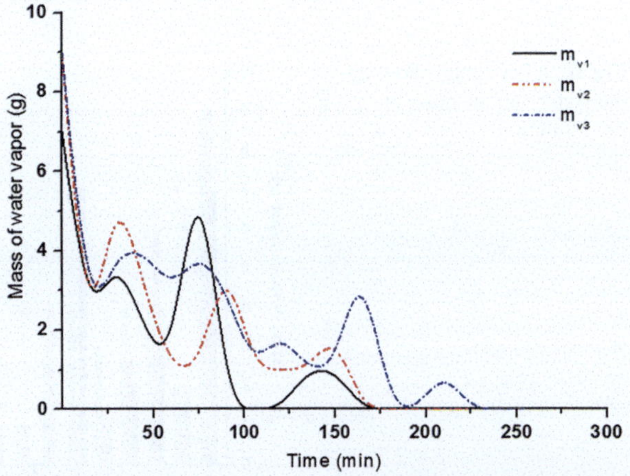

Fig. 4 A line graph of the mass of water vapour in time function

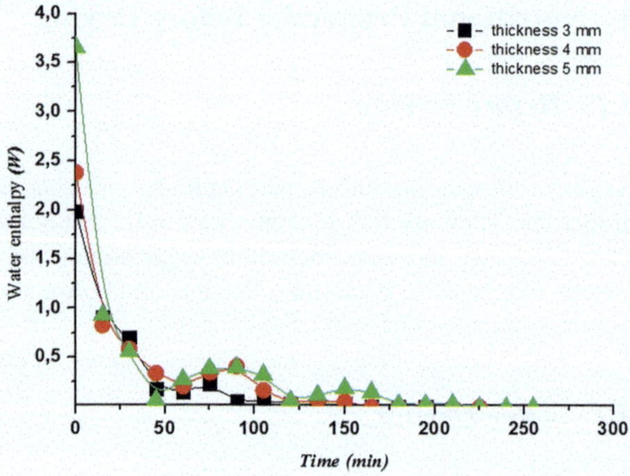

Fig. 5 A line graph that shows the water enthalpy in time function

The water enthalpy of the 3 mm-thick product begins at 2 w and fluctuates until 110 min, at which point it turns null. Then, at 120 min, 4 mm thickness begins at 2.4 W and fluctuates until it reaches zero. Finally, the thickness of 5 mm begins at 3.7 W and varies until 130 min, when it is null. These variations reflect the heat generated during drying within the chamber.

A bar graph of the mass of water vapour in time is revealed in Fig. 6. Where we observe that the first product has a minimum value of 1 W at three different times and 7.1 W at 75 min. The second product has a maximum weight of 9 W at the beginning and a minimum of 1 W six times. The third product has a minimum value of 1 W at four distinct points and a maximum weight of 9 W at the beginning.

In the line graph of the product mass in time function that appears in Fig. 7. We can see that Product One's mass starts at 30 g and decreases to 5 g after 195 min. Product Two's mass begins at 31 g and is reduced to 1 g after 210 min. The third product's mass begins at 45 g and shrinks to 6 g after 225 min. 13%. These values

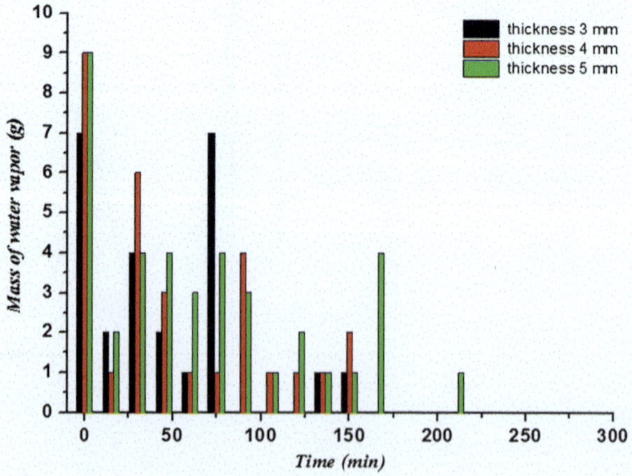

Fig. 6 A bar graph showing the water vapour mass in time function

give us the percentage of the product's weight loss: Tray 1, Tray 2, and Tray 3 were 13%, 3%, and 13%, respectively.

The temperature of the product is depicted in Fig. 8. As a function of time, and we can see that the first product has a maximum value of 57 °C at 160 min and a low value of 26 °C at the beginning. The second product, meantime, has the same minimum value as the first product and a maximum value of 53 °C at 190 min. The third product has a minimum value of 26.5 °C at the beginning and a maximum value of 54 °C at 225 min.

The temperature in time function is displayed in Fig. 9. And it is evident that the ambient temperature is almost constant between 28 and 29 °C. The temperature within the chamber fluctuates between 49.5 and 62.3 °C which is the highest temperature achieved during the drying. The overall temperature ranges from 38 to 42 °C.

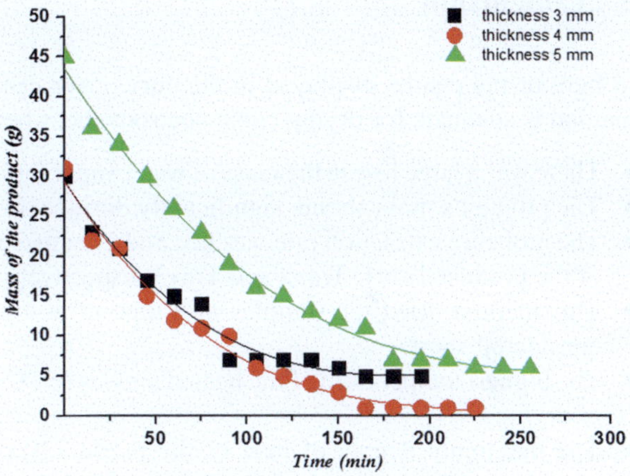

Fig. 7 A line graph that shows the product mass in time function

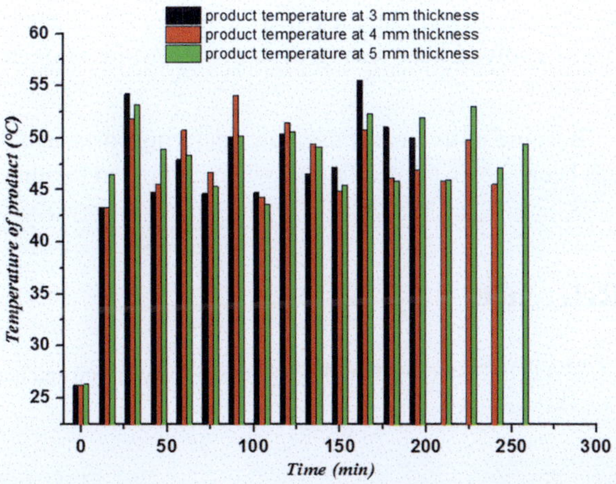

Fig. 8 A bar graph that shows the temperature of the product

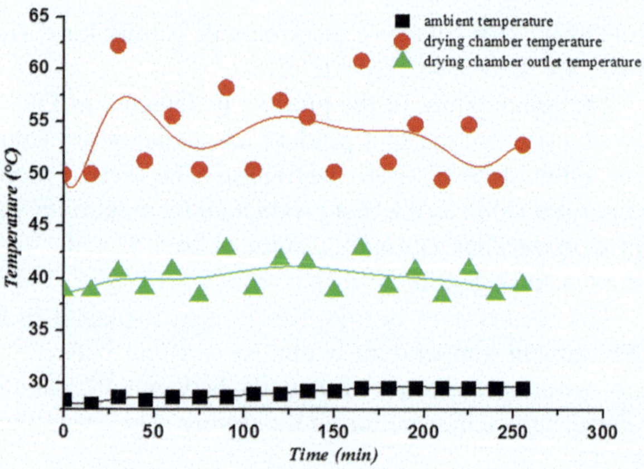

Fig. 9 A line graph that shows the ambient and chamber temperature plus the total one

5 Conclusion

In light of the results displayed in the curves, we came to the conclusion that this method is adequate for drying cubic beetroot slices because:

- There was a large loss in the mass of water vapor throughout the entire experiment.
- The product's mass shrank significantly during a brief period of roughly 4 h.
- The moisture extraction rate was fast and effective considering the loss in weight of the product Tray1, Tray2 and Tray3 respectively 13%, 3% and 13%.
- This method dried the samples in a time of 3–4 h to completely dry which is considered short.
- The highest temperature of the heated air was 63 °C obtained after almost an hour of drying.
- The maximum degree of heat of the product Tray 1, Tray 2 and Tray 3 was obtained at 57 °C, 53 °C and 54 °C, respectively at 160 min, 190 min and 225 min respectively.
- The total water enthalpy was decreased significantly to almost zero which means the evolved heat during drying was strong enough to release all of the product water.

Because of its ease of use, low operating costs, and practical design, this approach has been shown to be effective when compared to other vegetable drying techniques.

References

1. Ceclu L, Nistor O (2020) Red beetroot: composition and health effects—a review. J Nutr Med Diet Care 6(1):1–9

2. Ben Haj Koubaier H et al (2014) Betalain and phenolic compositions, antioxidant activity of Tunisian red beet (*Beta vulgaris* L. conditiva) roots and stems extracts. Int J Food Properties 17(9):1934–1945
3. Guldiken B et al (2016) Home-processed red beetroot (*Beta vulgaris* L.) products: changes in antioxidant properties and bioaccessibility. Int J Mole Sci 17(6):858
4. Pandharipande S, Bele B (2013) Osmotic drying rate estimation for dehydration of beetroot slices using artificial neural network. Int J Comput Appl 975:8887
5. Kaur K, Singh A (2014) Drying kinetics and quality characteristics of beetroot slices under hot air followed by microwave finish drying. Afr J Agric Res 9(12):1036–1044
6. Lech K et al (2015) Drying kinetics and bioactivity of beetroot slices pretreated in concentrated chokeberry juice and dried with vacuum microwaves. Drying Technol 33(13):1644–1653
7. Singh B, Hathan BS (2017) Process optimization of spray drying of beetroot juice. J Food Sci Technol 54(8):2241–2250
8. Vallespir F et al (2018) Improvement of mass transfer by freezing pre-treatment and ultrasound application on the convective drying of beetroot (*Beta vulgaris* L.). Food Bioprocess Technol 11(1):72–83
9. Szadzińska J et al (2019) Ultrasound-and microwave-assisted intermittent drying of red beetroot. Drying Technol
10. Dhiman A et al (2021) Status of beetroot processing and processed products: thermal and emerging technologies intervention. Trends Food Sci Technol 114:443–458
11. Malakar S, Alam M, Arora VK (2022) Evacuated tube solar and sun drying of beetroot slices: comparative assessment of thermal performance, drying kinetics, and quality analysis. Sol Energy 233:246–258
12. Liu Y, Duan Z, Sabadash S (2020) Effect of hot air drying temperatures on drying characteristics and physicochemical properties of beetroot (*Beta vulgaris*) slices. In: IOP conference series: earth and environmental science. IOP Publishing
13. Chaudhary V, Kumar V (2020) Study on drying and rehydration characteristics of tray dried beetroot (*Beta vulgaris* L.) and functional properties of its powder. Chem Sci Rev Lett 9(33):98–108
14. Tomic N et al (2020) Comparison of supercritical CO_2-drying, freeze-drying and frying on sensory properties of beetroot. Foods 9(9):1201
15. Hamid MG, Mohamed Nour AAA (2018) Effect of different drying methods on quality attributes of beetroot (*Beta vulgaris*) slices. World J Sci Technol Sustain Dev 15(3):287–298

Effect of Dip-Coating Seed Layer on the Morphological and Optical Properties of ZnO Nanorods Growth by Hydrothermal Method

S. A. Aldhehabi and B. E. Belkerk

Abstract This paper presents the fabrication and characteristic analysis of ZnO nanorods grown on glass substrates using a two-step approach involving dip-coating for the seed layer and hydrothermal growth for ZnO nanorods. The study systematically investigates the effects of different dip-coating cycles using advanced characterization techniques to investigate the impact on the morphology, crystal structure, and properties of the nanorods, Incorporating field-emission scanning electron microscopy (FE-SEM), X-ray diffraction (XRD) examination, and UV–VIS double-beam spectrometry. The ZnO nanorods exhibit a hexagonal morphology, with average diameters ranging from 60 to 450 nm through control of the dip-coating cycles. X-ray diffraction analysis reveals improved crystallinity with increasing dip-coating cycles. The UV transmittance spectra show characteristic absorption peaks in the UV range (360–380 nm) for all samples, with a decreasing band gap observed as the number of dip-coating cycles increases. These findings provide valuable insights into tailored ZnO nanorod synthesis and their potential applications in areas such as photocatalysis and optoelectronics. The study contributes to the fundamental understanding of nanomaterial properties, offering pathways for optimizing ZnO nanorod-based devices and technologies.

Keywords Dip-coating · Hydrothermal · ZnO · Nanorods

S. A. Aldhehabi (✉) · B. E. Belkerk
Laboratory Microsystems and Instrumentation (LMI), Faculty of Technology Sciences, Mentouri Brothers University Constantine 1, Route de Ain El Bey, 25017 Constantine, Algeria
e-mail: samer.aldhehabi@doc.umc.edu.dz

Biotechnology Research Center (CRBt), PB E73/UV 03 Ali Mendjeli, Nouvelle Ville, 25000 Constantine, Algeria

S. Hamamda et al. (eds.), *Proceedings of the 2nd International Conference of Nanotechnology for Environmental Protection and Clean Energy Production*, Springer Proceedings in Materials 45, https://doi.org/10.1007/978-981-97-1916-7_11

1 Introduction

Zinc oxide (ZnO) has garnered significant attention in scientific and technological circles owing to its remarkable properties as a semiconductor material [1]. These attributes have stimulated extensive research, elevating ZnO's importance in nanotechnology [2], optoelectronics [3], and catalysis [4], among other disciplines. The crystalline structure of ZnO, predominantly wurtzite [5], bestows a multitude of advantageous traits upon the material. Notably, ZnO's direct bandgap endows it with exceptional light absorption and emission capabilities [6], rendering it an exemplary candidate for optoelectronic applications [7]. With an approximate bandgap energy of 3.37 eV, ZnO exhibits pronounced sensitivity to ultraviolet (UV) light [8], amplifying its value in UV photodetectors, sensors, and related devices. Additionally, ZnO displays a significant exciton binding energy of approximately 60 meV [9], greatly influencing its optical emission and charge carrier dynamics. The multifaceted attributes of ZnO offer boundless prospects for pioneering advancements in diverse technological applications.

The crystalline quality of ZnO films obtained through chemical and physical deposition processes exhibits significant variability, contingent upon the employed synthesis conditions and growth mechanisms [10]. Typically, these synthesized ZnO materials manifest polycrystalline structures, characterized by a diverse range of nanostructures with distinct optical and electrical properties [11].

In recent times, A variety of ZnO nanostructure forms have demonstrated promising optical and electrical characteristics, ZnO nanostructures have garnered remarkable prominence in the realm of miniaturized and high-performance semiconductor devices, primarily owing to their exceptional optical performance. Moreover, the utilization of diverse thin film deposition methods has facilitated the fabrication of these ZnO nanostructures, unlocking a plethora of applications encompassing light-emitting diodes [12], lasers [13], gas sensors [14], and UV photodetector [15].

In this study, we present a comprehensive investigation of ZnO nanorods fabricated on seed layers using varying dip-coating cycles (from 3 to 10) followed by hydrothermal growth. The morphological, structural, and optical characteristics of the produced ZnO nanorods will be systematically analyzed to gain deeper insights into the influence of deposition cycles on both the development and properties of the nanorods. The outcomes of this research hold great potential for advancing the understanding and application of ZnO nanorods in cutting-edge technologies.

2 Experimental

2.1 *Sample Preparation and Deposition of ZnO Seed Layer*

The ZnO thin seed layers were cultivated on the glass slides through a precise dip-coating process. For this purpose, a solution was prepared using Zinc acetate dehydrate $Zn(CH_3COO)_2 2H_2O$ as the precursor and ethanol C_2H_6O as the solvent. A 0.025 M zinc acetate dehydrate solution was meticulously prepared by magnetic stirring in 90 ml of ethanol at room temperature for 2 h, followed by additional stirring for 60 min at 60 °C. Once a transparent solution was obtained, the dip-coating process was conducted under experimentally optimized growth conditions, Involving factors such as the concentration of the chemical precursor solution, withdrawal speed, and dipping speed.

Multiple deposition cycles were carried out to produce four distinct samples, denoted as 3 dip, 5 dip, 7 dip, and 10 dip, each possessing ZnO seed layers. To promote crystallization and structural integrity, all samples (3 dip, 5 dip, 7 dip, and 10 dip) underwent annealing in a furnace at 300 °C for a duration of 30 min. this precisely regulated thermal treatment guaranteed the development of sturdy and precisely delineated ZnO seed layers on the glass substrates.

2.2 *Growth of ZnO Nanorods*

ZnO nanorods were meticulously cultivated via a hydrothermal process, building upon a well-established ZnO seed layer. The solution of the chemical precursor, an amalgamation of zinc acetate dihydrate ($Zn(CH_3COO)_2 2H_2O$) and hexamethylenetetramine [$(CH)_2 6N_4$, HMT], was dissolved in distilled water at a precise equimolar ratio of 0.025 M.

To foster homogeneity and reactivity, the solution underwent meticulous magnetic stirring at ambient temperature for a duration of 2 h, followed by an additional hour within the reactor at 60 °C. The pursuit of a transparent and harmonious solution was relentlessly pursued, marking a pivotal milestone in the process.

The ZnO seed layer, previously germinated, became the foundation for the subsequent hydrothermal process. Submerging the samples into the hydrothermal reactor at precisely 80 °C for an extended duration of 20 h initiated the controlled growth of ZnO nanorods. This carefully tailored hydrothermal environment nurtured the nanorods, promoting their desired morphology and crystalline structure.

To culminate this intricate synthesis, the samples were cautiously retrieved from the reactor and subjected to annealing at a precisely maintained temperature of 400 °C for a crucial duration of 4 h. This thermal treatment, meticulously administered, ushered in the pinnacle of quality, furnishing high-quality ZnO nanorods with exceptional properties and structural integrity.

3 Results and Discussion

3.1 Morphology and Structural Analysis

Scanning electron microscopy (SEM) images of ZnO nanorods fabricated on glass substrates at different cycle dip-coating times are depicted in Fig. 1a. The nanorods exhibit a uniform distribution and a homogenous, hexagonal structure (inset of Fig. 1), with vertical alignment on the glass substrate. Top-view images (Fig. 1a.1–a.4) reveal an average diameter ranging from 60 to 450 nm for the ZnO nanorods.

Notably, the cycle dip-coating process induces significant alterations in the surface morphology of the ZnO nanorods, particularly in terms of their diameter. Remarkably dense ZnO nanorod arrays, consistent with previous studies [16], are observed.

Figure 1b demonstrates the approximate diameter scanned within a 1 μm area for each sample. With an increase in the number of dip-coating cycles, a prominent growth in the diameter of ZnO nanorods is evident. This observation can be attributed to an increase in the number of seeds available on the substrate, facilitating the formation of nanorods with larger diameters.

Intriguingly, the sample subjected to 5 dip-coating cycles exhibits the largest diameter of 450 nm, while those subjected to 3, 7, and 10 cycles display diameters of 90 nm, 100 nm, and 60 nm, respectively.

The X-ray diffraction (XRD) profiles of ZnO nanorods, produced using a hydrothermal method with different cycle dip-coating, are depicted in Fig. 3. These patterns exhibit a polycrystalline nature, demonstrating the characteristic hexagonal wurtzite structure for all four ZnO nanorod samples [17].

Notably, Fig. 2a illustrates the absence of any peaks corresponding to other elements in the XRD patterns which denotes that there are no impurities present in the sample, confirming the purity and exclusive presence of ZnO in the samples.

Table 1 provides a visual representation of the influence of cycle dip-coating on the intensity of the 002 peak, along with the corresponding tabulated results, which

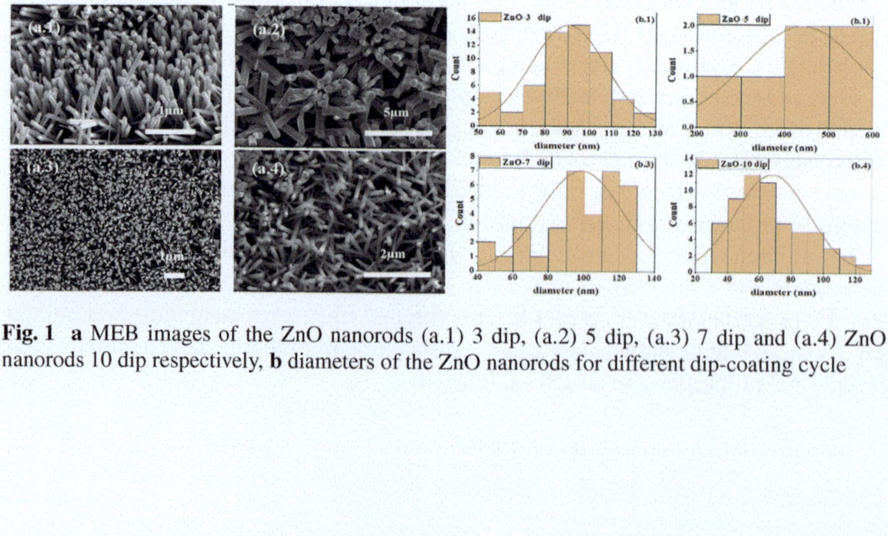

Fig. 1 **a** MEB images of the ZnO nanorods (a.1) 3 dip, (a.2) 5 dip, (a.3) 7 dip and (a.4) ZnO nanorods 10 dip respectively, **b** diameters of the ZnO nanorods for different dip-coating cycle

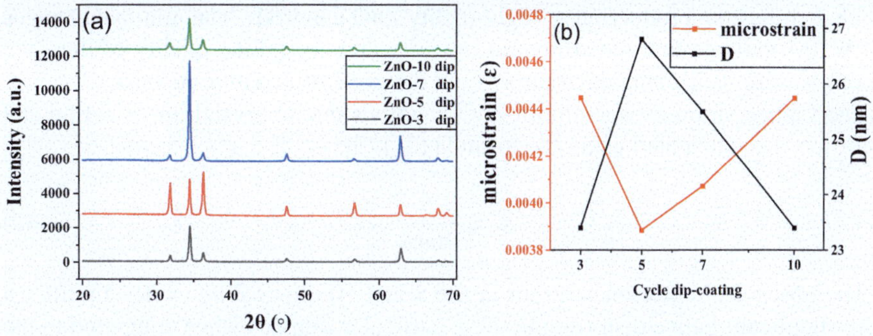

Fig. 2 The X-ray diffraction analyses with, **a** pattern of ZnO nanorods at different times of cycle dip-coating and **b** variation of microstrain and crystallite size with different times of cycle dip-coating calculated using Debye Scherrer's equations

are presented in Table 1. The observed trend reveals The intensity of the 002 peak demonstrates a gradual increase with the escalation in cycle dip-coating [18].

Specifically, the XRD peak associated with the seven dip-coating cycle demonstrates the highest intensity, suggesting an enhanced degree of crystallinity at this particular coating stage. However, as the number of coating cycles is further increased to ten, a significant reduction in peak intensity is observed, indicating a diminishing level of crystallinity. The crystallite size (D) was determined using the Debye–Scherrer formula (1) [19] and is presented in Table 1:

$$D = \frac{\kappa\lambda}{\beta \mathrm{COS}\theta} \tag{1}$$

D denotes the crystallite size, λ (1.5418 nm) represents the wavelength of the incident X-ray beam (CuKa), β signifies the full width at half-maximum (FWHM), and θ corresponds to the Bragg diffraction angle.

The crystallite sizes of ZnO exhibited a range from 23.4 to 26.8 Å, and their variation was influenced by the cycle dip-coating process. With an increase in the number of cycle dip-coating, the crystallite sizes demonstrated an initial increase followed by a subsequent decrease. Figure 2b visually depicts this trend, notably, the Full Width at Half Maximum (FWHM) values showed only slight variations,

Table 1 Structural parameters of ZnO nanorods as determined from the diffraction peak (002)

Sample	$2\theta_{(002)}$	I (a. u.)	d (Å)	FWMH
ZnO-3 dip	34.49022	2030	2.60105	0.31654
ZnO-5 dip	34.47203	2102	2.60255	0.27597
ZnO-7 dip	34.51252	5900	2.5996	0.29
ZnO-10 dip	34.49022	1845	2.60105	0.31651

which signifies the production of high-quality ZnO nanorods. The minimal changes in FWHM indicate that the structural integrity and crystalline quality of the ZnO nanorods were well-maintained throughout the cycle dip-coating process.

The Lattice strain broadening in ZnO nanorods due to crystal imperfection and distortion was calculated using the formula (2) [20]:

$$\varepsilon = \frac{\beta}{4\tan\theta} \tag{2}$$

The observed higher microstrain in the 3 and 10 dip-coating cycles signifies a more significant lattice distortion in these particular samples when compared to the others. Moreover, as the number of cycle dip-coating increased, there was a consistent and progressive increase in the lattice strain values, furthermore, the systematic rise in lattice strain values with an increasing number of cycle dip-coating suggests a cumulative effect of the coating process on the lattice arrangement and integrity of the ZnO nanorods.

3.2 *Optical Properties*

Figure 3a illustrates the transmission spectra of ZnO nanorods at various cycle dip-coating times. Evidently, the transmittance of ZnO thin films diminishes with the progressive increase in cycle dip-coating [21], and a shift towards lower UV wavelengths is observed [22]. Furthermore, the absorption peak at 375 nm, 376 nm, 382 nm, and 383 nm for 3, 5, 7, and 10 cycle dip-coating, respectively, indicates ZnO nanoparticle absorption. Notably, all ZnO nanorod samples exhibit a pronounced intensity drop starting from 360 nm, corresponding to a sharp peak attributed to the free exciton. The observed light absorption within the spectral range of 360–380 nm suggests the excitation potential of ZnO nanorods by blue light, indicative of excellent optical quality. These findings are consistent with the existing literature [23].

The band gap is measured by extrapolating the linear portion curve of (αhν)2 against the photon energy (hν), using Tauc's plot [24] for direct transition semiconductor on the following Eq. (3):

$$(\alpha h\nu)^2 = A(h\nu - E_g)^n \tag{3}$$

In this context, α, hν, A, E_g, and n symbolize the absorption coefficient, photon energy, a constant, the optical band-gap energy, and a factor dependent on the transmission type (equivalent to 1/2 for allowed straight transmission), respectively. The absorption coefficient (α) is ascertainable from the transmittance through the application of the following Eq. (4):

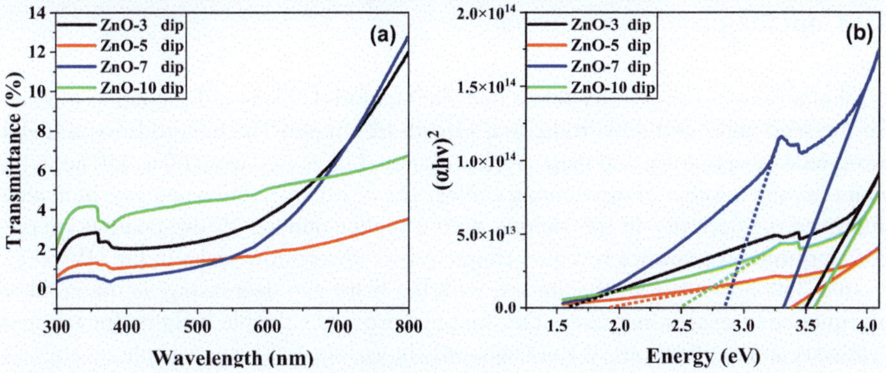

Fig. 3 **a** UV–VIS transmittance spectra for ZnO at different cycle dip-coating, **b** the plots of f (αhυ)2 as a function of photon energy with values of the band-gap for ZnO at different cycle dip-coating

$$\alpha = \frac{\ln(T)}{d} \tag{4}$$

where T is the transmittance, and d is the average length of ZnO nanorods.

The plot of $(\alpha h\nu)^2$ against hυ was utilized to determine the energy bandgap (E_g) of ZnO nanorods (refer to Fig. 3b). Two distinct absorption edges were observed, indicating the presence of two electronic transitions. The energy values for the first and second transitions are denoted as E_{gh} and E_{gL}, respectively (where $E_{gh} > E_{gL}$) [25].

The first gap, E_{gh}, , is referred to as the fundamental bandgap, responsible for the direct electron transition when ZnO nanorods are excited by blue light. This corresponds to the peak of absorption observed in the transmittance spectrum within the range of (360–380) nm (as shown in Fig. 3a). The second gap, E_{gL} is termed the optical bandgap or deep-level emission, arising from the recombination of electrons with deeply trapped holes in oxygen vacancies or can be linked to zinc ions in ZnO nanomaterials [26].

The values of E_{gh} were determined as 3.50, 3.38, 3.32, and 3.57 eV for 3, 5, 7, and 10 cycle dip-coating, respectively, and were obtained at higher photon energy. In contrast, the values of E_{gL} were found to be 1.60, 1.90, 2.76, and 2.50 eV for the respective cycle dip-coating conditions and were obtained at lower photon energy, notably, an inverse relationship was observed between the cycle dip-coating and the band gap. As the cycle dip-coating increased, the band gap decreased, approaching the ZnO band gap value of 3.37 eV.

4 Conclusion

In summary, we successfully fabricated ZnO nanorods using a dip-coating method for the seed layer and a hydrothermal growth technique. The nanorods exhibited a hexagonal morphology, and their average diameter increased from 60 to 450 nm with an increasing number of dip-coating cycles. The X-ray diffraction analysis indicated improved crystallinity in the sample with a higher number of dip-coating cycles. UV transmittance spectra revealed characteristic absorption peaks in the UV range around 360–380 nm for all samples, with the band gap decreasing as the number of dip-coating cycles increased. Our findings provide valuable insights for tailored synthesis and applications of ZnO nanorods in various fields such as photocatalysis and optoelectronics.

References

1. Barzinjy AA, Azeez HH (2020) Green synthesis and characterization of zinc oxide nanoparticles using *Eucalyptus globulus* Labill. leaf extract and zinc nitrate hexahydrate salt. SN Appl Sci 2:991
2. Moradpoor H, Safaei M, Mozaffari HR et al (2021) An overview of recent progress in dental applications of zinc oxide nanoparticles. RSC Adv 11:21189–21206
3. Kumari P, Misra KP, Chattopadhyay S, Samanta S (2021) A brief review on transition metal ion doped ZnO nanoparticles and its optoelectronic applications. Mater Today Proc 43:3297–3302
4. Shah AA, Bhatti MA, Tahira A et al (2020) Facile synthesis of copper doped ZnO nanorods for the efficient photo degradation of methylene blue and methyl orange. Ceram Int 46:9997–10005
5. Liu Q, Nie Y, Shang J et al (2021) Exceptional deformability of Wurtzite zinc oxide nanowires with growth axial stacking faults. Nano Lett 21:4327–4334
6. Li X, Liu X, Li Y et al (2021) Using novel semiconductor features to construct advanced ZnO nanowires-based ultraviolet photodetectors: a brief review. IEEE Access 9:11954–11973
7. Saadi H, Benzarti Z, Sanguino P et al (2022) Improving the optical, electrical and dielectric characteristics of ZnO nanoparticles through (Fe + Al) addition for optoelectronic applications. Appl Phys A 128:691
8. Abdulrahman AF, Ahmed SM, Barzinjy AA et al (2021) Fabrication and characterization of high-quality UV photodetectors based ZnO nanorods using traditional and modified chemical bath deposition methods. Nanomaterials 11:677
9. Kumar V, Gupta R, Bansal A (2021) Hydrothermal growth of ZnO nanorods for use in dye-sensitized solar cells. ACS Appl Nano Mater 4:6212–6222
10. López Guemez ADR, Cordero García A, Cervantes López JL et al (2023) Evaluation of the photocatalytic activity of ZnO nanorods and nanoflowers grown from seed layers deposited by spin coating. Boletín de la Sociedad Española de Cerámica y Vidrio S0366317523000304
11. Kamei R, Hosomi T, Kanao E et al (2021) Rational strategy for space-confined seeded growth of ZnO nanowires in meter-long microtubes. ACS Appl Mater Interfaces 13:16812–16819
12. Mohammad SM, Hassan Z, Abd-Alghafour NM et al (2020) Ultraviolet electroluminescence from flowers-like n-ZnO nanorods/p-GaN light-emitting diode fabricated by modified chemical bath deposition. J Lumin 226:117510
13. Fadzliana N, Samsuri SAM, Chan SY et al (2020) Influence of Al doping on random lasing in ZnO nanorods. Opt Laser Technol 124:106004
14. Moon D-B, Bag A, Lee H-B et al (2021) A stretchable, room-temperature operable, chemiresistive gas sensor using nanohybrids of reduced graphene oxide and zinc oxide nanorods. Sens Actuat B Chem 345:130373

15. Mohammad SM, Rajamanickam S, Hassan Z et al (2021) Self-powered UV photodetector performance optimization based on Ag nanoparticles-encapsulated-ZnO nanorods by photo-deposition method. Sens Actuat A 331:113032
16. Toe MZ, Pung S-Y, Yaacob KAB, Han SS (2021) Effect of dip-coating cycles on the structural and performance of ZnO thin film-based DSSC. Arab J Sci Eng 46:6741–6751
17. Alsultany FH, Hassan Z, Ahmed NM et al (2018) Effects of ZnO seed layer thickness on catalyst-free growth of ZnO nanostructures for enhanced UV photoresponse. Opt Laser Technol 98:344–353
18. Ali M, Sharif S, Anjum S et al (2020) Preparation of Co and Ni doped ZnO nanoparticles served as encouraging nano-catalytic application. Mater Res Express 6:1250d5
19. Khomutinnikova L, Evstropiev S, Meshkovskii I et al (2023) Ceramic ZnO–SnO_2–Fe_2O_3 powders and coatings—effective photogenerators of reactive oxygen species. Ceramics 6:886–897
20. Soliman TS, Rashad AM, Ali IA et al (2020) Investigation of linear optical parameters and dielectric properties of polyvinyl alcohol/ZnO nanocomposite films. Physica Status Solidi (A) 217:2000321
21. Sedky A, Amin SA, Mohamed M (2019) Electrical, photoluminescence and ferromagnetic characterization of pure and doped ZnO nanostructures. Appl Phys A 125:308
22. Ridwan J, Yunas J, Umar AA et al (2022) Vertically aligned Cu-doped ZnO nanorods for photocatalytic activity enhancement. Int J Electrochem Sci 17:220813
23. Pandi DV, Saraswathi V, Muthukumarasamy N et al (2021) C-axis oriented ZnO nanorods based quantum dot solar cells. Opt Mater 112:110774
24. Idiawati R, Mufti N, Taufiq A et al (2017) Effect of growth time on the characteristics of ZnO nanorods. IOP Conf Ser Mater Sci Eng 202:012050
25. Sedky A, Afify N, Almohammedi A et al (2023) Structural, optical, photoluminescence and magnetic investigation of doped and Co-doped ZnO nanoparticles. Opt Quant Electron 55:456
26. Gill R, Ghosh S, Sharma A et al (2020) Vertically aligned ZnO nanorods for photoelectrochemical water splitting application. Mater Lett 277:128295

Elaboration of a Reduced Graphene Oxide–Chitosan Biocomposite

Abdallah Cheraitia, Maroua Mezrag, and Roquiya Laghrib

Abstract In this work, to develop a reduced graphene oxide-chitosan (rGO-Cs) biocomposite by solution intercalation and cyclic voltammetry, we used reduced graphene oxide (rGO) and chitosan. We have synthesized reduced graphene oxide (rGO), from recycled, processed and activated graphite extracted from batteries or carbon from electric motors, by electrochemical process using KOH as electrolyte. We have extracted Chitosan from shrimp shells chemically in two steps: Extraction of chitin by demineralization, deproteinization and discoloration of shrimp shells. Then by deacetylation we obtain chitosan of off-white color, soluble in acetic acid and non-conductive and with a viscosity of 3.267 Pa s with a good yield and a DDA of 72%, which indicates the good quality of the chitosan produced; Using elaborated chitosan and commercial chitosan, reduced graphene oxide-chitosan (rGO-Cs) composites were synthesized; We characterized the produced samples by FTIR and UV–Vis spectroscopy, by cyclic voltammetry and by rheology in addition to conductivity and solubility tests.

Keywords Biocomposite · Reduced graphene oxide · Chitosan · Exfoliation

1 Introduction

Chitosan is a natural polymer derived from chitin, which is the main component of the exoskeleton of crustaceans such as shrimp, crabs, and lobsters. It was used in traditional medicine for its wound-healing properties [1]. Chitosan has a variety of

A. Cheraitia (✉) · M. Mezrag
Process Engineering Department, Industrial Analysis and Materials Engineering Laboratory (LAIGM) , University 8 May 1945, Guelma, Algeria
e-mail: cheraitiaabdellah@univ-guelma.dz

R. Laghrib
Higher Technical Teaching Normal School-Skikda, Laboratory of Silicates, Polymers and Nanocomposites (LSPN), University 8 May 1945, Guelma, Algeria

S. Hamamda et al. (eds.), *Proceedings of the 2nd International Conference of Nanotechnology for Environmental Protection and Clean Energy Production*, Springer Proceedings in Materials 45, https://doi.org/10.1007/978-981-97-1916-7_12

properties that make it a versatile material with a wide range of potential applications. It is biodegradable, biocompatible, and non-toxic, and it has antimicrobial and antioxidant properties.

Pure chitosan is not a good conductor of electricity; we can modified chemically it by introducing functional groups that can interact with electrons witch increase the electrical conductivity of chitosan [2, 3].

Graphene-chitosan composites are emerging materials that combine the properties of graphene and chitosan. The graphene is a very strong and conductive material [4, 5], with chitosan, a natural polymer with safe biocompatible and biodegradable properties [6]. This combination of properties makes graphene-chitosan composites highly promising for a variety of applications [7], including electronics (sensors and displays…etc.), biomedical devices (implants, wound dressings, and drug delivery systems), and environmental remediation to remove pollutants from water and air [8–10].

2 Experimental Procedure

2.1 Electrochemical Exfoliation of Graphite

We have synthesized black powder of reduced graphene oxide (rGO) by electrochemical exfoliation of graphite in an electrolytic KOH solution by applying a voltage (10 V) between the graphite electrodes (Fig. 1).

The proposed mechanism for electrochemical exfoliation [6] is based on the polarization of graphene layers by OH^- ions. The reactions that occur at the electrodes are as follows:

- At the cathode: $2H_2O + 2e^- \rightarrow H_2(g) + 2OH^-$
- At the anode: $4OH^- \rightarrow O_2(g) + 2H_2O + 4e^-$

Fig. 1 Electrochemical exfoliation steps of graphite to rGO in KOH solution

The intercalation of graphitic layers is caused by the presence of OH^- ions, the formation during the electrochemical process of O_2 leads to the exfoliation of graphite by expansion of space between layers of graphite and partial oxidation of graphene.

2.2 *Extraction of Chitosan*

We have extracted chitosan from shrimp shells in four main steps (Fig. 2).

First, we wash, dry, and grind shrimp shells into a fine powder. Next, we dissolve minerals such as $CaCO_3$, the main mineral compound in the shell, in an HCl acid solution.

Finally, we remove proteins by an alkaline treatment using concentrated NaOH solution at a high temperature of 90–100 °C (Fig. 3).

After bleaching with an oxidant such as H_2O_2, we obtain white chitin. The next step is to extract chitosan by deacetylating chitin. This is a delicate step that involves

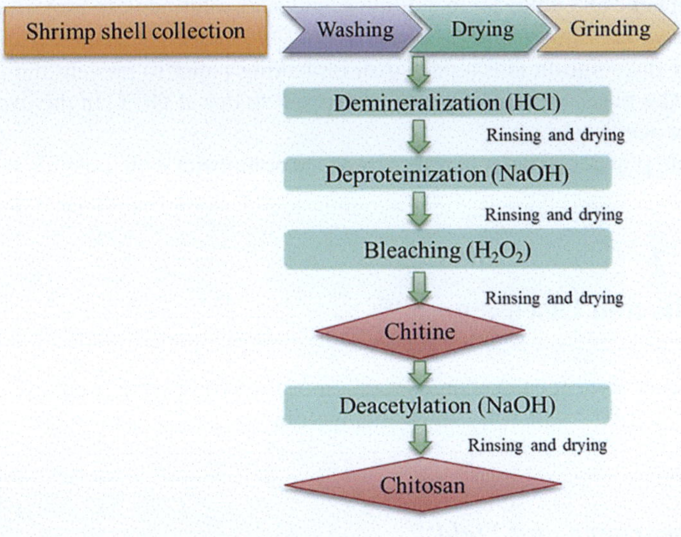

Fig. 2 Chitosan extraction steps from shrimp shells

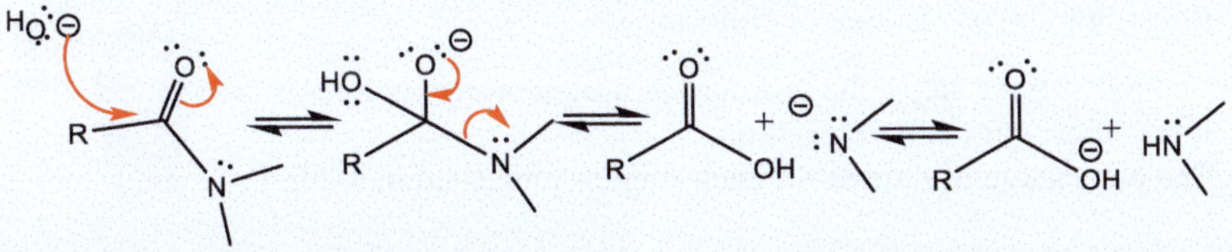

Fig. 3 Mechanism of deproteinization

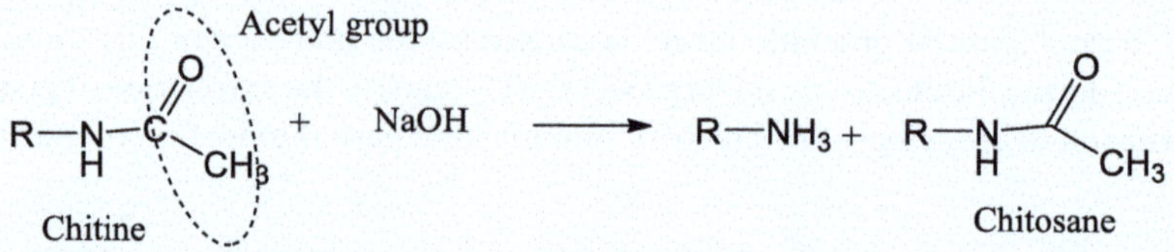

Fig. 4 Deacetylation of chitin reaction

substituting a maximum and sufficient number of acetyl groups, which are hydrolyzed to amine groups in an alkaline medium (Fig. 4).

Finally, we purify the extracted chitosan by filtration and washing.

2.3 Elaboration of rGO-Chitosan Composite

We synthesize the rGO-Chitosan composite (rGO-Cs) thin layer and powder by mixing graphene and chitosan in a dilute acetic acid solution with proportions rGO/Cs 50/50 w/w and liquid/solid proportion 1/10 v/w, under vigorous stirring for 6 h.

We pour the solution into a petri dish on baking paper to prevent the film from sticking to the glass of the petri dish and leave it to dry at 90 °C in the oven for 6 h to eliminate acetic acid.

We obtain powder by scraping the film of the composite with a spatula in stainless steel.

3 Results and Discussion

The techniques used for the characterization of the synthesized products are: FTIR tests of conductivity, solubility and voltamperometry.

3.1 Appearance and Yield

The rGO is obtained in the form of a black powder.

Regarding the production of chitosan:

- In the first three steps, the powder has a light orange color.
- It becomes white after bleaching.
- The percentage of returns for each stage is presented in Table 1.

Table 1 Appearance and yield of products in different steps of extraction of chitosan

	Preparation	Demineralization	Deproteinization	Bleaching	Deacetylation
Appearance					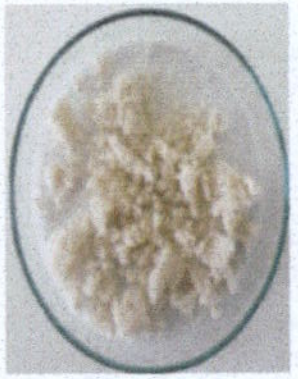
Color	Light orange	Light orange	Light orange	White	Off-white color
Yield % w	/	70.37	85.53	92.80	23.80

3.2 Conductivity and Solubility

Reduced graphene oxide (rGO) is a good conductor because of the existence of the conjugation of the structure and the weak presence of oxygen functions, while chitosan and chitin are not conductors because they are organic compounds. Chitosan is soluble in acetic acid, while rGO is poorly soluble. The elaborated rGO-chitosan (rGO-Cs) composites are conductors and insoluble in acetic acid Table 2.

3.3 FTIR Characterization

The infrared spectrum of the reduced graphene oxide–chitosan composite in powder form shows the same characteristic peaks of chitosan, namely:

- The amide group peaks at 3455 cm^{-1} (N–H and O–H stretching), 1633 cm^{-1} (C=O stretching), and 1560 cm^{-1} (N–H bending)
- The C–H stretching peak at 2883 cm^{-1}
- The C–O–C stretching peak at 1020 cm^{-1}

In addition to these peaks, the composite spectrum shows a peak at 2100 cm^{-1}, attributed to the skeletal vibrations of the graphite domains (the aromatic Sp^2 C=C) of graphene.

A broad peak is also observed between 2900 and 3450 cm^{-1}, which may be due to the presence of the OH of acetic acid, used for the synthesis of the composite. This acid contributes to the formation of hydrogen bonds between the chitosan chains and graphene, which strengthens the mechanical properties of the composite (Fig. 5).

3.4 Voltamperometry Tests

Deposition of chitosan on reduced graphene oxide electrodes is a promising method for improving their electrochemical properties. The process is reversible and can be repeated multiple times (Fig. 6). The addition of KCl to the electrolyte increases the conductivity of the solution, which results in an increase in the current intensity.

The deposition of CS on reduced graphene oxide (rGO) electrodes was carried out in a three-electrode cell. The working electrode was the rGO electrode, the counter electrode was a platinum electrode, and the reference electrode was an Ag/AgCl electrode. The electrolyte was a solution of CS in 0.75 M acetic acid. In the second case, 1 M KCl was added to the electrolyte to increase its conductivity.

The potential sweep was carried out between − 600 and + 600 mV versus Ag/AgCl. A peak of oxidation was observed at 200 mV, which decreased with the number of cycles.

Table 2 Appearance solubility and conductivity rGO-Cs composites

	Appearance		Solubility		Conductivity	
rGO-Cs film		Black	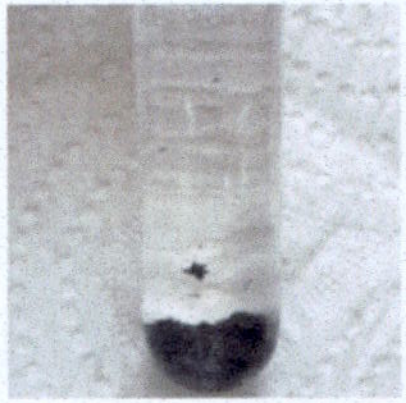	Not soluble		Conductor
rGO-Cs powder		Black	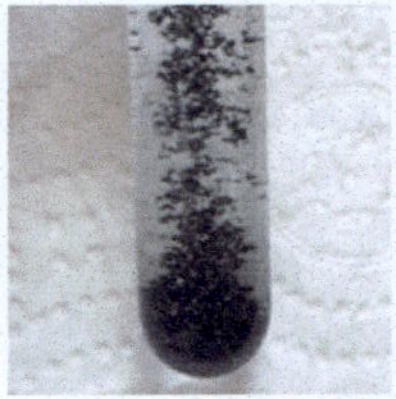	Not soluble		Conductor

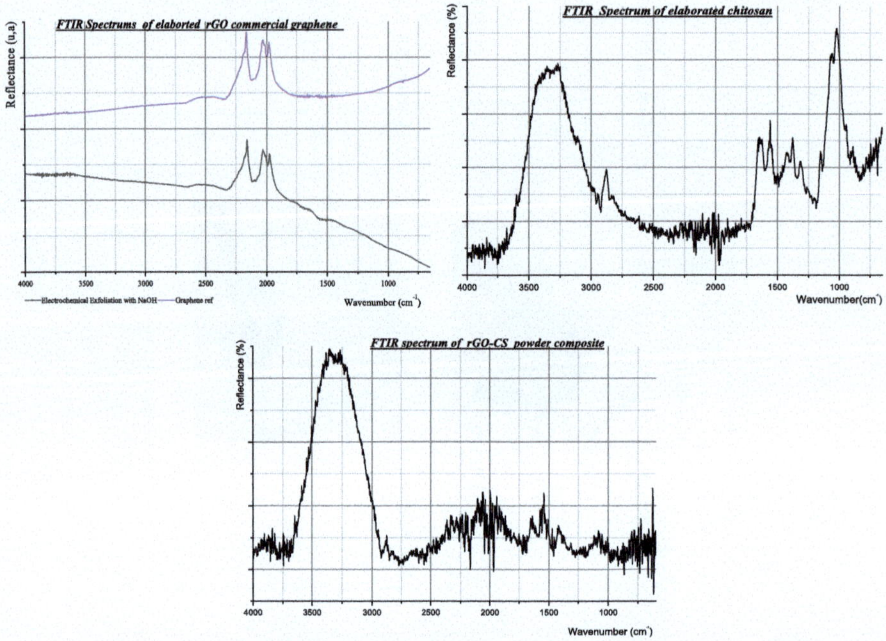

Fig. 5 FTIR spectrums for prepared rGO/graphene ref, chitosan and rGO-Cs composite

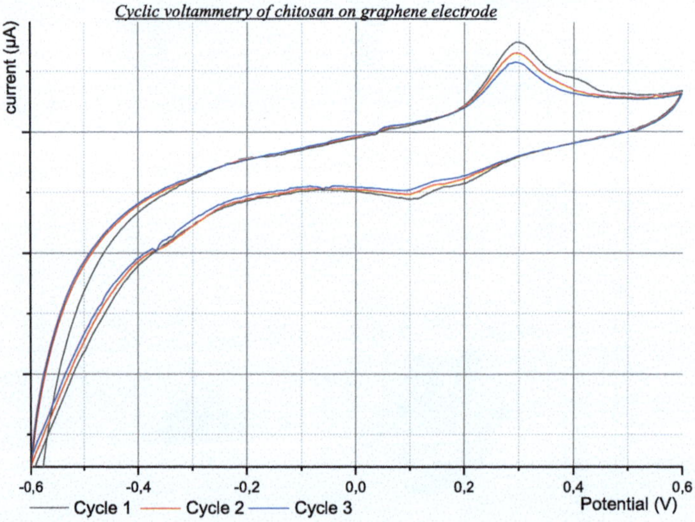

Fig. 6 Voltamperometry tests of deposition of chitosan on rGO electrodes

Voltammogram obtained from the deposition of CS on rGO electrodes (Fig. 6) show a peak of oxidation at 0.3 V and a peak of reduction at 0.1 V. The current intensity decreases with the number of cycles, which suggests that the deposition process is reversible.

Deposition of chitosan (CS) on reduced graphene oxide (rGO) electrodes is a simple and effective method for improving their electrochemical properties, such as conductivity and capacitance. This makes them promising candidates for use in a variety of electrochemical applications, such as energy storage devices, sensors, and actuators.

Tests on the $[Fe(CN)_6]^{4-} \rightleftharpoons [Fe(CN)_6]^{3-}$ system were performed on the elaborated rGO-Cs electrode, the voltammograms show two peaks: the oxidation peak at 0.3 V/Ag–AgCl and the reduction peak at 0.128 V/Ag–AgCl.

4 Conclusion

We have synthesized rGO by electrochemical exfoliation and extracted chitosan from shrimp shells. We then synthesized rGO-Cs, a composite material with promising electrochemical properties, using a simple chemical method. The method involves mixing rGO and chitosan in a diluted acetic acid solution.

We also used an electrochemical method to synthesize rGO-Cs. This method involves the deposition of chitosan on a rGO electrode.

Our results showed that adding KCl to the electrolyte increased the conductivity of the solution, which resulted in an increase in the current intensity. The deposition process is reversible and can be repeated multiple times.

The rGO-Cs is a promising material for electrochemical applications, such as energy storage devices, sensors, and actuators. These studies could focus on optimizing the deposition conditions to further improve the electrochemical properties of rGO-Cs.

References

1. Gopi S, Thoma S, Pius A (2020) Chitin and chitosan preparation and properties. In: Gupta MN (ed) Handbook of chitin and chitosan (vol 1). Wiley-VCH, Weinheim, Germany
2. Kumar S et al (2023) Electrical conductivity enhancement of chitosan composites with graphene nanoplatelets by chemical modification. Carbohydr Polym 278:118985–118993
3. Liu J et al (2022) Graphene-chitosan composites for electrical energy storage devices. Compos Sci Technol 220:111084–111093
4. Adetayo A, Runsewe D (2019) Synthesis and fabrication of graphene and graphene oxide: a review. J Compos Mater 53(9):1449–1468
5. Velazquez-Meraz O et al (2019) Improvement of physical and mechanical properties on biopolymer matrix composites using morphed graphene. Compos Sci Technol 184:107836
6. Hu H, Wang X, Jia J, Chen J (2010) Electrochemical exfoliation of graphite flakes for scalable production of graphene. ACS Nano 4(6):3131–3137

7. Cao N et al (2017) Facile synthesis of fluorinated polydopamine/chitosan/reduced graphene oxide composite aerogel for efficient oil/water separation. Chem Eng J 326:17–28
8. Kumar S, Tiwari VK (2021) Chitosan-based composites for biomedical applications: a review of electrical conductivity and biocompatibility. Carbohyd Polym 264:117927–117946
9. Zhang M et al (2022) Chitosan-carbon nanotube composites with high electrical conductivity and biocompatibility for flexible electronics. Mater Sci Eng C 124:112118–112126
10. Ganjali MR et al (2021) Chitosan composites: preparation and applications in removing water pollutants. In: Vijay Thakur VK et al (eds) Handbook of composites from renewable materials. Scrivener Publishing LLC, pp 562–577

Structural and Hydrophobic Properties of Cobalt Doped ZnO Thin Films Prepared by Electrodeposition Method

Zehira Belamri

Abstract In this work, thin films of cobalt-doped ZnO were prepared by thermal oxidation method of electrodeposited zinc layer on aluminum substrates for fabrication of hydrophobic coatings. XRD, Raman spectroscopy, and FEG-SEM methods used for characterization of studied samples. By measuring the contact angle between the film's surface and a deposited water drop, the wettability characteristics of the synthesized films were evaluated (WCA). The manufactured ZnO thin films were found to be polycrystalline with Wurtzite hexagonal structures, growing predominantly along the (101) plan, according to structural characterisation by XRD. No secondary phase for the cobalt-doped sample; this is confirmed by Raman spectroscopy. FEG-SEM images reveal that the grains shape is spherical for cobalt-doped ZnO thin film, however the undoped sample exhibit a compact grains with nanostructured pyramidal shape. Incorporation of cobalt into zinc ions sites leads to the decrease of crystallites size from 39 to 31 nm and enhanced the hydrophobic property of studied ZnO thin film.

Keywords Co-doped ZnO · Nanostructure · Thermal oxidation · Hydrophobic

1 Introduction

Zinc Oxide (ZnO) stays a principal nanostructured technological Group II-VI material. It is a semiconductor with a direct band gap of 3.2–3.4 eV at room temperature. Varying the value of the conductivity by doping processes, the gap can be changed. ZnO exhibits a natural n-type electrical conductivity which is due to the presence of the interstitial zinc atoms. The dopants usually used for ZnO generally belong to groups III and IV of Mendeleïev's table. In this case, they will replace the zinc atoms

Z. Belamri (✉)
Phase Transformation Laboratory, Frères Mentouri_Constantine 1 University, 25000 Constantine, Algeria
e-mail: belamri.zehira@umc.edu.dz

S. Hamamda et al. (eds.), *Proceedings of the 2nd International Conference of Nanotechnology for Environmental Protection and Clean Energy Production*, Springer Proceedings in Materials 45, https://doi.org/10.1007/978-981-97-1916-7_13

by occupying their atomic sites. Doping can also be done using elements from group VII of the periodic table.

In recent years, ZnO has attracted considerable attention; it's used in the fabrication of many devices, such as gas sensors [1], solar cells and photovoltaic [2, 3] fine hydrophobic property [4]. In order to create self-cleaning glass, Rabeel et al. [5] developed ZnO thin films that were deposited on glass substrates using a spray pyrolysis technique. The concept of preparing these surfaces presents enormous opportunities in the field of corrosion inhibition of metals and alloys.

The main objective of this present work is the fabrication of hydrophobic coating based on ZnO, using the thermal oxidation method of electrodeposited zinc thin film on metallic substrates. The study is devoted to the effect of doping by Co on the physical and wettability properties of these thin layers.

2 Materials and Methods

Before deposition, the substrates undergo a mechanical polishing until the desired surface shape and thickness of 2 mm is obtained. They are cleaned in ultrasound baths for 15 min with distilled water and methanol, respectively.

The Zn layers are electrodeposited at room temperature from an aqueous zinc acetate precursor solution of 0.2 mol/L, with agitation under − 10 V for 15 min. this is with respect to the distance between the aluminum substrate (cathode) and the platinum (anode) of 1.5 cm as represent in Fig. 1. ZnO thin films were obtained by thermal oxidation at 500 °C for 2 h of electrodeposited Zn layers. The concentrations of doping cobalt element are chosen of 0.01% (the doped sample is noted CZnO).

The structural characterizations were carried out using a PANALYTICAL Empyrean diffractometer ($\lambda_{cu} = 1.540$ Å) and HORIBA labRAM HR evolution type Raman spectrometer at room temperature with a monochromatic radiation source of 473 nm. For the morphological study, a Field Emission Gun Scanning Electron Microscope (SEM, Jeol FEG JSM-7100 F) equipped with an energy dispersive X-ray spectrometer (EDX) was used. the hydrophobicity was measured by contact

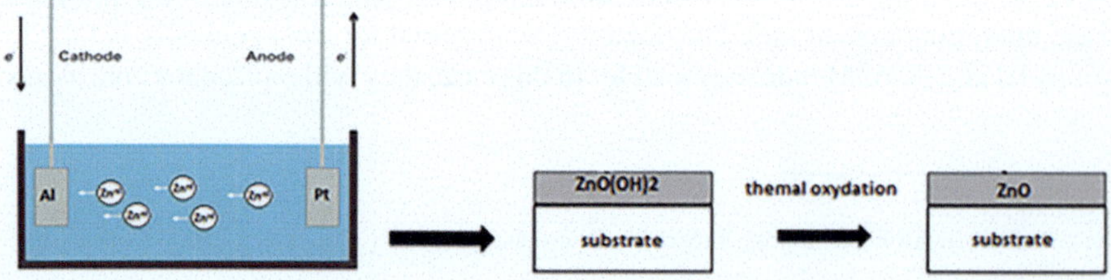

Fig. 1 Image represents the steps of the experiments

angle measurements at room temperature using an optical system composed of a lamp delivering white light and projecting the image of the drop deposited on the sample [leybold type light source (6 V, 30 W)], and a projection lens allowing the enlargement of the image of the drop projected on a screen of size (30 * 30 cm^2).

3 Results and Discussion

3.1 X-Ray Diffraction Study

The XRD spectra (Fig. 2) reveal that undoped and CZnO thin films were polycrystalline. They show multiple peaks corresponding to (100), (002), (101), (102), (110), (103), (112), (201) and (202) planes related to the hexagonal Wurtzite crystal structure (according to ICDD file N° 01-070-2551) with a preferential orientation along (101). No new peaks associated with Co oxides appeared in CZnO sample. This confirms the successful substitution replacement of Co ions in Zn sites in the ZnO lattice. Zinc phase indicate by circle and the aluminum phase due to substrate indicate by star in Fig. 2.

The three principal peaks (100), (002), (101) in XRD spectrum of CZnO sample, shift towards smaller angles (Fig. 3) without changing the Wurtzite structure of ZnO, this is evidence of cobalt ions incorporation into the ZnO crystal lattice where the ionic radius of Co (0.058 nm) is less then to that of Zn (0.074 nm) which leads to

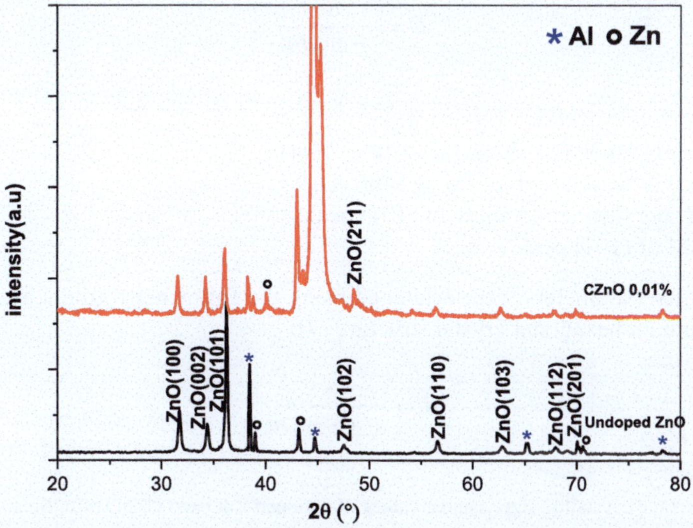

Fig. 2 XRD spectra of undoped and CZnO thin films treated 2 h at 500 °C

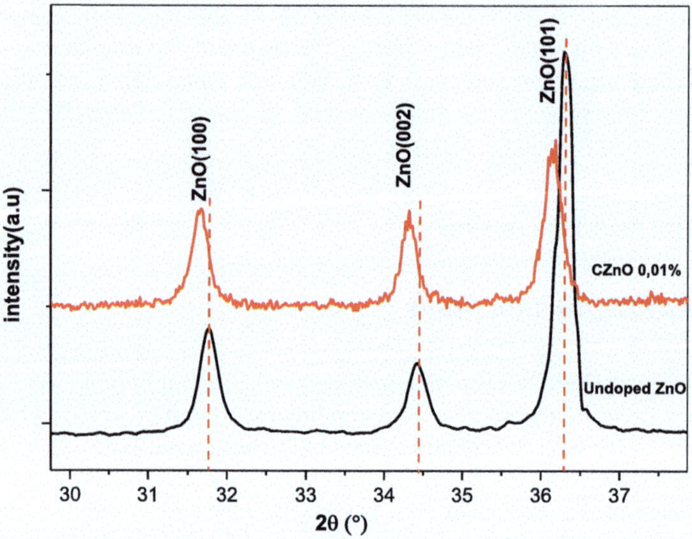

Fig. 3 XRD spectra of undoped and CZnO thin films treated 2 h at 500 °C show the shift of ZnO principal peaks towards smaller angles for CZnO sample

decrease of crystallite size, this is confirmed by lattice parameters values *a* and c (Table 1).

Average crystallite size calculated using Scherrer formula [5, 6]:

$$D = \frac{K\lambda}{\beta \cos \theta_B} \tag{1}$$

where

K a constant known as shape factor (k = 0.94),
λ The X-ray wavelength of Cu (1.5406 Å),
θ_B The Bragg diffraction angle of (101) peak,
β FWHM of (101) peak.

The lattice parameters were evaluated using the following formula available in the literature on hexagonal crystal structure [7]:

$$a = \frac{\lambda}{\sqrt{3} * \sin \theta} \tag{2}$$

Table 1 Average crystallite size, a and c values of undoped ZnO and CZnO thin films

Sample	$2\theta_{101}$ (°)	FWHM (°)	D (nm)	c (Å)	a (Å)
ZnO	36.299	0.225	39	5.206	3.250
CZnO	36.142	0.281	31	5.222	3.015

$$c = \frac{\lambda}{\sin\theta} \tag{3}$$

where

λ The wavelength of X-rays (1.5406 Å).
θ The diffraction angle of (100) for a parameter and the peak (002) for c parameter.
c_0 5.207 Å.

3.2 *Raman Spectroscopy Study*

Vibration properties of undoped and cobalt-doped ZnO thin films (CZnO) were investigated by Raman scattering technique in this work. Figure 4 shows the Raman spectra of studied samples, they consist of four modes: E_2^{Low}, $E_2^{(high)}$–$E_2^{(Low)}$, E_2^{high} and A_1(LO)/E_1(LO) of the hexagonal ZnO structure phonon. Also, the polar mode of A1(LO)/E1(LO) at around 570 cm^{-1} has been presented more intense for doped sample. The peaks intensity for the doped sample is greater than that of the undoped sample especially the peak E_2^{high}. This linked to the incorporation of Co ions the ZnO nanostructure. The appearance of A_1(TO) mode on the spectrum of CZnO sample, reflects the crystallization quality of the ZnO thin films of the hexagonal Wurtzite structure which confirms the XRD analysis results.

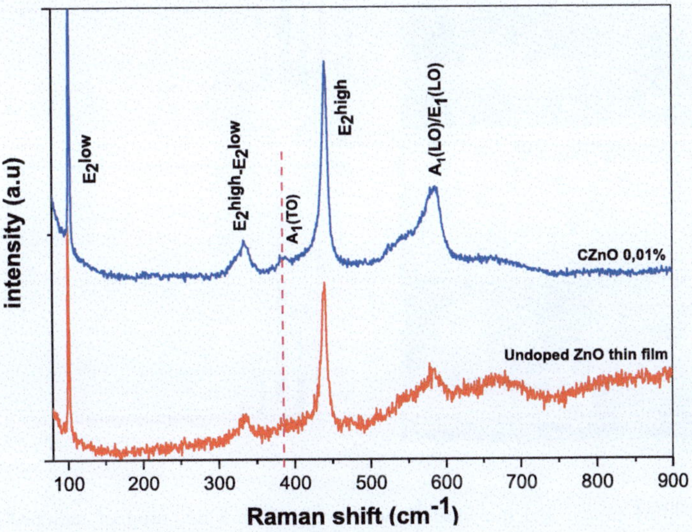

Fig. 4 Raman spectra of undoped and cobalt-doped ZnO thin films

3.3 Morphological and Wettability Characterizations

Figure 5a shows that the undoped ZnO film exhibits a compact morphology with pyramid-shape nanostructure attached to each other with a well-defined grain boundary and few nanorods. However, the surface morphology changed for cobalt doped sample; FEG-SEM image of cobalt-doped ZnO thin film (Fig. 5b) reveal more uniform and compact morphology with homogeneously dispersed spherical ZnO nanostructure. EDS analysis show the presence of zinc and oxygen in ZnO thin film (Fig. 5c) and no Co are detected in doped sample (Fig. 5d), which can be due to the low concentration of Co. The wetting properties of studied layers were analyzed by measuring the water contact angle of these layers. Measurements show that the obtained contact angle is 120.91° for undoped ZnO (inserted image in Fig. 5a) which indicates that the undoped ZnO film structure is inherently hydrophobic. The shape of the droplet on the surface of CZnO sample was more specific (inset of Fig. 5b) with contact angle of 128°, this indicates that Co improved the hydrophobic property of ZnO thin film.

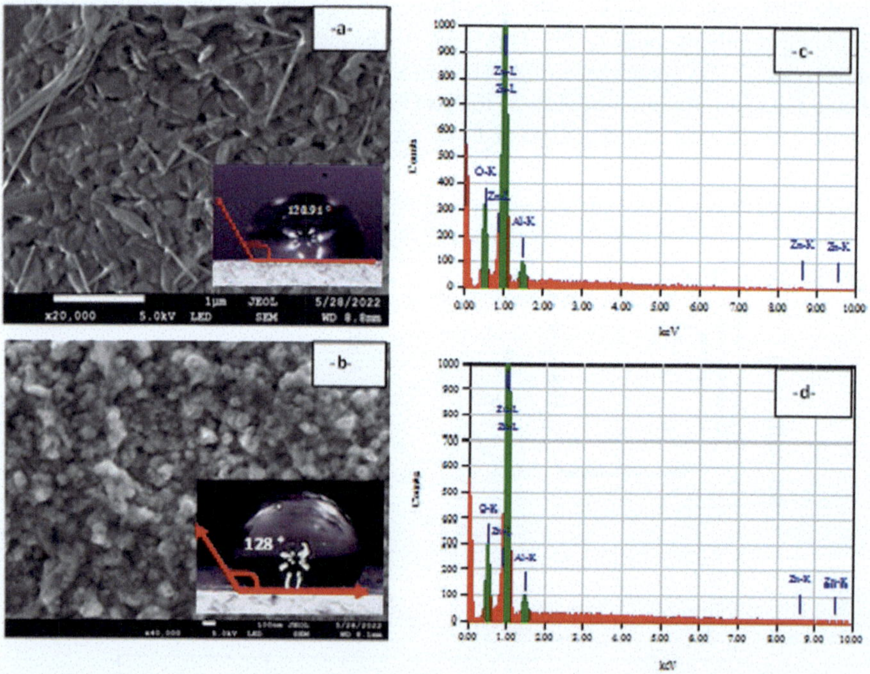

Fig. 5 FEG-SEM image and the inset show the image of water drop of undoped ZnO thin film (**a**) and of cobalt-doped ZnO thin films (**b**). (**c**) and (**d**) corresponding EDS with suitable indexing

4 Conclusion

In this work, the study is devoted to the cobalt-doping effect on the structural and morphological properties of the ZnO thin film which in turn affects the hydrophobicity of their surface. A low concentration of cobalt leads to a decrease in the average crystallite size and improves the structural state of the ZnO thin film. The change in the surface morphology after cobalt-doping enhances the hydrophobicity of the studied ZnO layer.

References

1. Kang Y, Yu F, Zhang L, Wang W, Chen L, Li Y (2021) Review of ZnO-based nanomaterials in gas sensors. Solid State Ionics 360:115544
2. Guermat N, Daranfed W, Bouchama I, Bouarissa N (2021) Investigation of structural, morphological, optical and electrical properties of Co/Ni co-doped ZnO thin films. J Mol Struct 1225:129134
3. Darenfad W, Guermat N, Mirouh K (2021) A comparative study on the optoelectronic performance of undoped, Mg-doped and F/Mg co-doped ZnO nanocrystalline thin films for solar cell applications. J Nano Electron Phys 13:06016
4. Belamri Z, Darenfad W, Guermat N (2023) Impact of annealing temperature on surface reactivity of ZnO nanostructured thin films deposited on aluminum substrate. J Nano Electron Phys 15:02026
5. Rabeel M, Javed S, Khan R, Aftab AM, Rehman S, Kim D, Khan MF (2022) Controlling the wettability of ZnO thin films by spray pyrolysis for photocatalytic applications. Materials 15:3364
6. Patra S, Sarkar S, Bera SK, Ghosh R, Paul GK (2009) Hydrophobic self-cleaning surfaces of ZnO thin films synthesized by sol–gel technique. J Phys D Appl Phys 42:075301
7. Zhao J, Ni J, Zhao X, Xiong Y (2011) Preparation and characterization of transparent conductive zinc doped tin oxide thin films prepared by radio-frequency magnetron sputtering. Univ Technol Mater Sci 26:388–392

Design of a New Photo-Diode Based on (α-PbO)/(α-SnO) Lateral Heterostructure

Ali Bakhtatou and Ali Hamidani

Abstract α-SnO and α-PbO monolayers are two stable nanostructures that are separately synthesized by epitaxial growth on the same substrate (SiO_2/Si) with superior functionalities, making them industrially important material for many applications. We have theoretically predicted a new lateral heterostructures $(\alpha\text{-PbO})_n$/$(\alpha\text{-SnO})_n$ formed by periodically repeating narrow SnO and PbO strips joined along their adjacent interface and investigated the structural, mechanical and electronic properties by using first-principles calculations. We demonstrate that creating these lateral heterostructures by α-PbO and by α-SnO monolayer materials will allow for adjusting various physical and chemical parameters such as tunable band gap, which are required to fit wide spectral range in optoelectronics applications. We obtained that the $(\alpha\text{-PbO})_7$/$(\alpha\text{-SnO})_7$ heterostructure shows an Anderson type-II band alignment where CBM is located at α-PbO domain while VBM is located at α-SnO domain. Consequently, the photon-generated electron-hole pairs are spatially separated, resulting in much longer exciton lifetime compared to that of the two monolayers separately which needs to be confirmed by further experimental measurements. This new structure can be used as a forward p-n diode and offer an interesting optoelectronic properties.

Keywords Two-dimensional materials · α-PbO monolayer · α-SnO monolayer · Lateral heterostructures · Band alignment

1 Introduction

Alpha lead-oxide (α-PbO) and alpha tin-oxide (α-SnO) 2D sheets were demonstrated to offer the possibility for oxide opto-electronics [1, 2].

Bulk α-PbO is a semiconductor with an indirect band gap of about ~ 1.9 eV [3]. Monolayer and multilayered α-PbO have been successfully synthesized [1].

A. Bakhtatou (✉) · A. Hamidani
8 Mai 1945 Guelma University, BP401, 24000 Guelma, Algeria
e-mail: bakhtatou.ali@univ-guelma.dz

S. Hamamda et al. (eds.), *Proceedings of the 2nd International Conference of Nanotechnology for Environmental Protection and Clean Energy Production*, Springer Proceedings in Materials 45, https://doi.org/10.1007/978-981-97-1916-7_14

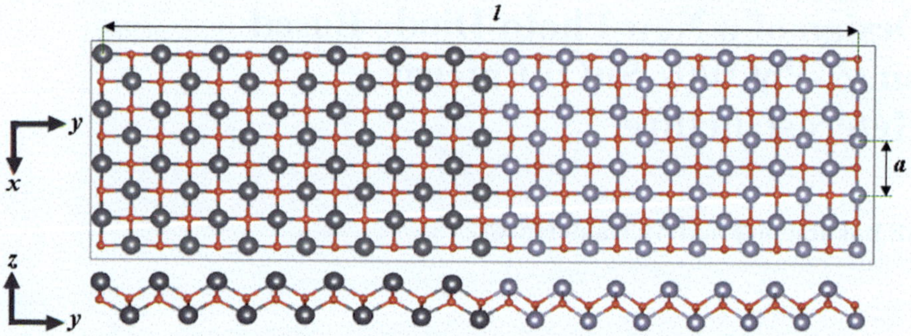

Fig. 1 Schematic representation of (α PbO)$_7$/(α SnO)$_7$ relaxed heterostructure cell with n = 7. Pb, Sn and O atoms are represented by grey, violet and small red balls respectively

Experimental results have suggested that these atomic sheets show hydrophobicity, microwave stability, thermal robustness, anti-corrosive behavior and acid resistance [1]. Moreover, α-PbO monolayer can be exploited in many applications in different fields; electronics, optoelectronics, photocatalysis, sensors and energy storage [1].

Bulk α-SnO has a small indirect band gap of about ~ 0.7 eV [4]. SnO monolayer was successfully synthesized and can be exploited in sensing as well as catalysts [5].

Here, we design theoretically novel electronic materials by stacking PbO and SnO horizontally to form lateral heterostructures (LHSs). These LHSs are labeled (α-PbO)$_n$/(α-SnO)$_n$, where n is the number of primitive cells in the stripe of α-PbO or α-SnO, n ranges from 1 to 10. The atomic representation of the heterostructure (α-PbO)$_7$/(α-SnO)$_7$ is presented in Fig. 1. We study theoretically the structural and electronic properties of these LHSs by first-principles calculations and prove that (α-PbO)$_7$/(α-SnO)$_7$ lateral heterostructure could function as a photodiode leading to device applications with remarkable functionalities.

2 Computation Details

The calculations reported in this paper work are performed within the projector augmented wave formalism (PAW) [6]. The generalized gradient approximation (GGA) of Perdew-Burke-Ernzerhof (PBE) [7] is used to approximate the exchange and correlation part as implemented in the Vienna ab initio simulation package (VASP) code [8]. A plane-wave basis set is take on with a kinetic-energy cutoff of 570 eV. For the electronic properties, the number of k-points in the reciprocal directions was chosen roughly in proportion to the size of the reciprocal lattice. 13 × 13 × 1 k-grid was under taken for α-PbO and α-SnO mono-layers. For the heterostructures we have used 13 × 6 × 1. The convergence criterion of energy value was chosen as 10^{-6} eV between two consecutive steps. Allowed maximum Hellmann–Feynman force convergence, on each atom in the cell, was set to 10^{-2} eV $Å^{-1}$. All calculations were performed with van der Waals correction (vdW) [9].

3 Structural Properties

2D α-PbO and α-SnO are isostructural with the space group P4/nmm. At room temperature, they crystallize in the tetragonal structure (Fig. 2a). The structure is multilayered (sheets) form parallel to the xOy plane (Fig. 2b). α-PbO monolayer (as well as α-SnO monolayer) is formed of a planar of oxygen atoms intercalated between two planes of Pb (Sn) atoms. (Fig. 2c, d). A lead or tin atoms are surrounded by four oxygen atoms at the corners of a tetrahedron (Fig. 2e, f). Our calculated lattice parameter of α-PbO (4.013 Å) and α-SnO (3.811 Å) are in good agreement with experiment values [1, 5].

The primitive cell of α-PbO (or α-SnO) is construct from 2 lead atoms (or 2 tin atoms) and 2 oxygen atoms. Lattice mismatch between α-PbO and α-SnO is 4.74%. $(\alpha\text{-PbO})_n/(\alpha\text{-SnO})_n$ contains 2n Pb atoms, 2n Sn atoms and 4n O atoms. Lateral heterostructure is extends in the y direction which give the width of the heterostructure. As well as, the lattice constant, *a,* is the interface length on the x side (Fig. 1). The calculated fundamental lattice parameters of 2D α-PbO and 2D α-SnO, the optimized lattice constants *a* and *l* of the heterostructures are shown in Table1.

4 Energetic Stability

The energetic stability of an interface can be characterized by its formation energy E_{form}. Defined as the energy required to create it from its constituent materials. The energy cost per primitive cell per unit length along the interface [10]:

$$E_{form} = (E_T - n(E_{PbO} + E_{SnO}))/(2a.n) \tag{1}$$

where E_T is the total energy of the heterostructure $(\alpha\text{-PbO})_n/(\alpha\text{-SnO})_n$, n is the number of primitive cells in the 2D strip of α-PbO (or α-SnO). E_{PbO} and E_{SnO} are the energy per primitive cell for α-PbO and α-SnO monolayers. *a* is the periodic length along the interface; factor 2 for the two interfaces within a supercell. According to this definition, negative E_{form} favors the creation of the interface. The calculated E_{form} values are shown in Table 1 for takes the values from 1 to 10. The calculated E_{form} values are negative for n = 1 to n = 8 and mean that the formation of these heterostructures is feasible. The interface is unstable for n = 9 and n = 10.

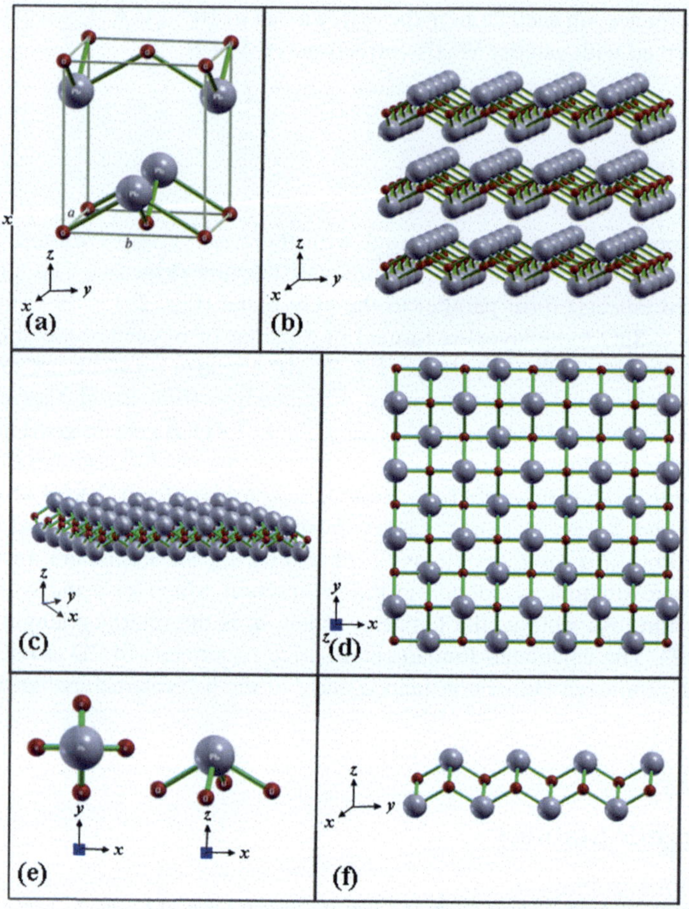

Fig. 2 **a** Tetragonal lattice structure of α-PbO bulk. **b** Layered structure of α-PbO. **c** Front view of a α-PbO monolayer. **d** Top view of a α-PbO monolayer. **e** Square pyramid of the lead atom with the 4 oxygen atoms. **f** Zigzag Pb–O bonds in the Oy direction (same shape in the Ox direction). *Remark* α-SnO bulk or monolayer have the same structures as the α-PbO structures above

5 Mechanical Stability

To investigate the mechanical stability of the heterostructures, we have calculated their independent four elastic constants such as C_{11}, C_{12}, C_{22} and C_{66} for α-PbO, α-SnO structures and eight $(\alpha\text{-PbO})_n/(\alpha\text{-SnO})_n$ heterostructures. The obtained elastic constants are shown in Table 2. The Born criteria stability [11]: $C_{11} > 0$, $C_{66} > 0$ and $C_{11} > |C_{12}|$ for α-PbO and α-SnO monolayers, and the three elastic stability criteria $C_{11} > 0$, $C_{66} > 0$ and $C_{11}.C_{22} > C_{12}.C_{21}$ for the heterostructures are met which confirms their mechanics stability.

Table 1 Optimized lattice parameters, formation energy (E_{form}) and Band gaps E_g of 2D PbO, 2D SnO and $(\alpha\text{-PbO})_n/(\alpha\text{-SnO})_n$ heterostructures, n ranges from 1 to 10

	a (Å)	l (Å)	E_{form} eV/nm	E_g eV
2DPbO	4.013			2.57 (d)
2DSnO	3.811			3.02 (i)
$n = 1$	3.966	7.696	– 0.103	2.45 (i)
$n = 2$	3.942	15.492	– 0.055	2.41 (d)
$n = 3$	3.940	23.299	– 0.035	2.28 (d)
$n = 4$	3.939	31.089	– 0.023	2.22 (d)
$n = 5$	3.935	38.896	– 0.016	2.17(d)
$n = 6$	3.932	46.710	– 0.010	2.14 (d)
$n = 7$	3.932	54.472	– 0.006	2.14 (d)
$n = 8$	3.932	62.264	– 0.003	2.12 (d)
$n = 9$	3.929	70.098	+ 0.001	
$n = 10$	3.930	77.880	+ 0.001	

(d) and (i) indicate direct and indirect band gaps respectively

Table 2 Elastic constants C_{11}, C_{12}, C_{22} and C_{66} of bare 2D α-PbO, α-SnO and $(\alpha\text{-PbO})_m/(\alpha\text{-SnO})_m$ heterostructures, n ranging from1 to 8

	$C11$ N/m	$C12$ N/m	$C22$ N/m	$C66$ N/m
2DPbO	42.4	28.2	42.4	28.9
2DSnO	49.1	40.4	49.1	39.7
$n = 1$	44.1	33.3	45.7	33.0
$n = 2$	43.6	33.0	46.2	32.9
$n = 3$	43.7	32.8	45.7	32.8
$n = 4$	43.6	32.7	45.7	32.7
$n = 5$	43.5	32.6	45.5	33.0
$n = 6$	43.5	32.6	45.5	32.9
$n = 7$	42.6	32.8	46.0	32.9
$n = 8$	42.8	33.1	45.4	33.1

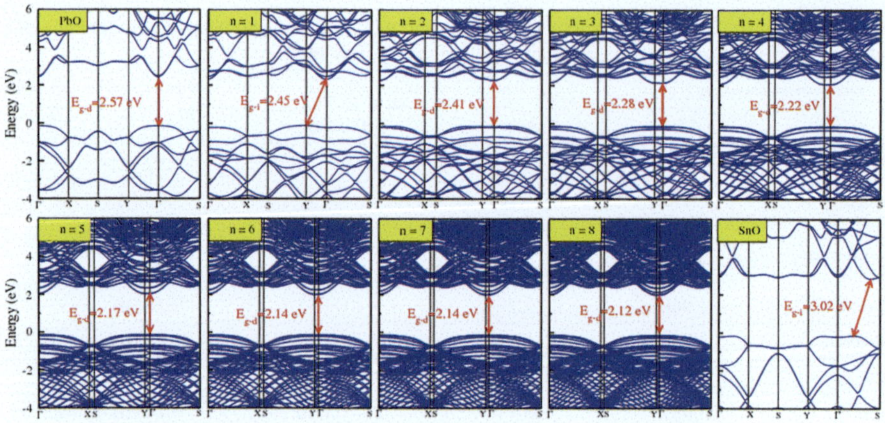

Fig. 3 Band structure of α-PbO and α-SnO monolayers, and $(\alpha\text{-PbO})_n/(\alpha\text{-SnO})_n$ lateral heterostructures along the high symmetry directions of BZ for n ranging from 1 to 8. E_{g-d} and E_{g-i} indicate direct and indirect band gaps respectively

6 Electronic Structure

6.1 Band Structure

In this part, we calculated the electronic properties of the α-PbO and α-SnO monolayers, as well as the eight $(\alpha\text{-PbO})_n/(\alpha\text{-SnO})_n$ heterostructures. Figure 3 shows the calculated electronic band structures. We found that 2D α-PbO is a direct band gap semiconductor of $E_{g-d} = 2.57$ eV while 2D α-SnO is a large indirect gap semiconductor of $E_{g-i} = 3.02$ eV which is consistent with previous theoretical results [12, 13]. We also note that among these lateral heterostructures, only the $(\alpha\text{-PbO})_1/(\alpha\text{-SnO})_1$ heterostructure has an indirect band gap of 2.45 eV, and its minimum direct band gap occurs at the Γ point with a value of 2.55 eV. The other heterostructures show a direct band gap at the Γ point.

The calculated band gap values are listed in Table 1. We found that the band gaps of the $(\alpha\text{-PbO})_n/(\alpha\text{-SnO})_n$ lateral heterostructures are smaller than the band gap values of PbO and SnO monolayers and also than their average value of 2.80 eV. As well as, all the band gaps of $(\alpha\text{-PbO})_n/(\alpha\text{-SnO})_n$ LHSs exhibit in visible band which is rare among semiconducting 2D materials [1].

6.2 1D Quantum Well Structure

We have also calculated the planar averaged potential and the macroscopic-averaged potential $V(y)$ in y direction as shown in Fig. 4. The maximum potential value between PbO and SnO sides is 0.876 V.

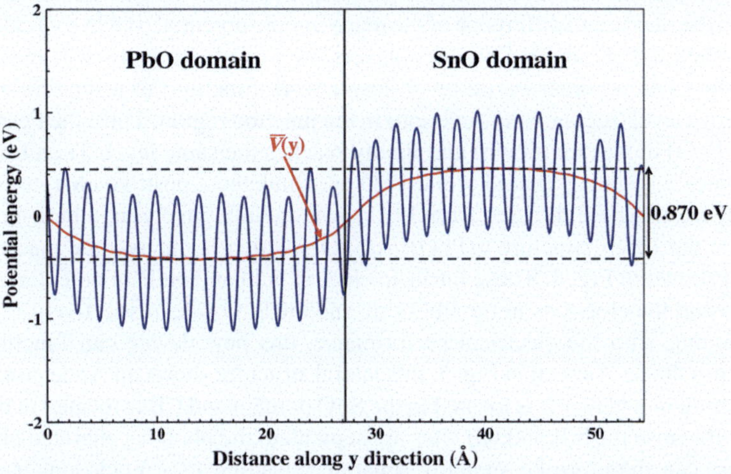

Fig. 4 Average electronic potential energy V(y) along y direction in $(\alpha PbO)_7/(\alpha SnO)_7$ heterostructure

In Fig. 5, we plotted valence band maximum (VB), conduction band minimum (CB) and Fermi level (E_F) of α-PbO and α-SnO monolayers before and after junction of $(\alpha\text{-PbO})_7/(\alpha\text{-SnO})_7$ heterostructure. The work function is the Fermi level counted from vacuum which is positioned at 0 eV. It can be seen that the work function of α-PbO and α-SnO monolayers and heterostructures are – 4.438 eV, – 5.258 eV and – 4.642 eV, respectively.

The work function difference between the two monolayers is equal to their Fermi level difference. For α-PbO and α-SnO, this difference is 0.820 eV, which is very

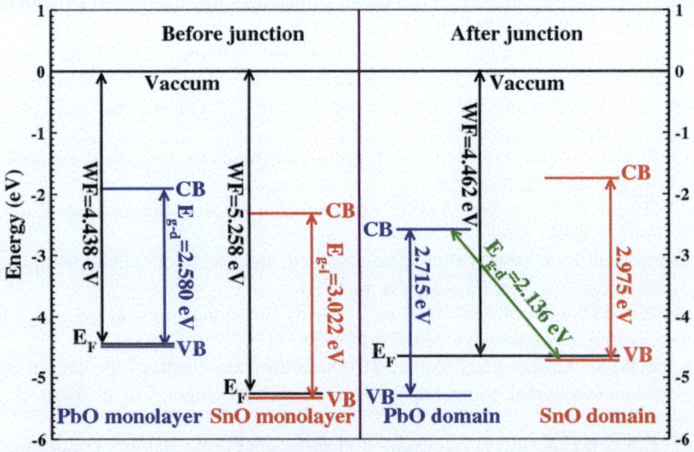

Fig. 5 Band alignment of α-PbO and α-SnO monolayers, and the $(\alpha\text{-PbO})_7/(\alpha\text{-SnO})_7$ heterostructure

similar to the maximum difference of the macroscopic potential (0.876 eV) calculated above. When α-PbO and α-SnO monolayers are joined, electrons will move from n-type PbO with a higher Fermi level (lower work function) to p-type SnO with a lower Fermi level (higher work function) in the junction region. Thus, the Fermi level of SnO (PbO) moves up (down) and finally reaches the same level. Therefore, SnO accumulates a negative charge while PbO accumulates a positive charge, forming a built in electric field directed from PbO towards SnO, involving a built-in potential inside the heterostructure $\alpha(PbO)_7/(\alpha SnO)_7$ which is the average macroscopic potential shown in Fig. 4. These intrinsic electric field and built-in potential confine the holes and the electrons in the PbO side and the SnO side, respectively.

In keeping with the electrical performance, this new device can function as a forward p-n diode. As seen in Fig. 5, this lateral structure shows an Anderson type-II band alignment where VB is located in the SnO domain and CB is located in the PbO domain. Because the VB and CB sites are separated, the photon-generated electron-hole pairs can therefore be spatially separated, leading to a much longer exciton lifetime and the possible presence of exciton condensation between the layers, which may be beneficial for image development, diodes, photoelectric devices and lasers.

References

1. Kumar P, Liu J, Ranjan P, Hu Y, Yamijala SS, Pati SK, Irudayaraj J, Cheng GJ (2018) Alpha lead oxide (a-PbO): a new 2d material with visible light sensitivity. Small 14(12):1703346
2. Miller SA, Gorai P, Aydemir U, Mason TO, Stevanovi V, Toberer ES, Snyder GJ (2017) SnO as a potential oxide thermoelectric candidate. J Mater Chem C5:8854–8861
3. Keezer RC, Bowman DL, Becker JH (1968) Electrical and optical properties of lead oxide single crystals. J Appl Phys 39:2062
4. Hiramatsu H, Nomura K, Yanagi H, Kamiya T, Kimura M, Hirano M, Hosono H (2009) Tin monoxide as an s-orbital-based p-type oxide semiconductor: electronic structures and TFT application. Phys Status Solidi B 206:2187–2191
5. Daeneke T, Atkin P, Orrell-Trigg R, Zavabeti A, Ahmed T, Walia S, Liu M, Tachibana Y, Javaid M, Greentree AD, Russo SP, Kaner RB, Kalantar-Zadeh K (2017) Wafer-scale synthesis of semiconducting SnO monolayers from interfacial oxide layers of metallic liquid tin. ACS Nano 11(11):10974–10983
6. Kresse G, Joubert D (1999) From ultrasoft pseudopotentials to the projector augmented-wave method. Phys Rev B 59:1758–1775
7. Perdew JP, Burke K, Ernzerhof M (1996) Generalized gradient approximation made simple. Phys Rev Lett 77:3865–3868
8. Kresse G, Furthmüller J (1996) Efficient iterative schemes for ab initio total-energy calculations using a plane-wave basis set. Phys Rev B 54:11169
9. Grimme S (2006) Semiempirical GGA-type density functional constructed with a long-range dispersion correction. J Comput Chem 27(15):1787–1799
10. Zhang J, Xie W, Xu X, Zhang S, Zhao J (2016) Structural and electronic properties of interfaces in graphene and hexagonal boron nitride lateral heterostructures. Chem Mater 28(14):5022–5028
11. Mouhat F, Coudert F-X (2014) Necessary and sufficient elastic stability conditions in various crystal systems. Phys Rev B 90:224104
12. Bakhtatou A, Ersan F (2019) Effects of the number of layers on the vibrational, electronic and optical properties of alpha lead oxide. Phys Chem Chem Phys 21:3868–3876

13. Abdullahi YZ, Bakhtatou A, Mogulkoc Y, Ersan F (2023) Giant band gap bowing in two dimensional $Pb_{1-x}Sn_xO$ alloys and Janus PbSnO monolayers. J Phys Chem Solids 175:111203

Numerical Investigation of Laminar Forced Convection Heat Transfer Nanofluids Flow Using Different Base Fluids

A. Laichi, A. Bouhezza, H. Köten, and O. Kholai

Abstract In this paper, Numerical investigation of laminar forced convection heat transfer nanofluids flow through a horizontal circular pipe under a constant heat flux using different base fluids. In the first part, it has been studied the effect of using several base fluids such as water, engine oil and mercury at different temperature ranging from 280 to 320 K and at different Reynolds number values ranging from 100 to 2100 on average Nusselt number and pressure drop. In the second part, it has been studied the effect of suspended of Al_2O_3 at 3% volume fraction, 295 K and Reynolds number equal 1460 on heat transfer coefficient and local Nusselt number. The governing equations have been modeled using finite volume method approach with SIMPLE algorithm helping with ANSYS FLUENT software. The results indicate that the average Nusselt number and pressure drop for all base fluids increase as Reynolds number increase and decrease as temperature decrease. Engine oil has a higher average Nusselt number and pressure drop while mercury has a small average Nusselt number and pressure drop. Also, mercury has a small change with temperature. Moreover, average Nusselt number is only function of Prandtl and Reynolds number but pressure drop varies with Reynolds number. Add of Al_2O_3 nanoparticle to the base fluids enhanced the heat transfer coefficient and local Nusselt number for all nanofluids. It has been found that mercury has a higher heat transfer coefficient and small Nusselt number due to higher thermal conductivity and small thermal boundary layer thickness.

Keywords Heat transfer · Fluid flow · Nanofluids · Transport phenomenon

---

A. Laichi (✉) · A. Bouhezza
Skikda University, 21000 Skikda, Algeria
e-mail: a.laichi@univ-skikda.dz

H. Köten
Medeniyet University, Istanbul, Turkey

O. Kholai
Constantine-1-University, Constantine, Algeria

S. Hamamda et al. (eds.), *Proceedings of the 2nd International Conference of Nanotechnology for Environmental Protection and Clean Energy Production*, Springer Proceedings in Materials 45, https://doi.org/10.1007/978-981-97-1916-7_15

1 Introduction

The enhancement of heat transfer characteristic of base fluid using in industries such as, heat exchangers, electronics and energy storage and transportation is one of important research. There are two conventional methods using for enhancement the heat transfer, the passive method using the extended surface and rough surface and the active method using external power. Two methods doesn't sufficient for find a high performance of heating and cooling systems. The use of ultra-fine particle such as Al_2O_3, TiO_2 and SiO_2 dispersed in water has been reported by Masuda [1]. The author showed that the thermal conductivity of Al_2O_3-Water and TiO_2-Water nanofluids increased with increasing of volume concentration and no change for SiO_2-Water nanofluid. Also, for the effective dynamic viscosity increased with increasing of volume concentration and decreased with increasing of temperature. Moreover, it observed that the suspension has a higher dispersion for spherical nanoparticle. A new class of fluid call Nanofluids contained a small size of nanoparticle less than 100 nm has been found by Choi [2]. It reported that the dispersion of Cu nanoparticle in water enhanced the heat transfer coefficient as the thermal conductivity and flow rate increased. Palm et al. [3] studied the laminar forced convective heat transfer enhancement using Al_2O_3-Water nanofluids in radial flow cooling systems considering temperature-dependent properties. It found that the average heat transfer coefficient at 4% volume concentration increased with 25% than water base fluid. However, the modeling using temperature-dependent nanofluids properties gave a better heat transfer enhancement predictions as well as lower wall shear stresses than when considering constant properties throughout the domain. Convective heat transfer for water based Alumina nanofluids in a single 1.02 mm tube have been studied by Lai et al. [4]. The results showed that the heat transfer coefficient increased with volume flow rate and volume fraction of nanofluids. In addition, in the developing region, the HTC enhancement decreased with increasing axial distance from the test section entrance and it was higher for large volume fraction and volume flow rate, the fully developed region showed a small enhancement of HTC because stabilized of boundary layer. Many experimental studies have been reported for this reason, Kim et al. [5]. It studied the effect of dispersion of Al_2O_3 in water on the convective heat transfer coefficient in laminar and turbulent flow condition and found that the maximum enhancement showed when used Al_2O_3 in turbulent flow condition. In this study, the first one we describe the effect of different base fluid such as water, engine oil and mercury, Reynolds number and temperature on enhancement of heat transfer and pressure drop, in the second part, we describe the effect of addition Al_2O_3 nanoparticle to water, engine oil and mercury at 3% volume fraction on local heat transfer coefficient. 2D Laminar forced convective heat transfer in horizontal circular pipe are investigated.

2 Mathematical Model

Figure 1 shows representation of geometry using in this study. The flow is steady, axisymmetric laminar forced convective in horizontal circular pipe has a 4.57 mm diameter and 2 m length, nanoparticle and base fluid are in thermal equilibrium and single phase model is applied, no-slip condition and constant heat flux at the wall 2089.56 W/m^2.

The properties of different base fluids at different temperatures used in this study presented by Bergman et al. [6]. Also, the properties of different nanofluids at 295 K and 3% volume concentration of Al_2O_3 show in Table 1.

2.1 Continuity

$$\frac{1}{r}\frac{\partial}{\partial x}\left(\rho_{nf} V_x\right) + \frac{\partial}{\partial x}\left(\rho_{nf} V_r\right) + \frac{\rho_{nf} V_r}{r} = 0 \tag{1}$$

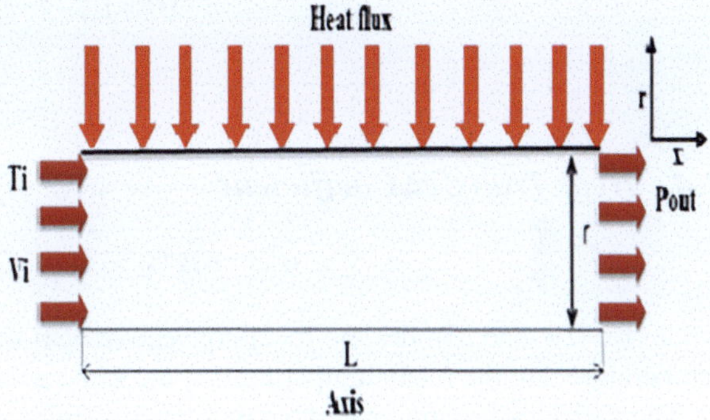

Fig. 1 Physical model configuration

Table 1 Nanofluids properties

Fluid	Density (kg/m^3)	Specific heat capacity (J/kg K)	Dynamic viscosity (N s/m^2). 10^{-6}	Thermal conductivity (W/m K). 10^3	Prandtl number
Al_2O_3 [7]	3970	765	/	40,000	/
Water-Al_2O_3	1087.16	3806.77	1275.18	659.6	7.3594
Engine oil-Al_2O_3	979.5385	1751.89	987,300	158.3	10,926.35
Mercury-Al_2O_3	13,254.06	145.12	2065	8902	0.03366

2.2 Momentum

$$Vr\frac{\partial Vx}{\partial r} + Vr\frac{\partial Vx}{\partial x} = -\frac{1}{\rho_{nf}}\frac{\partial P}{\partial x} + \upsilon_{nf}\left[\frac{1}{r}\frac{\partial}{\partial r}\left(r\frac{\partial V_x}{\partial r}\right) + \left(\frac{\partial^2 V_x}{\partial x^2}\right)\right] \tag{2}$$

$$Vr\frac{\partial Vr}{\partial r} + Vr\frac{\partial Vr}{\partial x} = -\frac{1}{\rho_{nf}}\frac{\partial P}{\partial r} + \upsilon_{nf}\left[\frac{1}{r}\frac{\partial}{\partial r}\left(r\frac{\partial V_r}{\partial r}\right) + \left(\frac{\partial^2 V_r}{\partial x^2}\right) - \frac{V_r}{r^2}\right] + g_r \tag{3}$$

2.3 Energy

$$V_r\frac{\partial T}{\partial r} + V_r\frac{\partial T}{\partial x} = \alpha_{nf}\left(\frac{1}{r}\frac{\partial}{\partial r}\left(r\frac{\partial T}{\partial r}\right) + \left(\frac{\partial^2 T}{\partial x^2}\right)\right) \tag{4}$$

where

$$\alpha_{nf} = \frac{K_{nf}}{(\rho c_p)_{nf}}$$

2.4 Heat Transfer Coefficient

$$q'' = h_x(T_w - T_m) \tag{5}$$

where hx, Tw and Tm are the local convective heat transfer coefficient, local wall temperature and the mean temperature at the same axial position respectively.

The mean temperature calculate with energy balance as:

$$T_m = \frac{2}{R^2 U_m}\int_0^R u(r)T(r)rdr \tag{6}$$

$$\phi = q''(\pi D_h L) = mc_p(T_{mi} - T_{mo}) \tag{7}$$

$$\phi = Sh_{ave}(T_w - T_m) \tag{8}$$

Sh_{ave}, T_w and T_m are the average convective heat transfer coefficient, average local wall temperature and the mean temperature of fluid respectively, ϕ is the rate of heat transfer.

2.5 Nusselt Number

Nusselt number is the ratio between the convection heat flux and conduction heat flux.

$$Nu = \frac{h}{k_{eff}} D_h = \frac{convection\,heat\,flux}{conduction\,heat\,flux} \tag{9}$$

where h, keff and Dh are the heat transfer coefficient, effective thermal conductivity and hydraulic diameter respectively. keff depend on the mean temperature of the fluid flow.

2.6 Nanofluids Properties

Thermo physical properties of nanofluids such as density, effective thermal conductivity, effective dynamic viscosity and heat specific capacity are calculated using different empirical model, which are:

Effective density of nanofluids is calculated using the model of Pak and Cho [8]:

$$\rho_{eff} = (1 - \varphi)\rho_f + \varphi\rho_s \tag{10}$$

where ρ_f, φ and ρ_s are density of base fluid, volume concentration and density of nanoparticle respectively.

Effective specific heat capacity is calculated using the model of Xuan and Roetzel's [9] and it is obtained in thermal equilibrium mixture of base fluid and nanoparticle.

$$C_{peff} = \frac{(1 - \varphi)(\rho C_p)_f + \varphi(\rho C_p)_s}{\rho_{eff}} \tag{11}$$

The effective thermal conductivity is calculated using the model of Hamilton and Crosser [10] for spherical nanoparticles:

$$\frac{k_{eff}}{k_f} = \frac{k_s + 2k_f - 2\varphi(k_f - k_s)}{k_s + 2k_f + \varphi(k_f - k_s)} \tag{12}$$

The effective dynamic viscosity is calculated using the model of Maïga et al. [11] for spherical nanoparticles:

$$\frac{\mu_{nf}}{\mu_f} = (1 + 7.3\varphi + 123\varphi^2) \tag{13}$$

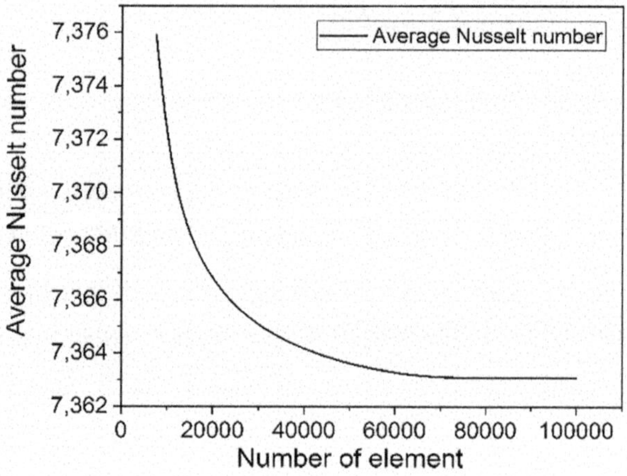

Fig. 2 Grid independency study

3 Numerical Method

We studied laminar forced convective heat transfer and fluid flow using different base fluid in the first step, after studied the effect of add nanoparticle on heat transfer enhancement and hydrodynamic behavior. ANSYS FLUENT 2020 R1 Computational fluid dynamic language used in this study. The steady state incompressible coupled non-linear partial differential equations have discretized using the finite volume method. A second order upwind scheme uses for pressure, momentum and energy while the SIMPLE algorithm applied for the velocity–pressure coupling. Converged solutions have been considered when the residual in each equation 10^{-6}. The non-uniform grid has been adopted in the radial direction, which is finer at an adjacent tube wall and uniform grid for axial direction. To find a grid independence solution, several element grids have studied at Reynolds number equal 1620.

For this study using different element grid 7500, 10,000, 15,000, 30,000, 60,000, 80,000, 100,000 and the average Nusselt number error between 60,000 grid and 80,000 less than 1%, finally selected 60,000 elements grid, which is captured the maximum variation of parameters studied. The variation of Nusselt number function of number of element shows in Fig. 2.

4 Results and Discussion

For illustrate the accuracy of results Fig. 3. It Shows a comparison between different theoretical and numerical results with the present model of local Nusselt number function of dimensionless axial position with Shah and London equation [12] and numerical result of Javad Bayat [13] at Reynolds number 1620 for water at 295 K.

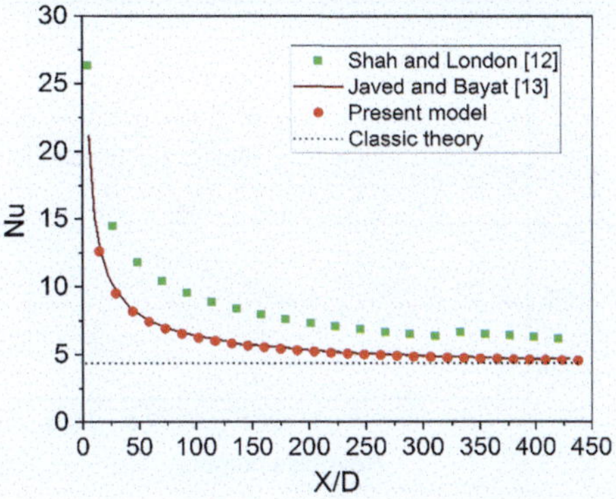

Fig. 3 Axial variation of the local Nusselt number at several studies

4.1 *Shah and London Equation*

We have two equations of Shah and London [12], Eq. (14) using for thermal developing region and Eq. (15) using for thermal fully developed region.

$$Nu(x) = 1.953\left(RePr\frac{D_h}{x}\right)^{1/3} \quad \text{For } RePr\frac{D_h}{x} \geq 33.3 \tag{14}$$

$$Nu(x) = 4.364 + 0.0722\left(RePr\frac{D_h}{x}\right)^{1/3} \quad \text{For } RePr\frac{D_h}{x} < 33.3 \tag{15}$$

4.2 *Effect of Temperature on the Average Nusselt Number for Different Base Fluids*

Figure 4 shows the variation of the average Nusselt number of water base fluid at three different temperatures and different Reynolds numbers. It observes linear increasing of average Nusselt number with increasing of Reynolds number because the increase of Reynolds number decrease the thickness of boundary layer due to enhance the heat transfer. Average Nusselt number decrease with increasing of temperature because temperature affects the thermo physical properties of water, this effect shows in Prandtl number when decrease 2.72 times in temperatures ranging from 280 to 320 K and when increase Reynolds number between 100 and 2100. This results confirmed that the average Nusselt number change function of Reynolds and Prandtl number (thermo-physical properties especially dynamic viscosity). Figure 5 shows the variation of average Nusselt number of engine oil base fluid at three different temperatures and different Reynolds numbers. The same variation of engine oil showed with water but it is not linear, average Nusselt number of engine oil at 2100 and 280 K higher than.

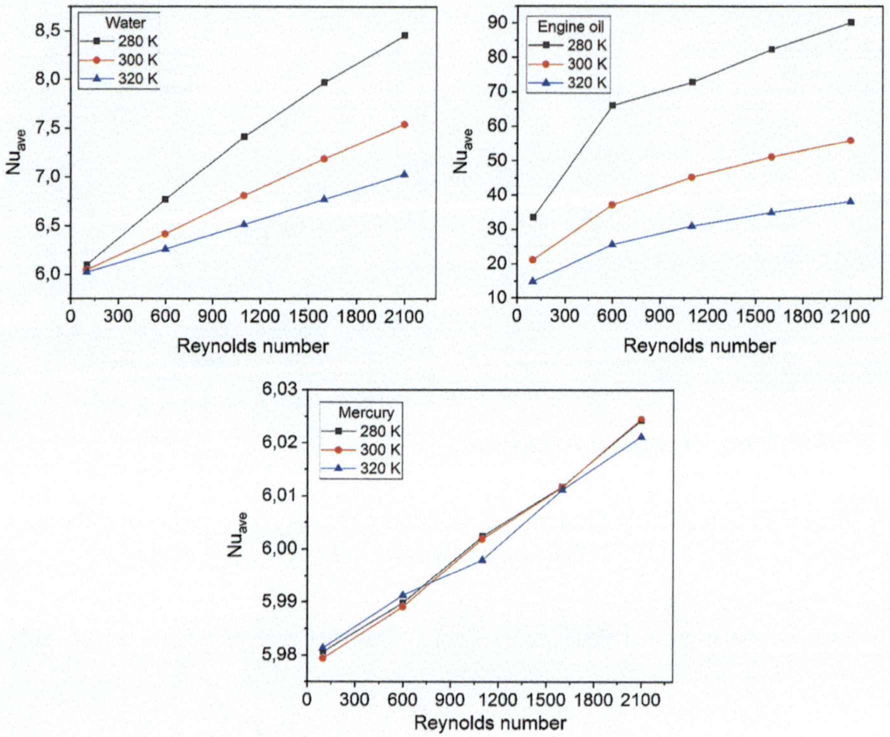

Fig. 4 The average Nusselt number of different base fluids at different Reynolds numbers and different temperatures

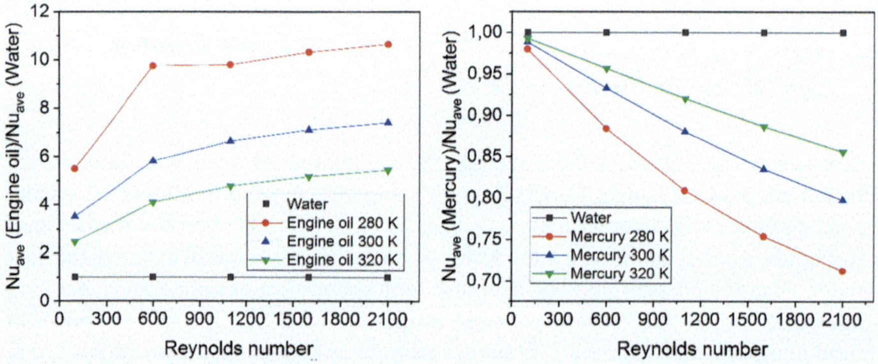

Fig. 5 The average Nusselt number ratio of engine oil and mercury compared with water at different Reynolds numbers and temperatures

Water with 10 times, because higher Prandtl number of engine oil 27,500 compare with Prandtl number of water 10.26. Prandtl number affects the thermal boundary layer thickness. Figure 6 shows the variation of average Nusselt number of mercury base fluid at three different temperature and different Reynolds numbers. Mercury has inverse behaviour with temperature and Reynolds number, average Nusselt number is great at high temperature and low Reynolds number, low temperature and high Reynolds number, but this variation is not large because of small Prandtl number and this cause due to small average Nusselt number of mercury than water and engine oil.

Figure 5 shows the ratio of average Nusselt number of engine oil and mercury respectively compared with water base fluid at different temperatures. It observed that at Reynolds number 2100 and temperature 280 K average Nusselt number of engine oil higher that water with 10 times, mercury is less with 25% than water at the same conditions of Reynolds number and temperature. The thermo physical properties such as dynamic viscosity affect the heat transfer behavior and it has high sensibility with temperature.

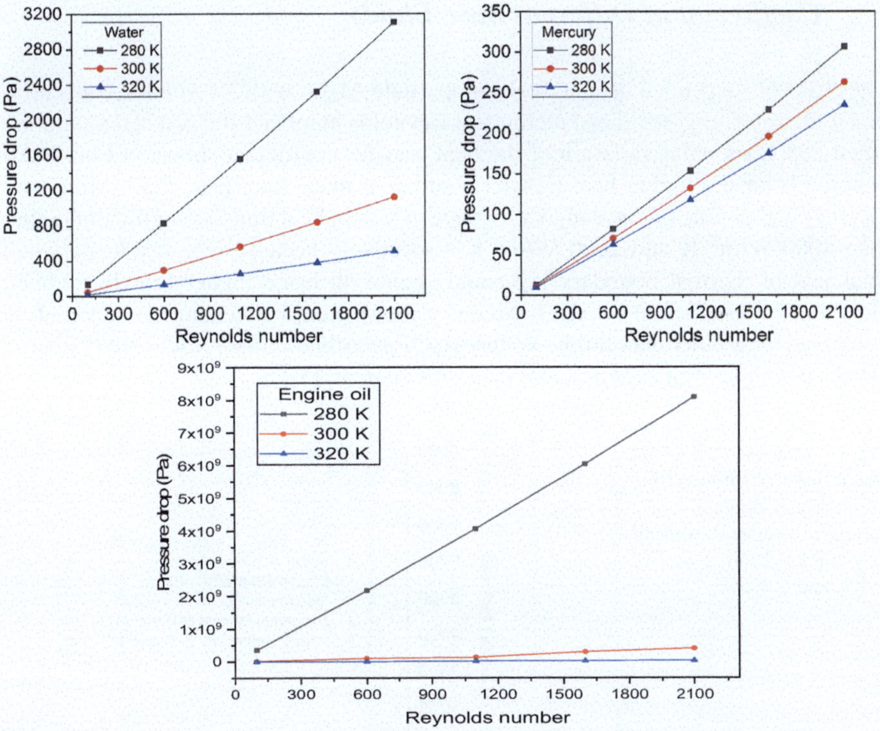

Fig. 6 Pressure drop of water, engine oil and mercury at different temperature and Reynolds numbers

4.3 Effect of Temperature on Pressure Drop for Different Base Fluids

The better base fluid whose has a higher heat transfer and lower pressure drop, in this reason we present compared study of pressure drop of different base fluids at different temperatures. Figure 6 illustrate the effect of temperature on pressure drop for several base fluids water, engine oil and mercury respectively, as a function of Reynolds. Pressure drop increase with increasing of Reynolds number and decrease with increasing of temperature for all base fluids. Temperature affects especially the dynamic viscosity, density with small change, it mean Prandtl number. It shows that higher pressure drop due to higher Prandtl (dynamic viscosity) equivalent to engine oil base fluid, it has a higher pressure drop than water and mercury. The effect of temperature shows especially for engine oil and water but it is small for mercury.

4.4 Effect of Nanoparticle on Local Heat Transfer Coefficient of Different Base Fluids

The effect of suspended of spherical nanoparticle Al_2O_3 with 3% volume concentration with water, engine oil and mercury at Reynolds number 1460 and at temperature equal 295 K on axial variation of the heat transfer coefficient shows in Fig. 7. All nanofluids have a higher heat transfer coefficient than water base fluid. Moreover, mercury and engine oil have higher average values of heat transfer coefficient higher than 8000 W/m^2 K and 2000 W/m^2 K respectively because mercury has a small thickness of thermal boundary layer and engine oil has a higher Prandtl number. Heat transfer coefficient of Al_2O_3-Mercury shows a uniform change state variation compared with other nanofluids. Water-Al_2O_3 nanofluid in the present study gives a good agreement with experimental data of Kim et al. [5].

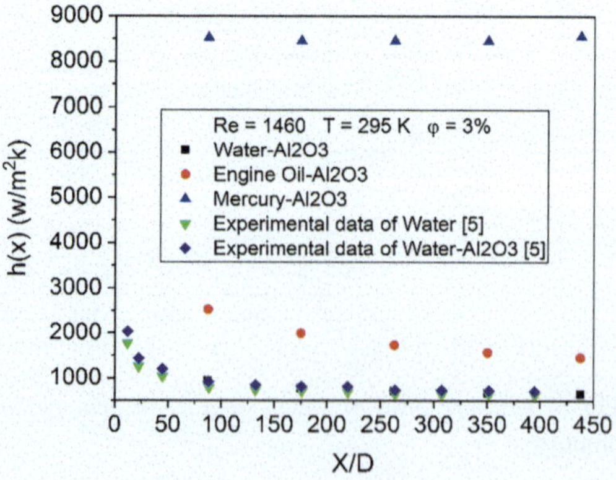

Fig. 7 Axial variation of the heat transfer coefficient of different nanofluids compared with experimental data at Re = 1460, T = 295 K and $\varphi = 3\%$

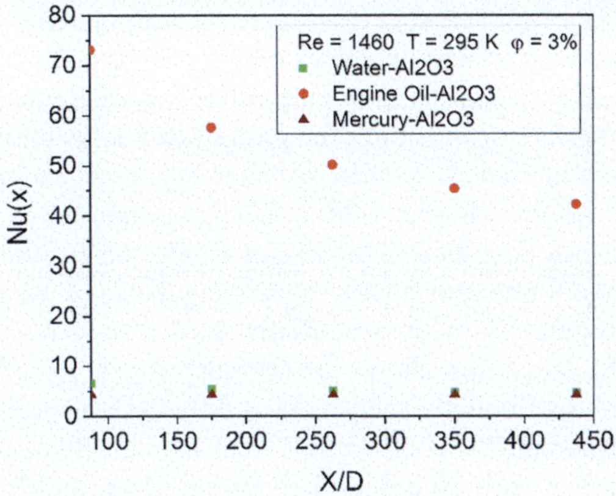

Fig. 8 Axial variation of the Local Nusselt number of different nanofluids at Re = 1460, T = 295 K and φ = 3%

4.5 *Effect of Nanoparticle on Local Nusselt Number of Different Base Fluids*

Figure 8 present the effect of different nanofluids water-Al_2O_3, engine oil-Al_2O_3 and mercury-Al_2O_3 at 3% volume concentration, temperature equal 295 K and Reynolds number equal 1460 on the axial local Nusselt number. Engine oil-Al_2O_3 nanofluid give a higher estimation of local Nusselt number with 10 order than Water-Al_2O_3 and Mercury-Al_2O_3 nanofluids because of higher convection heat flux than conduction heat flux at the surface of the tube. Mercury-Al_2O_3 nanofluid has a higher thermal conductivity about 8.902 W/m K so; the conduction heat flux is favorable in this case.

5 Conclusion

To investigate of 2D laminar forced convective heat transfer nanofluids flow in circular pipe under a constant heat flux we use a three different base fluids water, engine oil and mercury at different temperatures and different Reynolds numbers in the first step. Second step the same studied using Al_2O_3 nanoparticle at 3% volume concentration and at Reynolds number and temperature equal 1460 and 295 K respectively. Average Nusselt number increase with increasing of Reynolds number because decrease of thickness of dynamic boundary layer whiles its decrease with increasing of temperature because decrease of Prandtl number (decrease of dynamic viscosity and increase of thermal conductivity), this results showed for water and engine oil while does not show a large effect for mercury. Engine oil has a higher average Nusselt number than water than mercury because a higher Prandtl number. It illustrate that average Nusselt number change with Reynolds and Prandtl number Nu =

f (Pr, Re). Pressure drop increase with increasing of Reynolds number and decrease with increasing of temperature for all base fluids. However, pressure drop has higher values when using engine oil at higher Reynolds number and small temperature. Pressure drop is only function of Reynolds number. In the second part using Al_2O_3 nanoparticle suspended in water, engine oil and mercury at 3% volume concentration, Reynolds number 1460 and temperature 295 K. Add nanoparticle increase the heat transfer coefficient for all nanofluids, this results confirmed with compared of Water-Al_2O_3 nanofluids with experimental data of water Kim et al. [5]. Mercury-Al_2O_3 and Engine oil-Al_2O_3 nanofluids show a large heat transfer coefficient than water and the maximum shows for Mercury-Al_2O_3 nanofluid because has small thickness of thermal boundary layer. Mercury-Al_2O_3 nanofluid has a uniform evolution of axial heat transfer coefficient. Local Nusselt number of Engine oil-Al_2O_3 nanofluid is higher with 10 orders than Water-Al_2O_3 and Mercury-Al_2O_3 nanofluids and it has maximum in developing region and decrease in fully developed region because the thermal boundary layer is constant. Also, Prandtl number affect the local Nusselt number, higher evolution local Nusselt number shows in higher Prandtl number of Engine oil-Al_2O_3 nanofluid, small local Nusselt number shows for Mercury-Al_2O_3 nanofluid because it has higher conductivity. Temperature affects the heat transfer and hydrodynamic behavior. Nusselt number and heat transfer coefficient are function only with Reynolds number and Prandtl number. Dynamic viscosity has a higher effect on pressure drop for all base fluids. The use of nanoparticle enhances the heat transfer coefficient and it is change with base fluids. The use of nanofluids based on its heat transfer condition, base fluid application and environmental.

References

1. Masuda H, Ebata A, Teramae K, Hishinuma N (1993) Alteration of thermal conductivity and viscosity of liquid by dispersing ultra-fine particles. Netsu Bussei 7(4):227–233
2. Choi SUS, Eastman JA (1995) Enhancing thermal conductivity of fluids with nanoparticles (vol 231/MD 66). ASME Publications FED, pp 99–105
3. Palm SJ, Roy G, Nguyen CT (2006) Heat transfer enhancement with the use of nanofluids in radial flow cooling systems considering temperature-dependent properties. Appl Therm Eng 26(17):2209–2218
4. Lai WY, Vinod S, Phelan PE, Prasher R (2009) Convective heat transfer for water-based alumina nanofluids in a single 1.02-mm tube. J Heat Transf 131:112401
5. Kim D et al (2009) Convective heat transfer characteristics of nanofluids under laminar and turbulent flow conditions. Curr Appl Phys 9(2, suppl):e119–e123
6. Bergman TL, Lavine AS, Incropera FP, DeWitt DP (2011) Introduction to heat transfer, 6th edn. Wiley
7. Hu Y-P, Li Y-R, Lu L, Mao Y-J, Li M-H (2020) Natural convection of water-based nanofluids near the density maximum in an annulus. Int J Therm Sci 152:106309
8. Pak BC, Cho YI (1998) Hydrodynamic and heat transfer study of dispersed fluids with submicron metallic oxide particles. Exp Heat Transf 11(2):151–170
9. Xuan Y, Roetzel W (2000) Conceptions for heat transfer correlation of nanofluids. Int J Heat Mass Transf 43(19):3701–3707
10. Hamilton DRL, Crosser OK (1962) Thermal conductivity of heterogeneous two-component systems. Ind Eng Chem Fundam 1(3):187–191

11. Maïga SEB, Nguyen CT, Galanis N, Roy G (2004) Heat transfer behaviours of nanofluids in a uniformly heated tube. Superlattices Microstruct 35(3):543–557
12. Shah RK, London AL (1978) Laminar flow forced convection in ducts: a source book for compact heat exchanger analytical data. Academic Press
13. Bayat J, Nikseresht AH (2011) Investigation of the different base fluid effects on the nanofluids heat transfer and pressure drop. Heat Mass Transf 47(9):1089–1099

Electronic Properties and Magnetic Stability in Binary Iron Pnictide

Nabila Benmeddah and Fatma Temmar

Abstract In this work, we presented a first-principle study of the structural, electronic and magnetic properties of iron selenide in its pure state. The calculations were performed by the Pseudo-Potential method which is based on the density functional formalism (DFT), using the local spin density approximation (LSDA). The lattice parameters and total energy of FeSe were calculated for three different cases of magnetism in order to examine magnetic stability and electronic properties. We discussed the most stable state as well as the one closest to the creation of the phenomenon of superconductivity in this binary iron pnictide.

Keywords Magnetism · Superconductivity · DFT study · Fe-pnictides · Band structures · Fermi surfaces

1 Introduction

Superconductivity was discovered in 2008, by the team of Japan Hideo Hosono at the Tokyo Institute of Technology, in iron-based compounds (Fe-pnictides), at temperatures almost as in cuprates (T_c up to 55 K, [1–5]) which attracted enormous interest. It has been induced in RFeAsO by doping with fluorine, cobalt [1], thorium [2], strontium [3], oxygen vacancies [4], or by applying high pressure [5, 6].

All of the parent compound (RFeAsO, R = La, Ce, Nd, Pr, Sm, Gd) display a similar antiferromagnetic phase transition from tetragonal to orthorhombic structural distortion on cooling from 250 to 100 K [7, 8]. When these materials are doped, the structural and magnetic transitions are suppressed and superconductivity appears [9–13].

N. Benmeddah (✉)
Materials Computational Physics Laboratory, Djillali Liabès University, 22000 Sidi Bel Abbès, Algeria
e-mail: nphysique@yahoo.fr; nabila.benmeddah@univ-saida.dz

F. Temmar
Physico-chemical Studies Laboratory, Dr. Moulay Taher University, 20000 Saida, Algeria

S. Hamamda et al. (eds.), *Proceedings of the 2nd International Conference of Nanotechnology for Environmental Protection and Clean Energy Production*, Springer Proceedings in Materials 45, https://doi.org/10.1007/978-981-97-1916-7_16

FeSe, another iron-based superconductor, reaches a critical temperature above "100 K" by growing a thin layer on a $SrTiO_3$ substrate, thus surpassing all other iron-based superconductors [14, 15] and opening the way to superconductors in thin layers as well as to the synthesis of complex materials.

In short, the discovery of new iron-based superconductors has opened a new horizon for superconductivity at high critical temperature. In this work, we focus the study on one of the parent of the binary compound FeSe type iron pnictides, we discuss the magnetic stability and the electronic structure, in particular his band structures and Fermi surfaces while relying on the FP-LAPW method based on the DFT.

2 Computational Details

We compute the **FeSe** electronic band structure in the framework of Density Functional Theory implemented in the Wien2K full-potential APW+lo package [16, 17], we considered 200 **k** points in the BZ and checked that all the properties presented in this paper were converged with this mesh. The results that we present here, have been obtained within the LDA to the exchange-correlation potential [18, 19].

Our calculations of electronic properties were done by using the muffin-tin radii R_{MT} for both of atoms Fe, Se initialized to be 2.0 a.u. [20].

The electronic states of the atoms in the crystal have been chosen with the valence configurations of

$$\mathrm{Fe}^{26} : 1S^2 \;\; 2S^2 \;\; 2P^6 \;\; 3S^2 \;\; 3P^6 \;\; 3d^6 \;\; 4S^2.$$

$$\mathrm{Se}^{34} : 1S^2 \;\; 2S^2 \;\; 2P^6 \;\; 3S^2 \;\; 3P^6 \;\; 3d^{10} \;\; 4S^2 \;\; 4P^4.$$

To ensure convergence, the LAPW basis set was defined by the cutoff $R_{MT}K_{\mathrm{max}}$ = 7 (the minimum of R_{MT} and the maximum plane-wave cut-off parameter K_{max}). 6 $\times$ 6 $\times$ 6 k-point grid was used to perform the Brillouin Zone integration [21, 22] in the self-consistent calculations with charge convergence until 10^{-3} eV.

Iron-based superconductor FeSe belongs to tetragonal structure with space group P4/nmm (#129). The equilibrium lattice parameters of FeSe have a value of a = 3.762 Å and c = 4.420 Å [20].

Iron selenide has the simplest crystal structure of all iron-based superconductors. FeSe is, like pnictides, an almost two-dimensional material, being only a stack of iron-selenium planes. It is stand out by the fact that it does not contain any interplanar atoms and, obviously, that it contains only two atomic species. By minimizing the total energy, the lattice constants and atomic positions will optimized.

Figure 1 shows the optimized structures of FeSe, Fe, Se atoms arrange in parallel planes in a supercell of 24 atoms according to the direction (1, 1, 2) thus, the distribution of the two Fe, Se atoms in the conventional mesh is presented on the right of the figure.

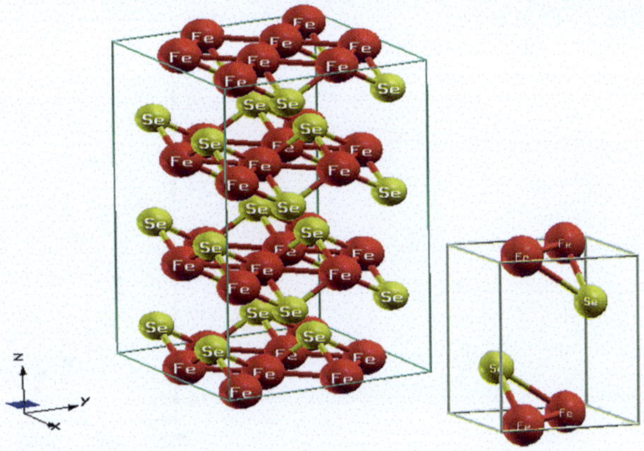

Fig. 1 FeSe crystal structure with P4/nmm space group

Table 1 Structural parameters of FeSe with space group: P4/nmm #129

Compound	Structural parameter (Å)	Atomic positions
FeSe	$a = 3.762^{a}$	Fe $\left(\frac{1}{4}, \frac{3}{4}, 0\right)$ [20]
	$c = 4.420^{a}$	Se $\left(\frac{1}{4}, \frac{1}{4}, z\right)$ [20]

[a]Experimental value, with Z = 0.2402 [20]

The experimental structural parameters and atomic positions used in the calculation are shown in Table 1.

3 Results and Discussions

3.1 *Magnetic Stability*

In order to study the stability of FeSe compound we performed the optimization calculations. In Fig. 2, we represent the variation of the total energy as a function of the volume for our material in three different cases (non-magnetic, ferromagnetic and finally antiferromagnetic) in the tetragonal structure.

From Fig. 2, we find that the most stable phase for the compound is the non-magnetic (without spin polarized) phase (NM) thus coinciding with the volume closest to the associated experimental study.

The structural parameters such as the lattice constant, the modulus of compressibility as well as the minimum energy obtained during the calculations are given in Table 2.

By analyzing Table 2 we find that the lattice parameters decrease, the total energy E and the modulus of compressibility B increase, when changing from the magnetic state to the no magnetic state for the FeSe compound. However, the experimental results are slightly overestimated compared to our results, in particular the parameter

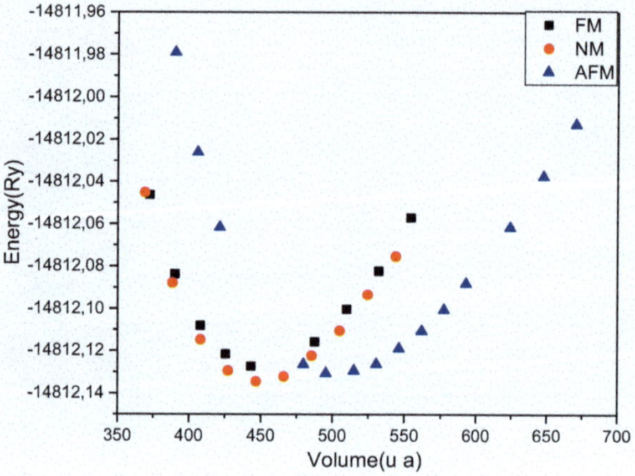

Fig. 2 Total energy variation with volume in 3 cases of magnetism

Table 2 Structural parameters and total energy calculated in magnetic and nonmagnetic states

Compound	Lattice parameters	Results	Expérimental results
FeSe with SP	a (Å) c (Å)	3.6588 5.2988	3.765 5.518 [23]
	B (GPa)	99.9142	
	B′	4.7328	
	Emin (Ry)	− 14,812.133918	
FeSe without SP	a (Å) c (Å)	3.6349 4.3753	3.775 5.512 [24]
	B (GPa)	123.9174	
	B′	4.8190	
	Emin (Ry)	− 14,812.134479	

c. The most stable state is that of the no magnetic FeSe, which we will prove again in the next section. We have relaxed our material using the experimental structural parameters [20], to be able to compare with the literature and to reproduce better the correct electronic structure of the studied system.

3.2 *Electronic Properties*

In this section, we present the results for the bi-atomic FeSe compound from the family of the iron-based superconductors. The band structure for the compound have been presented in Fig. 3. The letters Γ, X, M, Z, A, R represent the points of high symmetry of this structure. Two hole-like bands are centered at the Γ point and two electron-like bands at the M, R points. The calculated band structure along the Γ–Z line suggests that, because the dispersion of the bands is linear in kz-direction this system can be treated as a two-dimensional one. This result agrees qualitatively with the previous calculations.

To get an idea of the origin of the energy bands, we calculated the total and partial density of states of the FeSe compound (Fig. 4).

Fe atom has the greatest density near the Fermi level, as the results of the previous figure have shown.

The density of states at the Fermi level is largely dominated by iron d electrons, while the other contributions at this level are marginal in comparison, also Se contribution exceeds Fe contribution at – 15 eV further E_F.

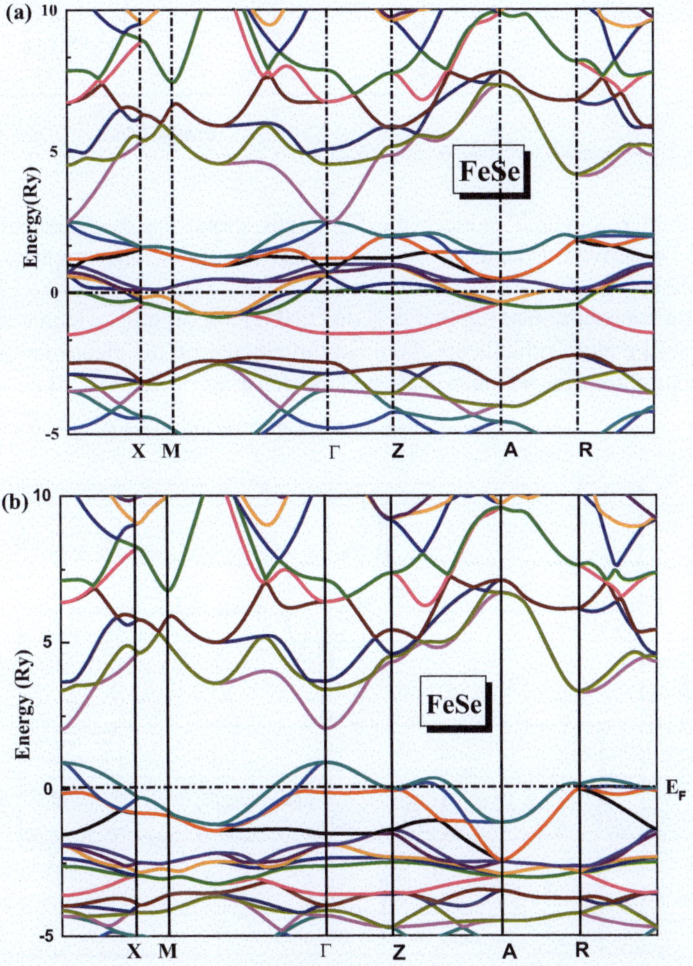

Fig. 3 Band structure of FeSe, in the tetragonal structure, calculated with the functional DFT. **a** Non-magnetic state, **b** antiferromagnetic state

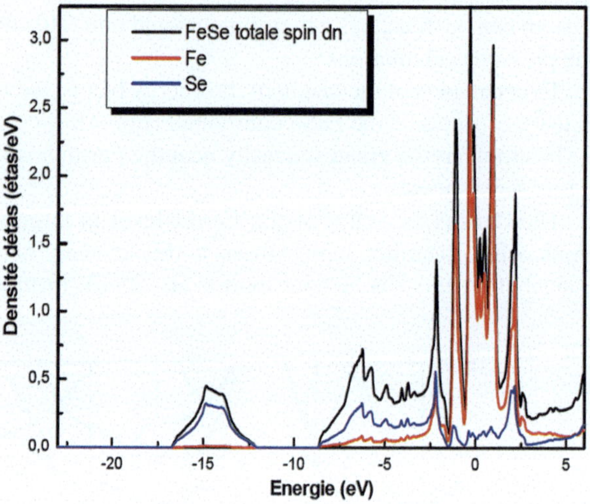

Fig. 4 The total (black, solid) DOS of tetragonal FeSe obtained using experimental lattice parameters and theoretically relaxed z_{Se}

Theoretical z_{Se}, which reduces the Se-height above the Fe plane, affects the individual Fe-derived d-bands around E_F differently due to the orientation of the d-orbitals [23] (Fig. 5).

The main contribution of the PDOS is due to the ***Fe-d*** orbitals, which are the most dominant so, Fe atom orbitals are the most influencing on the electronic properties of the structure, for this, we have evaluated the densities of states of all ***Fe-d*** orbitals (Fig. 6).

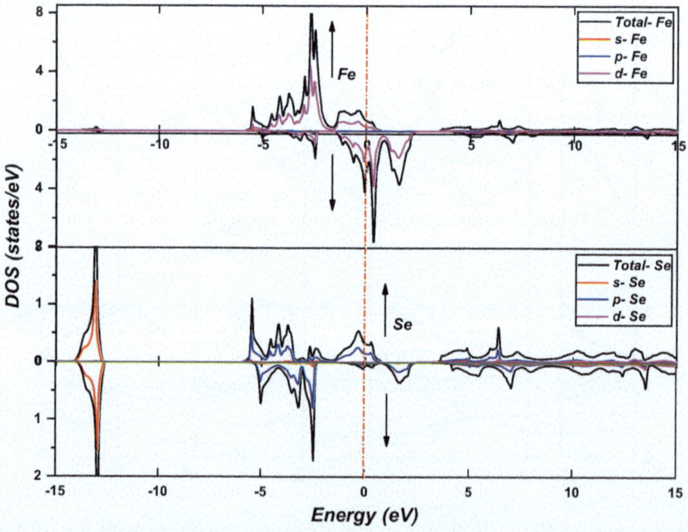

Fig. 5 Total and partial DOS of Fe, Se atoms in FeSe compound obtained by functional DFT-LSDA

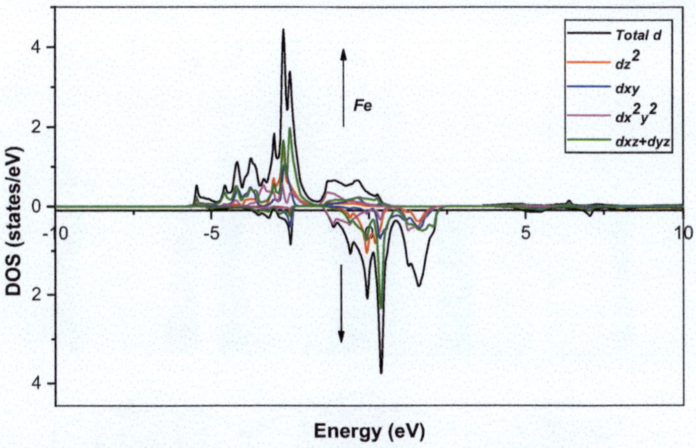

Fig. 6 Density of states projected into the 3d orbitals of the iron atom

The orbital $d_{XZ} + d_{YZ}$ represents the maximum density at Fermi level thus, the other orbitals all present intense peaks near the Fermi level, in particular the $d_Z{}^2$ and the d_{XY}, which suggests an attenuation limited by the behavior of iron electrons.

3.3 *Fermi Surface of the FeSe*

The Fermi surface of FeSe is illustrated in Fig. 7. Two bands intersect the Fermi level, giving rise to two distinct cylindrical surfaces parallel to the k_z axis. The curvature of the surfaces along this axis is low due to the almost two-dimensional nature of the material. The surface is plotted on a two-dimensional Brillouin Zone in Fig. 7. The two cylinders become two pockets of very similar areas, composed of two holes in the center of the Brillouin zone. The other cylinders are resulting from electrons contribution in non-magnetic state.

The superconducting FeSe obtained in thin layers, whose Tc is much higher attending 100 K, has a different Fermi surface. Generally, the true electronic structure of Iron-based superconductors has 4 (or 5) Fermi surfaces: 2 (or 3) hole (cylinders) in the center of the first Brillouin zone, as it was presented for the FeSe antiferromagnetic Fig. 7b, and 2 other zone-corner showing electrons contribution at Fermi surfaces, as it was presented for the FeSe no-magnetic.

Superconductivity coincides with the non-magnetic state when we consider the number of hole and electron contribution at Fermi level.

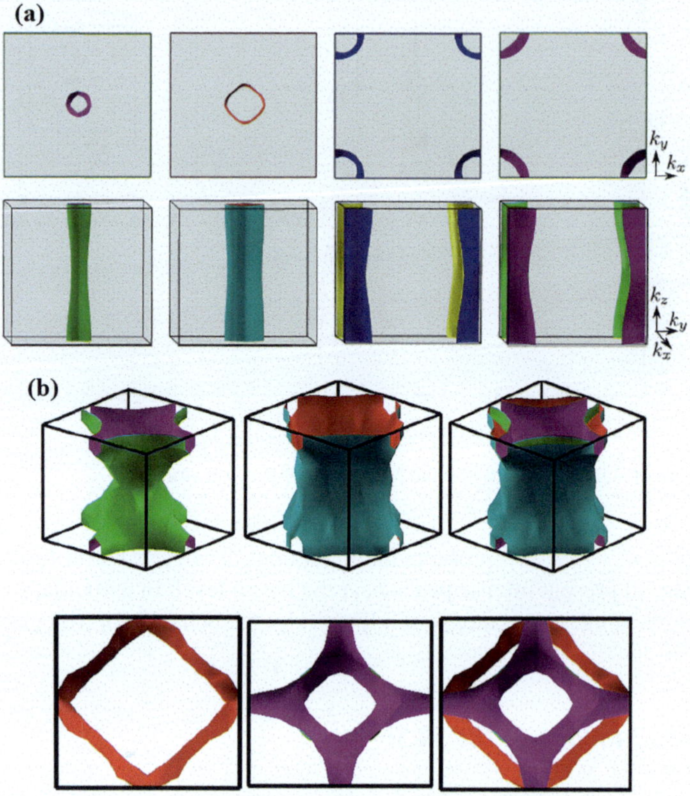

Fig. 7 Fermi surface: **a** non-magnetic state **b** antiferromagnetic state

4 Conclusions

In summary, FeSe with its critical temperature $T_C = 8$ K undergoes a structural transition from the tetragonal to the orthorhombic phase at $T_s = 87$ K with non-magnetic order, but with emerging nematic electronic structure. On the other hand, the calculations based on density functional theory (DFT) predict magnetic states for FeSe [25–28]. Recently, exotic states with hidden magnetic order were proposed. Our calculations based on LSDA allowed us to study the electronic and magnetic properties based on purely experimental parameters, we find that lattice constants (a, c) decrease from magnetic to non-magnetic state thus, the total energy increase proving the non-magnetic character of FeSe at low temperature. A complement study was added to compare different properties when the probability of antiferromagnetic state is present. The band structure in the nonmagnetic state is much more charged at the Fermi level than in the antiferromagnetic state. In both cases of magnetism, the density of state marks a remarkable contribution of the iron atom in particular

Fe-d where the orbital $d_{XZ} + d_{YZ}$, dz^2 and d_{XY} represent the big contribution at E_F successively.

Since the number of distinct pockets forming the Fermi surface gives the maximum number of different gaps that can exist on this surface, the non-magnetic state reinforced quantitatively the superconducting state while the magnetic state shows the same quality of the pockets.

References

1. Sefat AS, Huq A, McGuire MA, Jin R, Sales BC, Mandrus D (2008) Phys Rev B 78:104505
2. Wang C, Li L, Chi S, Zhu Z, Ren Z, Li Y, Wang Y, Lin X, Luo Y, Jiang S, Xu X, Cao G, Xu Z (2008) Europhys Lett 83:67006
3. Kasperkiewicz K, Bos JWG, Fitch AN, Prassides K, Margadonna S (preprint). arxiv:0809.1755
4. Ren Z, Che G, Dong X, Yang J, Lu W, Yi W, Shen X, Li Z, Sun L, Zhou F, Zhao Z (2008) Europhys Lett 83:17002
5. Kamihara Y, Watanabe T, Hirano M, Hosono H (2008) J Am Chem Soc 130:3296
6. Bos JWG, Penny GBS, Rodgers JA, Sokolov DA, Huxley AD, Attfield JP (2008) Chem Commun 3634
7. Dong J, Zhang HJ, Xu G, Li Z, Li G, Hu WZ, Wu D, Chen GF, Dai X, Luo JL, Fang Z, Wang NL (2008) Europhys Lett 83:27006
8. Drew J, Pratt FL, Lancaster T, Blundell SJ, Baker PJ, Liu RH, Wu G, Chen XK, Watanabe I, Malik VK, Dubroka A, Kim KW, Roessle M, Bernhard C. arxiv:0805.1042
9. Cruz C, Huang Q, Lynn JW, Li J, Ratcliff W II, Zarestky JL, Mook HA, Chen GF, Luo JL, Wang NL, Dai P (2008) Nature 453:899
10. McGuire MA, Christianson AD, Sefat AS, Sales BC, Lumsden MD, Jin R, Payzant EA, Mandrus D, Luan Y, Keppens V, Varadarajan V, Brill JW, Hermann RP, Sougrati MT, Grandjean F, Long GJ (2008) Phys Rev B 78:094517
11. Zhao J, Huang Q, de la Cruz C, Li S, Lynn JW, Chen Y, Green MA, Chen GF, Li G, Li Z, Luo JL, Wang NL, Dai P (preprint). arxiv:0806.2528, https://doi.org/10.1038/nmat2315
12. Zhao J, Huang Q, de la Cruz C, Lynn JW, Lumsden MD, Ren ZA, Yang J, Shen X, Dong X, Zhao Z, Dai P (2008) Phys Rev B 78:132504
13. Qiu Y, Bao W, Huang Q, Yildirim T, Simmons J, Lynn JW, Gasparovic YC, Li J, Green M, Wu T, Wu G, Chen X (preprint). arxiv:0806.2195
14. Song YJ, Hong JB, Min BH, Lee KJ (2011) The pseudogap behavior in the stoichiometric FeSe superconductor (Tc ~ 9.4 K). J Korean Phys Soc 59(2):312
15. Ge J-F, Liu Z-L, Liu C, Gao C-L (2014) Superconductivity above 100 K in single-layer FeSe films on doped $SrTiO_3$. Nat Mater 14(3):285–289
16. Blaha P, Schwarz K, Madsen GKH, Kvasnicka D, Luitz J (2001) WIEN2K, an augmented plane wave plus local orbitals program for calculating crystal properties. Vienna University of Technology, Vienna
17. Sjöstedt LN, Singh DJ (2000) Solid State Commun 114:15
18. Perdew JP, Wang Y (1992) Phys Rev B 45:13244
19. Anisimov VI, Zaanen J, Andersen OK (1991) Phys Rev B 44:943
20. Hsu FC et al (2008) Proc Natl Acad Sci U S A 105:14262
21. Murnaghan FD (1944) Proc Natl Acad Sci U S A 30:5390
22. Monkhorst HJ, Pack JD (1976) Phys Rev B 13:5188
23. Singh PP. arXiv:0908.0303v3 [cond-mat.mtrl-sci] 22 Feb 2010
24. Yeh KW, Huang TW, Huang YL, Chen TK, Hsu FC, Wu PM, Wu MK (2008) Tellurium substitution effect on superconductivity of the α-phase iron selenide. Eur Phys Lett 84(4):37002
25. McQueen TM et al (2009) Phys Rev B 79:014522

26. McQueen TM, Williams A J, Stephens PW, Tao J, Zhu Y, Ksenofontov V, Casper F, Felser C, Cava RJ (2009) Phys Rev Lett 103057002
27. Watson MD et al (2015) Phys Rev B 91:155106
28. Li YF, Zhu LF, Guo SD, Xu YC, Liu BG (2009) J Phys Condens Matter 21:115701

Structural and Morphology Characterization of an EVA/ZnO Polymer in Thin Film Form for Solar Applications

Maifi Lyes, Ayadi Aicha, Hioual Ouided, Mazinz Zaina, Agroui Kamel, and Chari Abdelhamid

Abstract The objective of this work is to develop and study the structural properties, as well as the morphology of thin films of ZnO nano powders prepared by the sol-gel method, based on the fact that reducing grain size to the nanoscale introduces new physical characteristics and opens promising prospects for solar applications. The first part is an experimental study on the fabrication of ZnO. The second part is an experimental study on the organic polymer EVA. We used THF-soluble EVA and doped it with In_2O_3. Experimental analysis using XRD and SEM techniques is conducted to examine the doping rate, efficiency, and performance of the materials.

Keywords EVA · ZnO · Morphology · Structural

1 Introduction

Nanoparticles have always existed in nature, but they also exist as a result of human activity. Currently, research on nanoparticles is focused on improving and developing new properties to be used in various products in our daily lives. Nanoparticles are defined as synthetic particles with one or more dimensions measuring less than 100 nm. However, the terminology regarding different aspects of nanotubes is still subject to discussion. These particles are synthesized due to their unique properties specific to their dimensions [1–7].

M. Lyes (✉) · M. Zaina · A. Kamel
Research Center on Semiconductor Technology for Energetic TESE-CRTSE, Algiers, Algeria
e-mail: maifi.lyes@crtse.dz

M. Lyes · A. Aicha · H. Ouided · C. Abdelhamid
Department of Physics Thermodynamic Laboratory and Material Surface Treatment, University Frères Mentouri Constantine 1, Constantine, Algeria

H. Ouided
Department Computer Science, University Abbes Laghrour Khenchela, Khenchela, Algeria

S. Hamamda et al. (eds.), *Proceedings of the 2nd International Conference of Nanotechnology for Environmental Protection and Clean Energy Production*, Springer Proceedings in Materials 45, https://doi.org/10.1007/978-981-97-1916-7_17

There are natural nanoparticles (such as sea salt) and unintentionally anthropogenic artificial nanoparticles, such as domestic and industrial fumes, as well as intentionally manufactured ones with new properties, like oxides and polymers, for example [7–11].

Polymer materials are increasingly used in solar applications due to their interesting optical and mechanical properties. Ethylene-vinyl acetate (EVA) is an example of a polymer material widely used in the solar panel industry because of its transparency to sunlight and ability to adhere to other materials. However, to enhance the performance of solar panels, it is often necessary to incorporate zinc oxide (ZnO) nanoparticles into the polymer matrix. ZnO nanoparticles possess unique optical properties, such as a broad absorption spectrum and good electrical conductivity, which can improve solar panel efficiency [11–19].

Hence, the study and optical characterization of these EVA/ZnO polymer materials in the form of thin films are of great importance to understand their optical behavior and properties related to light transmission, absorption, and reflection. Commonly used techniques for characterizing EVA/ZnO polymer thin films include photoluminescence spectroscopy, X-ray diffraction (XRD), Fourier-transform infrared spectroscopy (FTIR), and scanning electron microscopy (SEM). These techniques allow for the measurement of material's optical properties like refractive index, light transmission, and reflection, as well as the morphology and structure of the thin films. This information is essential for optimizing solar cell design and improving their energy efficiency.

2 Experimental Procedure

We prepare a mass of zinc oxide weighed on an analytical balance (m = 12 g). After weighing it, we place it in a beaker and add a volume of water (v = 60 ml). We also prepare a mass of citric acid weighed on an analytical balance (m = 1 g). After weighing it, we place it in a beaker and add a volume of ethanol (v = 60 ml). To obtain homogeneous solutions, we place them on a stirrer for 20 min. After stirring, we mix the two quantities in a third beaker, which allows us to obtain a layer of ZnO gel. Then, we filter this layer using filter paper (Fig. 1).

Drying: The first step of the thermal treatment of our gel is to place the ZnO gel in an oven for 24 h to obtain nanoscale ZnO powder.

Annealing: The second step, which is also the most important in our thermal treatment, is the annealing process. After drying the ZnO gel, we obtain a white powder. This powder is amorphous, so we proceed with a 2-h annealing at a temperature of T = 450 °C to crystallize our powder into a Wurtzite-type hexagonal structure. After annealing, we obtain a crystallized white or gray powder, which corresponds to pure ZnO. These powders are shown in Fig. 2.

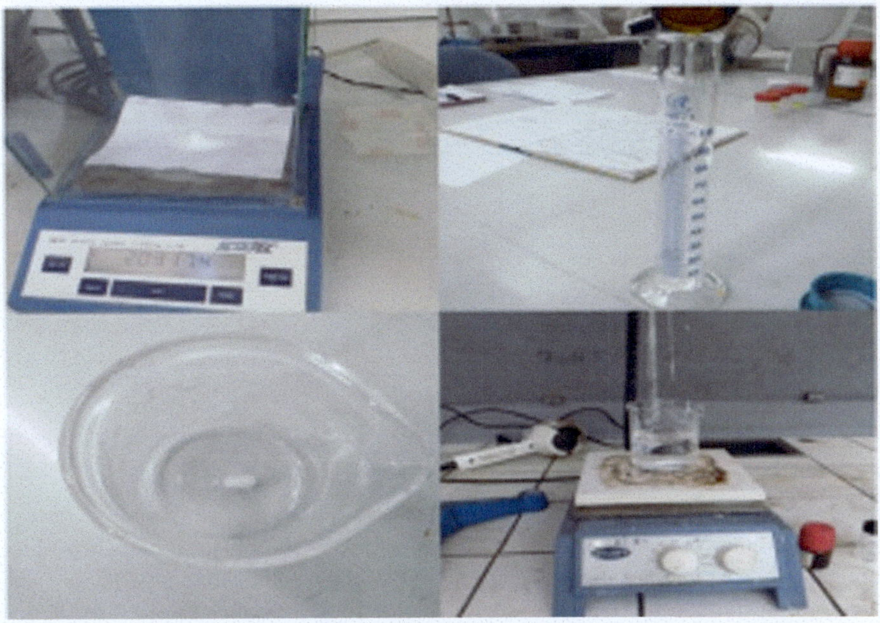

Fig. 1 Equipment used for sample preparation

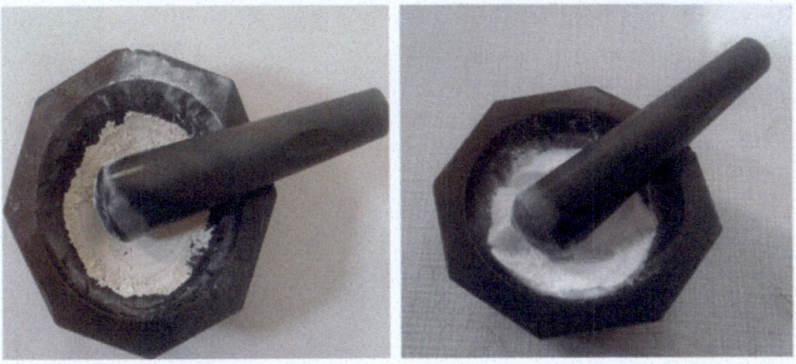

(a) (b)

Fig. 2 ZnO gel before and after annealing

On the other hand, we prepare a mass of EVA weighed on an analytical balance (m = 0.5 g). After weighing it, we place it in a beaker and add a volume of (V = 20 ml) of THF (Tetrahydrofuran) to obtain a viscous liquid after two hours of magnetic stirring.

We use ethyl vinyl acetate (EVA) containing 18% vinyl acetate (Fig. 3).

We introduced 0–30% ZnO/In_2O_3 into the EVA. At the end of this process, we obtain two samples that we will examine later: pure ZnO/EVA, and EVA/ZnO doped with 30% In_2O_3.

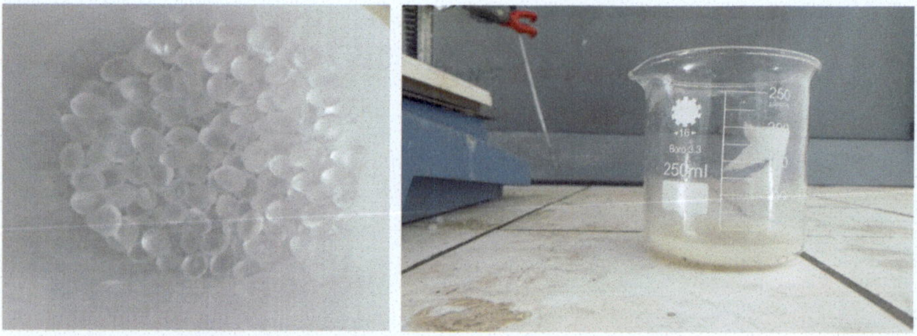

(a) (b)

Fig. 3 EVA pure and gel before and after

3 Results

This is the main part of the study, where we present the most significant results obtained when using various characterization methods at the University of Constantine1 and at the Algerian Semiconductor Technology Research Center for Energy (CRTSE). These techniques were applied to samples prepared from pure ZnO/EVA and doped with In_2O_3.

DRX characterization was performed using an Empyrean Panalytical diffractometer, where X-ray radiation emitted from cobalt Kα with a wavelength of $\lambda = 1.789$ Å was used (Fig. 4).

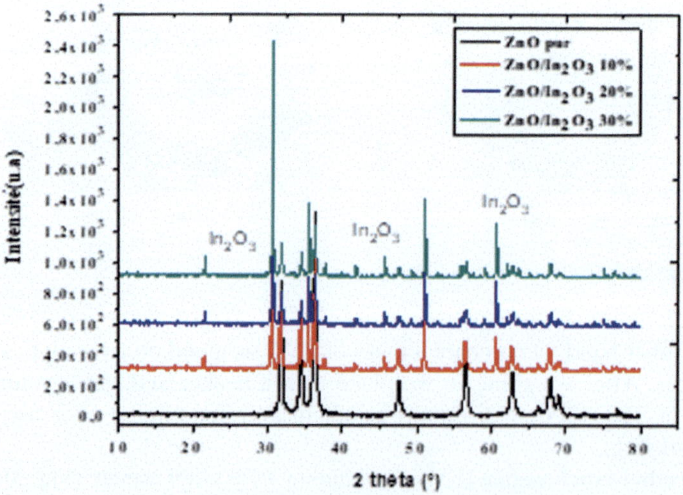

Fig. 4 X-ray diffraction analysis of pure and doped ZnO

(a) (b)

Fig. 5 Scanning electron microscopy (SEM) image of ZnO * 10,000. In_2O_3/ZnO (30%)

The first observation that can be made from the spectra of pure ZnO and doped ZnO is that we observe the peaks of the hexagonal Wurtzite structure of ZnO, whether it is pure or doped. This indicates a structure with the same structural characteristics as that of bulk ZnO, whether it is pure or doped with In_2O_3. As for the In_2O_3-doped layers, we can say that In_2O_3-doped ZnO up to a concentration of 30% forms solid solutions: the In_2O_3 atoms are placed in substitution or interstitial sites.

The second observation concerns the intensity of the peaks, where a decrease in the peaks is clearly evident, whether they are intense peaks or others. This is explained in terms of crystallinity, where the intensity of pure ZnO decreases significantly with increasing In_2O_3 concentration, while the intensity of In_2O_3 peaks gradually increases with increasing dopant concentration.

The morphology of the deposited ZnO/EVA layer was observed by scanning electron microscopy. Figure 5 depict the morphology of the undoped and In_2O_3-doped ZnO/EVA layer deposited on glass using the sol-gel technique.

In Fig. 5, a SEM image of ZnO and 10% In_2O_3-doped ZnO with a magnification of 10,000×, it can be observed that the powder exhibits a homogeneous particle size distribution. The grains are nano-sized particles with a size of 0.01 nm or smaller. In Fig. 5, a SEM image of 30% In_2O_3-doped ZnO with a magnification of 10,000×, it can be observed that the powder exhibits a uniform compact homogeneous particle size distribution with some irregularities. However, there are grains that are not nano-sized particles, possibly because the material was not ground properly due to the high amount of doping added compared to the initial amount.

4 Conclusion

The study using X-ray diffraction (XRD) has shown that our powders become crystalline with a Wurtzite hexagonal structure after being subjected to a thermal annealing at 450 °C. Additionally, doping with a high percentage has an impact on the average grain size. The morphological characterization conducted by scanning electron microscopy (SEM) has confirmed the structure observed by XRD and has also revealed the homogeneity of our powders, indicating a high lifetime. In conclusion of this study, ZnO emerges as an ideal semiconductor for solar applications as a thin film, potentially enhancing efficiency. The performance of ZnO can be enhanced by introducing highly thermally conductive semiconductor nanoparticles as dopants. In terms of dopants, we focused on the effect of In_2O_3 doping in ZnO. It was observed that for all doping levels, the absorbance increases with the doping rate.

References

1. Li J, Xue P, Ding W, Han J, Sun G (2009) Micro-encapsulated paraffin/high-density polyethylene/wood flour composite as form-stable phase change material for thermal energy storage. Sol Energy Mater Sol Cells 93(10):1761–1767
2. Alkan C, Sari A (2008) Fatty acid/poly (methyl methacrylate) (PMMA) blends as form-stable phase change materials for latent heat thermal energy storage. Sol Energy 82(2):118–124
3. Abhat A (1983) Low temperature latent heat thermal energy storage: heat storage materials. Sol Energy 30(4):313–332
4. Akrouche A (2011) Etude du Stockage et Déstockage d'Energie dans un Matériau à Changement de Phase, Mémoire de Magister, Université M'Hamed Bougara, Boumerdes, Ecole des mines de Nantes, France
5. Gharssallaoui A, Roudaut G, Chambin O, Voilley A, Saurel R (2007) Applications of spray-drying in microencapsulation of food ingredients: an overview. Food Res Int 40:1109–1121
6. Pielichowska K, Pielichowski K (2014) Phase change materials for thermal energy storage. Progr Mater Sci 65(0):67–123
7. Berreksi S, Latarsa S (2016) Simulation numérique du stockage thermique par chaleur latente dans une capsule sphérique, mémoire de master: systèmes. Thermique et Combustion, Université M'Hamed Bougara, Boumerdes
8. Himran NSYUK, Suwaono ARYA, Mansoori GA (1994) Characterization of alkanes and paraffin waxes for application as phase change energy storage medium. Energy Sour 16(1):117–128
9. Hadjieva M, Kanev S, Argirov J (1992) Thermophysical properties of some paraffins applicable to thermal energy storage. Solar Energy Mater Solar Cells 27(2):181–187
10. Sharma A, Tyagi VV, Chen CR, Buddhi D (2009) Review on thermal energy storage with phase change materials and applications. Renew Sustain Energy Rev 13(2):318–345
11. Alkan C (2006) Enthalpy of melting and solidification of sulfonated paraffins as phase change materials for thermal energy storage. Thermochim Acta 451(1–2):126–130
12. Hasnain SM (1998) Review on sustainable thermal energy storage technologies. Part I: Heat storage materials and techniques. Energy Convers Manage 39(11):1127–1138
13. Akgün M, Aydin O, Kaygusuz K (2007) Experimental study on melting/solidification characteristics of a paraffin as PCM. Energy Convers Manag 48(2):669–678
14. Gontard A, Gryska K, Lallemand L (2012) Amélioration de la conductivité thermique de matériau à changement de phase. Université de Pau des pays de l'Adour
15. Oró Prim E (2013) Thermal energy storage (TES) using phase change materials (PCM) for cold applications. University de Lleida
16. Bendkowska W, Tysiak J, Grabowski LJ, Blejzyk A (2005) Determining temperature regulating factor for apparel fabrics containing phase change material. Int J Clothing Sci Technol 17(3–4):209–214
17. Ying B, Kwok YL, Li Y, Yeung C-Y, Song Q-W (2004) Thermal regulating functional performance of PCM garments. Int J Clothing Sci Technol 16(1/2):84–96

18. Rupp J (1999) Interactive textiles regulate body temperature. Int Text Bull 45(1):58–59
19. Clerk Maxwell J (1892) A treatise on electricity and magnetism, 3rd edn, vol 2. Clarendon, Oxford, pp 68–73

Synthesis and Characterization of $Zn_{(1-x)}Ni_{(x=0.01-0.03)}O$ Nanofilms Under Specific Conditions: Enhancement of Red Emission, and Photodegradation Retarding Behavior

Abdelhalim Ouhaibi, Bachir Eddine Messaid, Rafik Serhane, Abdelhalim Zoukel, and Abderezak Zellagui

Abstract In this work, undoped and Nickel (1, 2, and 3%) doped ZnO nanofilms were synthesized on glass substrates by ultrasonic spray method. The structure was studied by X-ray diffraction (XRD). Optical properties of samples were analyzed by photoluminescence (PL) spectroscopy. Subsequently, sprayed nanofilms were used as catalysts in an experiment for the degradation of methylene blue (MB) under UV irradiation. XRD spectra confirmed the hexagonal 'wurtzite' structure with polycrystalline nature of undoped and Ni doped nanofilms, and revealed that no compound containing Nickel appeared confirming the good quality of the structures. Photoluminescence spectra showed that all samples exhibited strong UV emission with blue emission. The absence of green emission in all the spectra is obvious and unexpected. A regular red emission is strangely observed in the spectra of doped nanofilms only, which can be exploited by the development of photo or electroluminescent devices based on Ni doped ZnO. Phtodegradation of undoped ZnO exhibited high degradation rate activity by 87.1% compared to 61.4%, 49.7%, 72.7% of Ni doped ZnO

A. Ouhaibi (✉) · R. Serhane
Microelectronics and Nanotechnology Division, MEMS and Sensors Team, Center for Development of Advanced Technologies—CDTA, Baba Hassen, Algeria
e-mail: aouhaibi@cdta.dz

R. Serhane
e-mail: Rserhane@cdta.dz

B. E. Messaid · A. Zellagui
Department of Process Engineering, M'Hamed Bougara University, Boumerdes, Algeria
e-mail: b.messaid@univ-boumerdes.dz

A. Zellagui
e-mail: a.zellagui@univ-boumerdes.dz

A. Zoukel
Physico-Chemistry of Materials Laboratory, Laghouat University, Laghouat, Algeria
e-mail: ab.zoukel@lagh-univ.dz

S. Hamamda et al. (eds.), *Proceedings of the 2nd International Conference of Nanotechnology for Environmental Protection and Clean Energy Production*, Springer Proceedings in Materials 45, https://doi.org/10.1007/978-981-97-1916-7_18

(1%, 2%, and 3%) respectively. Therefore, Ni doping lowered the photocatalytic efficiency of ZnO nanofilms.

Keywords Spray · ZnO · Ni doping · Photoluminescence · Photodegradation

1 Introduction

ZnO is among the most interesting materials in its powder, thin films or nanofilms forms having multitude applications in several fields [1–7]. Particularly in recent years, many research have been focused on photodegradation of pollutant caused by chemical or biological products such as organic chemicals, bacteria, industrial dyes, and antibiotics that have a direct negative influence on the environment [8]. ZnO has a wide band gap of ~ 3.37 eV and high energy binding of excitons of 60 meV.

When irradiated by an energy greater than or equal to its gap, electron and hole (e^-/h^+) pairs energy will be created in conduction band (CB) and valence band (VB) respectively. The generated (e^-/h^+) pairs are the main actors in oxidation–reduction reaction of adsorbed organic molecules on the samples surface. The (e^-) react with O_2 molecules dissolved in the aqueous solution in forming superoxide $O^-{}_2$, while the (h^+) react with OH^- or H_2O to form hydroxyl radicals (·OH), which act in the degradation of the organic pollutant [9].

ZnO nanofilms can be nanostructured by many chemical and physical techniques namely magnetron thermal evaporation [10], sputtering [11], sol–gel [12], hydrothermal [13], chemical vapor deposition (CVD) [14]. In the present work, ultrasonic spray was used in the elaboration of samples due its simplicity and low cost [15]. The main goal of the study is to synthesize undoped and Ni-doped ZnO nanofilms in order to study the role of Nikel doping on the photocatalytic activity as well as on the generation of visible emissions, in particular the rare red emission in ZnO.

2 Experimental

Utrasonic spray technique was used to elaborate undoped and Nikel-doped ZnO nanofilms on glass substrates preheated to 400 °C. The starting solution of 0.3 M concentration of ZnO was prepared by dissolving Zinc acetate ($Zn(CH_3OO)_2, 2H_2O$) in 25 ml of Methanol. Nikel dichloride ($NiCl_2$, $6H_2O$) was used as Ni dopant. The molar ratio concentration Ni/Zn in the initial solution was varied to obtain Ni/ZnO atomic ratios of 1, 2 and 3%. Resulting aqueous solution was stirred for 1 h before use. Glass substrates have been cleaned as follow: 15 min in dilute acetone, 15 min in ethanol and 5 min of rinsing in deionized water before being dried with compressed air. The process of the ultrasonic spray technique used in this work is detailed in our old paper [16] except that the deposition operation of the present work was carried out in a clean room.

3 Characterization Details

The crystal structures of sprayed nanofilms were studied using X-ray diffractometer (XRD, BRUKER AXE D8 ADVANCE) with Cu Ka radiation (wavelength of 1.54 Å) in the range of diffraction angles between 20° and 85°. Photoluminescence characterization was carried out by FL3-DFX-iHR320 Photoluminescence equipment. Finally, the photodegradation activity was analyzed by measuring the decomposition rate of methylene blue (MB) solution exposed to UV irradiation in the presence of ZnO samples.

4 Result and Discussion

4.1 X-Ray Study

Nanoparticles X-Ray diffraction (XRD) patterns of ZnO are shown in Fig. 1. All the obtained peaks can be indexed to single phase hexagonal wurtzite structure of ZnO with p63mc space group and polycrystalline nature according to standard values (JCPDS Card Number 36-1451) [17, 18]. No signal of the metallic Zn is detected by XRD. In addition, there are no peaks corresponding with the Ni compounds, suggesting the incorporation of Ni atoms into ZnO. The average grain size of Ni-doped ZnO nanoparticles was estimated using Scherer's equation [19]

$$\mathrm{D(hkl)} = \frac{0.9\lambda}{\beta\,\mathrm{Cos}\theta} \tag{1}$$

The following relations calculated the lattice parameters 'a' and 'c' [20]:

$$\frac{1}{d_{hkl}^2} = \frac{4}{3}\left(\frac{h^2 + hk + k^2}{a^2}\right) + \frac{l^2}{c^2} \tag{2}$$

And

$$\mathrm{2d(hkl)sin}\theta = \mathrm{n}\lambda \tag{3}$$

where λ is the wavelength of Cu Kα radiation, β is the full width at half maximum (FWHM) of corresponding peak, and θ is the diffraction angle.

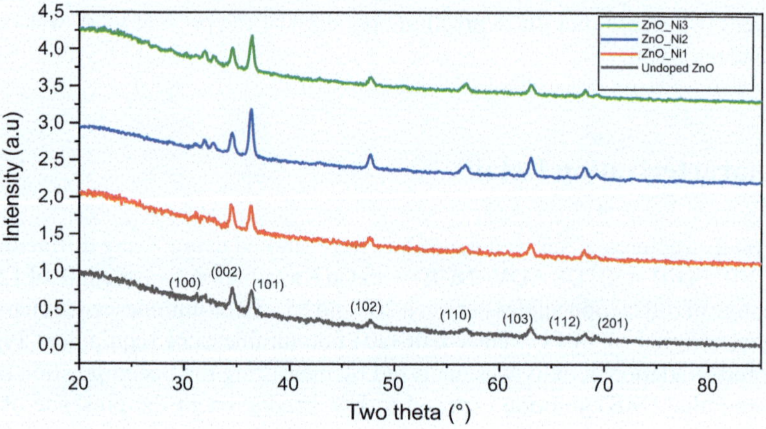

Fig. 1 XRD spectra of undoped and Ni-doped ZnO

Table 1 summarizes the calculated values of the different ZnO nanocrystallite sizes using the most intense (101) and (002) diffraction planes.

The average nanocrystallite size values increased slightly from 18.62 nm for pure ZnO to the maximum value of 21.26 nm for 1% Ni-doped ZnO, which is possibly due to the closeness of the ionic radius of Ni + 2 (0.69 A°) and Zn + 2 (0.74 A°).

So, the incorporation of Ni atoms with a maximum percentage of 3% did not modify the lattice architecture of ZnO. These results agree well with the literature [21].

Table 1 Grain size an lattice parameters of undoped and Ni doped ZnO Nanoparticles

Sample	(hkl) plane	a (Å)	c (Å)	<D> (nm)
ZnO	(002) (101)	3.239	5.189	18.62
ZnO_Ni1	(002) (101)	3.235	5.202	21.26
ZnO_Ni2	(002) (101)	3.239	5.189	19.08
ZnO_Ni3	(002) (101)	3.237	5.188	19.66
Reference (JCPDS card number 36-1451)	–	**3.249**	**5.206**	–

The definition of significance of bold in the table: Ref (JCPDS card number 36-1451) is the XRD reference data for ZnO wurtzite

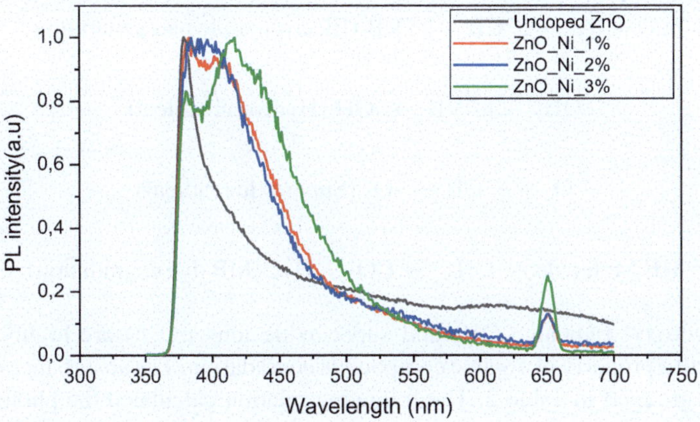

Fig. 2 Photoluminescence spectra of undoped and Ni doped ZnO

4.2 *Optical Measurement*

4.2.1 Photoluminescence Spectroscopy

Photoluminescence spectroscopy was used to obtain detailed insight into the effect of Ni dopant incorporation on the band gap, defects density as well as luminescence characteristics of ZnO. As shown in Fig. 2, normalized PL spectra of undoped and Ni doped ZnO nanofilms have a strong UV band peak around 379 nm (Eg ~ 3.27 eV). A narrow excitonic peak forms undoped ZnO PL spectrum indicating the good crystallinity of the structure. Besides, Ni doped ZnO PL spectra are composed of wide peaks ranging fromUV to visible region. The violet emission peaks located between 402 and 410 nm were attributed to the transition of electrons from a set of shallow levels below the conduction band to the valence band and interstitial oxygen atoms (O_i) [22]. The observed emission peak around 415 nm was assigned to the electronic transition energy from the conduction band to Zn vacancies [23]. The blue emission peaks between 441 and 480 nm are due to the transition of electrons from interstitial Zn_i zinc atoms to the top of the valence band. Beside, an important, rare, and uniform red emission appeared at 650 having oxygen interstitial (Oi) and oxygen vacancy (V_O) as origins [24, 25].

4.3 *Photocatalytic Activity*

The photocatalytic activity of undoped and Nikel-doped ZnO thin nanofilms was evaluated by the degradation of methylene blue (MB) dye solution as an organic pollutant under UV lamp. The degradation mechanism is described in the literature by the following chemical reactions [25]:

$$ZnO + hv \rightarrow e^{-}CB + h^{+}VB \text{ (Electron/hole pair generation)} \tag{4}$$

$$OH^{-} + h^{+}VB \rightarrow \cdot OH \text{ (Hydroxyl radical)} \tag{5}$$

$$O_2 + e^{-}CB \rightarrow \cdot O_2^{-} \text{ (Superoxide radical)} \tag{6}$$

$$\text{MB Molecule} + \cdot OH^{-} \rightarrow CO_2 + H_2O \text{(MB dye degradation)} \tag{7}$$

The hydroxyl radicals (·OH) and super-oxide ions ($\cdot O_2{}^{-}$) are highly reactive species with a property of organic compounds degradation. The absorbance measurements are grouped in Table 2. The following relation calculated the photocatalytic degradation percentages:

$$\text{Photodegradation}(\%) = (A_0 - A) \times 100\, A_0 \tag{8}$$

Where, A_0 and A are the initial absorbance of dye and the absorbance after UV irradiation respectively.

From Fig. 3, it is observed that undoped ZnO exhibited higher photocatalytic activity with 87.1% comparatively to Ni doped ZnO nano films 1%, 2% and 3% which reached 61.4%, 49.7%, 72.7% respectively. The high photocatalytic activity of undoped ZnO can be attributed mainly to the migration of charge carriers (e^{-} electrons and h^{+} holes) towards the catalytic surface during UV irradiation, as well as their participation in the redox reaction with the adsorbed dye, which is confirmed by the PL results. The photodegradation activity results revealed that Ni-doped ZnO restricted the degradation of MB under UV. Therefore, this result can be exploited in specific applications. The obtained result can be merged with some techniques in control theory [26–31].

Table.2 Absorbance values of the irradiated MB each 30 min for 210 min at $\lambda = 663$ nm

Time (mn)	0	30	60	90	120	150	180	210
Undoped ZnO	0.40	0.27	0.23	0.19	0.17	0.13	0.09	0.05
ZnO_Ni1	0.40	0.26	0.25	0.24	0.23	0.20	0.17	0.15
ZnO_Ni2	0.40	0.36	0.36	0.29	0.29	0.25	0.22	0.20
ZnO_Ni3	0.40	0.37	0.33	0.26	0.24	0.18	0.14	0.11

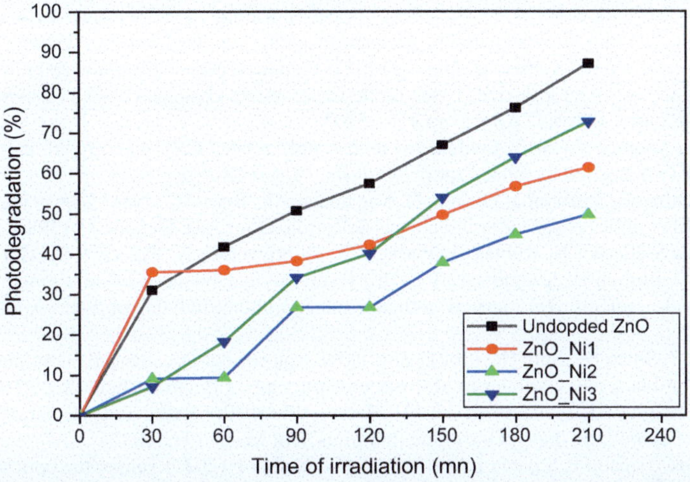

Fig. 3 MB degradation rate as a function of UV irradiation time for undoped and Ni-doped ZnO nanofilms

5 Conclusion

In the present work, Zinc oxide nanofilms were deposited by ultrasonic spray technique on glass substrates at 400 °C. The effects of doping by Nikel on structural, optical, and photocatalytic properties were analyzed. The X-Ray diffraction revealed the wurtzite polycrystalline structure of undoped and Ni doped nanofilms. The PL spectra of all sprayed ZnO nanofilms exhibited the presence of UV and visible light emission. The peaks located around 379 nm are attributed to the recombination of (electron/hole) in the valence band. Visible peaks from 402 to 480 nm were attributed to the electronic transitions from conduction band to Zn vacancies, interstitial atoms (Zn_I) to valence band and oxygen vacancies respectively. An important red emission appeared at 650 having oxygen interstitials (Oi) as origin. Ultimately, photocatalytic results showed that the best photocatalytic efficiency joined 87.1% by undoped ZnO nanofilms compared to Ni doped ZnO which restricted the degradation.

References

1. Chen D, Wang J, Liu Q, Xu Y, Li D, Liu Y (2011) Highly sensitive. ZnO thin film bulk acoustic resonator for hydrogen detection. J Micromech Microeng 21:115018
2. Zhang XY, Dhawan A, Wellenius P, Suresh A, Muth JF (2007) Planar ZnO ultraviolet modulator. Appl Phys Lett 91:071107
3. Shih W-C, Wang T-L, Pen Y-K (2012) Enhancement of characteristics of ZnO thin film surface acoustic wave device on glass substrate by introducing an alumina film interlayer. Appl Surf Sci 258:5424

4. Phan D-T, Chung G-S (2011) The effect of post-annealing on surface acoustic wave devices based on ZnO thin films prepared by magnetron sputtering. Appl Surf Sci 257:4339
5. Remashan K, Choi YS, Park SJ, Jang JH (2012) Enhanced capacitance ratio and low minimum capacitance of varactor devices based on depletion-mode Ga-doped ZnO TFTs with a drain-offset structure. J Phys D Appl Phys 45:435103
6. Ortel M, Wagner V (2013) Leidenfrost temperature related CVD-like growth mechanism in ZnO-TFTs deposited by pulsed spray pyrolysis. J Cryst Growth 363:185
7. Renganathan B, Sastikumar D, Gobi G, Yogamalar NR, Bose AC (2011) Nanocrystalline ZnO coated fiber optic sensor for ammonia gas detection. Opt Laser Technol 43:1398
8. Siwinska-Stefanska K, Kubiak A, Piasecki A, Dobrowolska A, Cza-czyk K, Motylenko M, Rafaja D, Ehrlich H, Jesionowski T (2019) Hydrothermal synthesis of multifunctional TiO_2-ZnO oxide systems with desired antibacterial and photocatalytic propertiesAppl. Surf Sci 463:791
9. Wang CC, Shieu FS, Shih HC (2021) Zinc oxide nanostructures enhanced photoluminescence by carbon-black nanoparticles in Moiré heterostructures. J Environ Chem Eng 9:104707
10. Abdulmunem OM, Mohammed Ali MJ, Hassan ES (2020) Optical and structural characterization of aluminium doped zinc oxide thin films. Opt Mater 109:110374
11. Singh SK, Tripathi P, Sulania I, Siva Kumar VV, Kumar P (2021) Tuning the optical and electrical properties of magnetron-sputtered Cu–ZnO thin films using low energy Ar ion irradiation. Opt Mater 114:110985
12. Kanabar N, Gadani K, Shrimali VG, Sagapariya K, Rathod KN, Udeshi B, Joseph J, Pandya DD, Solanki PS, Shah NA (2021) Structural and electrical properties of sol–gel grown nanostructured ZnO and $LaMnO_3$ particle-based nanocomposites. Appl Phys A 127:122
13. Abideen ZU, Shah AH, Teng F, Abro MI, Ullah K (2021) One-step hydrothermal synthesis of ZnO microtubes with an efficient. Micro Nano Lett 16:142
14. Ponja SD, Sathasivam S, Parkin IP, Carmalt CJ (2020) Highly conductive and transparent gallium doped zinc oxide thin films via chemical vapor deposition. Sci Rep 10:638
15. Saha JK, Bukke RN, Mude NN, Jang J (2020) Significant improvement of spray pyrolyzed ZnO thin film by precursor optimization for high mobility thin film transistors. Sci Rep 10:8999
16. Ouhaibi A, Ghamnia M, Dahamni MA, Heresanu V, Fauquet C, Tonneau D (2018) The effect of strontium doping on structural and morphological properties of ZnO nanofilms synthesized by ultrasonic spray pyrolysis method. J Sci Adv Mater Dev 3:29
17. Salavati-Niasari M, Davar F, Mazaheri M (2008) Preparation of ZnO nanoparticles from [bis(acetylacetonato)zinc(II)]-oleylamine complex by thermal decomposition. Mater Lett 62:1890–1892. https://doi.org/10.1016/j.matlet.2007.10.032
18. Salavati-Niasari M, Davar F, Fereshteh Z (2009) Synthesis and characterization of ZnO nanocrystals from thermolysis of new precursor. Chem Eng J 146:498–502. https://doi.org/10.1016/j.cej.2008.09.042
19. Boulila S, Ghamnia M, Boukhachem A, Ouhaibi A, Chakhoum MA, Fauquet C, Heresanu V, Tonneau D (2020) Photocatalytical properties of NiO nanofilms doped with Ba. Philosoph Mag Lett 100:6, 283–293. https://doi.org/10.1080/09500839.2020.1760389
20. Ouhaibi A, Saoula N, Ghamnia M, Dahamni MA, Guerbous L (2022) Effect of deposition temperature on morphological, optical, and photocatalytic properties of ZnO thin films synthesized by ultrasonic spray pyrolysis method. Cryst Res Technol: 2100224
21. Ahmad I, Aslam M, Jabeen U, Zafar MN, Malghani MNK, Alwadai N, Alshammari FH, Almuslem AS, Ullah Z (2022) ZnO and Ni-doped ZnO photocatalysts: Synthesis, characterization and improved visible light driven photocatalytic degradation of methylene blue Inorganica Chimica Acta 543:121167. ISSN 0020-1693. https://doi.org/10.1016/j.ica.2022.121167
22. Sowri Babu K (2018) Violet luminescence and its origin in ZnO nanoparticles prepared by co-precipitation method. AIP Conf Proc 1992:040012
23. Mahroug A, Boudjadar S, Hamrit S, Guerbous L (2014) G Structural, optical and photocurrent properties of undoped and Al-doped ZnO thin films deposited by sol–gel spin coating technique. Matter Lett 134:248

24. Sindhu HS, Rajendra BV, Kulkarni SD, Babu PD (2018) Defect induced white-light emission from Mn–doped ZnO films and its magnetic properties. Thin Solid Films 655:83
25. Kharroubi B, Bousmaha M, Bezzerrouk MA, Akriche A, Naceur R, Guezzoul M, Bensassi KB, Zahafi K, Zoukel A, Abdelkrim M, Bedrouni M, Bouslama M (2022) Photocatalytic efficiency of undoped and Cu-doped ZnO thin films coated inside transparent glass tube as one-piece photoreactor, Appl Surface Sci 601:154121. ISSN 0169-4332. https://doi.org/10.1016/j.apsusc.2022.154121
26. Bounemeur A, Chemachema M (2021) Adaptive fuzzy fault-tolerant control using Nussbaum-type function with state-dependent actuator failures. Neural Comput Appl 33:191–208
27. Abdelhamid B, Mouhamed C, Najib E (2017) Indirect robust adaptive fuzzy control of uncertain two link robot manipulator. In: Recent advances in electrical engineering and control applications 2017. Springer International Publishing, pp 125–139
28. Bounemeur A, Chemachema M, Essounbouli N (2018) Indirect adaptive fuzzy fault-tolerant tracking control for MIMO nonlinear systems with actuator and sensor failures. ISA Trans 1(79):45–61
29. Zahaf A, Bououden S, Chadli M, Chemachema M (2022) Robust fault tolerant optimal predictive control of hybrid actuators with time-varying delay for industrial robot arm. Asian J Control 24(1):1–5
30. Zahaf A, Bououden S, Chadli M, Boulkaibet I, Neji B, Khezami N (2023) Dynamic sensorless control approach for Markovian switching systems applied to PWM DC–DC converters with time-delay and partial input saturation. Sensors 23(15):6936
31. Abdelmalek Z, Abdelhamid B, Sofiane B, Boulkaibet I (2019) Fault diagnosis of uncertain hybrid actuators based model predictive control. In: Proceedings of the 4th international conference on electrical engineering and control applications: ICEECA 2019, 17–19 Dec 2019, Constantine, Algeria 2021. Springer, Singapore, pp 961–971

Synthesis and Characterization of Nickel Doped Fe_3O_4 Nanoparticles Using Solvothermal Method

E. Aliouat, N. Belmokhi, B. Mihoubi, H. Filali, R. Bouhroum, and N. Boukheit

Abstract Over the past 20 years, research on nanomaterials has made remarkable progress not only in the controlled synthesis and the study of novel properties, but also in the multiple applications of matter at the nanoscale. Currently, Fe_3O_4 magnetite nanoparticles continue to pose an enigma, as the understanding of certain fundamental properties remains unclear despite vigorous efforts and substantial progress in research. In fact, these materials have attracted considerable attention in the fields of new technology with different but complementary motivations using various types of experimental techniques and theoretical approaches. In this work, we present the results obtained from a thorough analysis of Fe_3O_4 nanoparticles with various Nickel concentrations synthesized through a simple solvothermal method. Characterization methods are mainly structural (XRD, SEM) and physical (UV–Vis, FTIR, …). The SEM images show that the prepared NPs exhibit spherical shape with nanometric size confirmed by X-ray diffraction. Furthermore, the Nickel substitution of iron cations into the crystalline structure had an impact on the average crystallite size. Raman spectra of Fe_3O_4 magnetite nanoparticles are very consistent with the literature.

Keywords Magnetite · Magnetic properties · Nanoparticles

E. Aliouat (✉) · N. Belmokhi · B. Mihoubi · H. Filali · N. Boukheit
Laboratory of Thermodynamic and Treatment of Materials Surface, Department of Physics, Faculty of Exact Sciences, Mentouri Brothers University, Constantine, Algeria
e-mail: esma.aliouat@doc.umc.edu.dz

R. Bouhroum
Laboratory of Innovative Techniques and Environmental Preservation, Mentouri Brothers University, Constantine, Algeria

S. Hamamda et al. (eds.), *Proceedings of the 2nd International Conference of Nanotechnology for Environmental Protection and Clean Energy Production*, Springer Proceedings in Materials 45, https://doi.org/10.1007/978-981-97-1916-7_19

1 Introduction

In recent years, the high demand for magnetic materials in biomedical applications has prompted researchers to explore innovative approaches for synthesizing iron oxide nanomaterials [1–3]. In addition to its very high magnetic properties and electronic conductivity [4, 5], magnetite's biocompatibility [6, 7] and low toxicity have made it the driving force behind research efforts aimed at a multidisciplinary field of study. These experimental studies have revealed that magnetite has made significant progress both in the controlled synthesis of nanoparticles and in the exploration of new properties, which is intensifying its use in advanced medical technology applications [8–11].

Several synthesis methods for iron-based nanoparticles have been developed, including co-precipitation [12], micro/nano-emulsion [13], green (biosynthesis) [14], and solvothermal methods [15–17]. Each synthesis method has a different yet significant impact on a specific biomedical application.

Despite the relatively low quantities of materials produced through the solvothermal route, this technique appears well-suited for synthesizing nanoparticles smaller than 25 nm, in various shapes, with superior performance. However, it is essential to note that stability and sensitivity to aggregation pose major challenges. Consequently, researchers have focused on surface coatings to preserve stability and properties of nanoparticles when suspended in physiological environments [18, 19], which is particularly important because magnetic nanoparticles in water- and protein-rich environments are exposed to different PH conditions during preparation and biomedical applications [20].

Furthermore, new perspectives on enhancing the magnetic properties of magnetite nanoparticles have intensified research activities through doping reactions with transition metals.

In this study, a simple solvothermal method was employed to synthesize nickel doped Fe_3O_4 nanoparticles. Advanced microstructural and magnetic analyses using usual characterization techniques (XRD, FTIR, Raman, VSM…), provided necessary insights for a better understanding of their properties.

2 Preparation Method

Synthesis of the $Ni_xFe_{3-x}O_4$ NPs was carried out based on solvothermal method, in which Ferric chloride (as iron source), Diethanolamine (DEA), Sodium hydroxide, Ammonium acetate and Ethylene glycol (as reducing agent) were used.

At first, iron chloride and nickel acetate were used by respecting the reports, 48 mmol of Sodium hydroxide and 240 mmol of Ammonium acetate were dissolved separately in 20 ml, 10 ml, and 40 ml of ethylene glycol respectively under magnetic stirring to form clear solutions. Then, 4 $cm^3$of DEA was mixed with the solutions and stirred for 2 h at room temperature. The obtained mixture was transferred to a

Teflon-lined autoclave and maintained at 200 °C for 12 h. Finally, the sediment was washed with distilled water and dried at 100 °C.

3 Results and Discussion

Analysis of the X-ray diffraction patterns of nickel-doped Fe_3O_4 nanoparticles (Fig. 1) shows good agreement with those of magnetite, which crystallizes in a spinel structure in the cubic space group Fd-3m with a lattice parameter of 0.8396 nm (JCPDS card no. 19-629). Thus, the solvothermal method enabled the synthesis of nickel-doped magnetite without second phase of Ni or one of its compounds. Moreover, the reaction precursors inducing magnetite formation were not detected by XRD. In view of this result, we can conclude that the Fe_3O_4 formation reaction is complete. This confirms the formation of a single-phase solid solution with no secondary phase [20].

Table 1 summarizes the influence of doping on the lattice parameter and crystallite size estimated, from the main reflection, by Scherer equation.

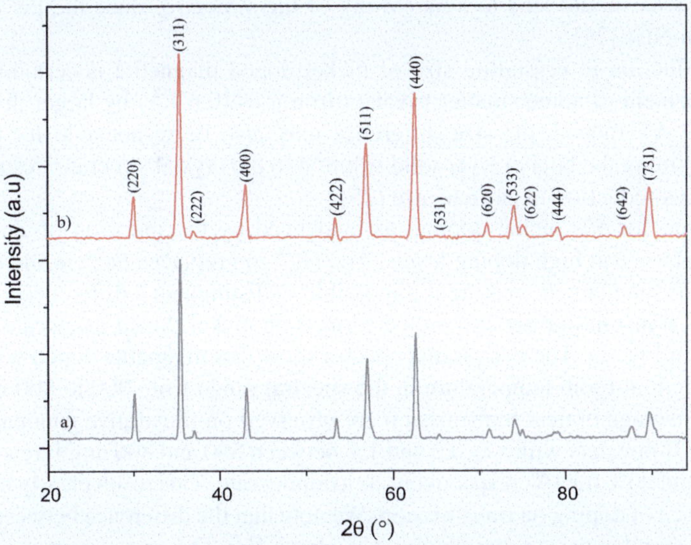

Fig. 1 XRD patterns of Fe_3O_4 nanoparticles doped with **a** 0.5% Ni; **b** 1% Ni

Table 1 Lattice parameter and crystallite size of nickel doped nanoparticles

Sample	Average grain size (nm)	Lattice parameter (Å)
0.5% Ni	22.62	8.385
1% Ni	17.59	8.392

The lattice parameter values of nickel-doped nanoparticles are very close to those of magnetite reported in the literature. Surface oxidation of Fe_3O_4 nanoparticles could explain the difference between the calculated and theoretical parameters. It has been reported that the thickness of the oxide layer can reach 3 nm [21]. What's more, the lattice parameter of the nanoparticles increases relatively slightly with the rate of nickel doping. This confirms the insertion of nickel into the magnetite lattice [22]. The main purpose of this insertion was to enhance the magnetic properties of magnetite nanoparticles. As the syntheses were carried out under similar conditions, the variation in the lattice parameter can be explained by the difference in the ionic radii of the two cations, but also by their preference for a specific coordination. On the other hand, studies of zinc doping of magnetite have reported that Zn^{2+} can occupy both octahedral and tetrahedral sites, and the sites that Zn^{2+} actually occupies depend on the amount of dopant (Zn) and the method of synthesis [23]. Thus, if we consider not only the ion radii (R Fe^{3+} = 64 pm, R Ni^{2+} = 0.69 pm) but also the size of the octahedral and tetrahedral sites of the spinel structure, the slight increase observed strongly suggests the substitution of ferric ions by nickel ions in the tetrahedral sites for rates of less than 3%. Indeed, the insertion of the Ni^{2+} cation, which is larger than the Fe^{3+}cation, into the tetrahedral site induces a slight broadening of the structure and thus a slight increase in the crystalline parameter. A substitution of ferrous ions by nickel ions should lead to a narrowing of the structure, since the Fe^{2+}cation is larger than Ni^{2+} [24].

The reduction in crystallite size of nickel doped magnetite is certainly due to the replacement of some smaller trivalent iron ions (0.63 Å) by larger divalent Ni ions (0.69 Å). Indeed, the size difference may also be responsible for the ionic forces confining the $Ni_xFe_{3-x}O_4$ solid solution to the crystal structure, resulting in a spontaneous reduction in particle size [25].

However, for 5% nickel doping, as evidenced by the insertion of Zn into the magnetite lattice at high doping levels. The Ni^{2+} ion replaces Fe^{2+}, while electrical neutrality is maintained, making it impossible to distinguish a difference.

Optical transmission spectra are performed on nickel-doped nanoscale powders as shown in Fig. 2a. The two similar spectra show that magnetite doped with nickel is transparent at room temperature in the spectral range from 300 to 800 nm. High values of average optical transmission are observed on illustrative nanoparticles of $NixFe_{3-x}O_4$ powders with x = 0.5 and 1% between 500 and 800 nm. It reaches 80% for 0.5% and 90% for 1%, respectively, as a doping rate. This result clearly highlights the influence of doping on transmission. We note that the difference between the two spectra is significant, approximately, it is about 5%. The transparent magnetite is very often explained by the inter-band electronic transition, due to decreasing of band gap energy [26].

As a precision tool compared with the tauc plot method, the technique based on the first derivative of the transmission with respect to energy is introduced. The latter method offers the applicability advantage in wide range of experimental conditions and validating its adequately compared with second method. The two methods' consistency leads to a good reliability of calculation [27]. The dT/dE $Ni_xFe_{3-x}O_4$ curves (Fig. 2b) with x = 0.5 and 1% clearly show that the difference in the powder

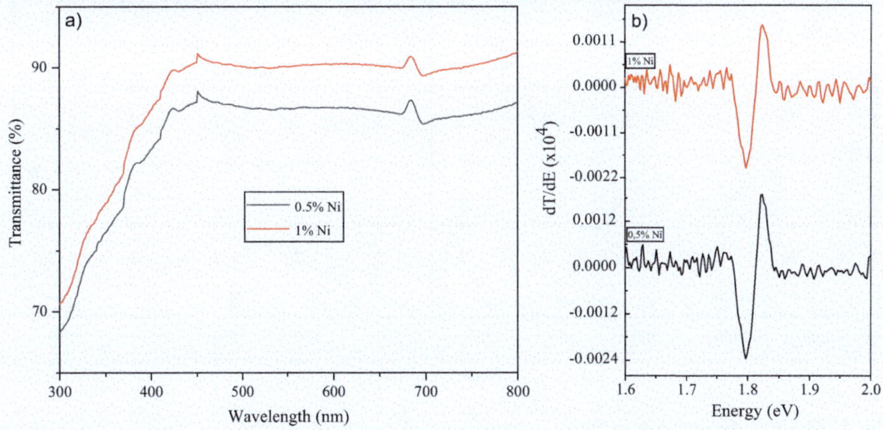

Fig. 2 Transmittance spectra (**a**) and dT/dE curves (**b**) of nickel-doped magnetite

gap is not significant, it is around 1.79 eV. The values of these gaps are important and confirm the semiconductor nature of our powders (Fig. 2).

Infrared spectroscopy analysis enables the determination of different vibrational bonds and the location of characteristic bands of chemical molecules present in metal oxide spinel systems. Figure 4 presents the FT-IR spectra of Fe_3O_4 nanoparticles doped with x = 0.5 and 1%, recorded in the range of 400–4000 cm^{-1}. The two

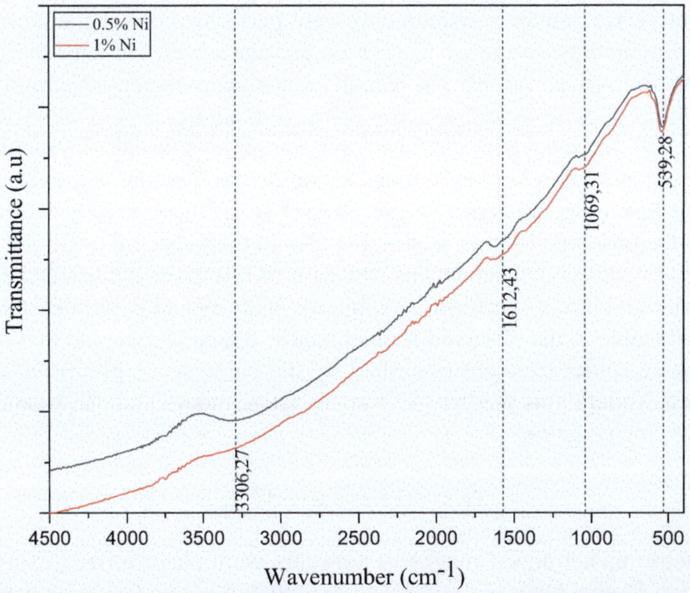

Fig. 3 FTIR spectra of nickel-doped magnetite

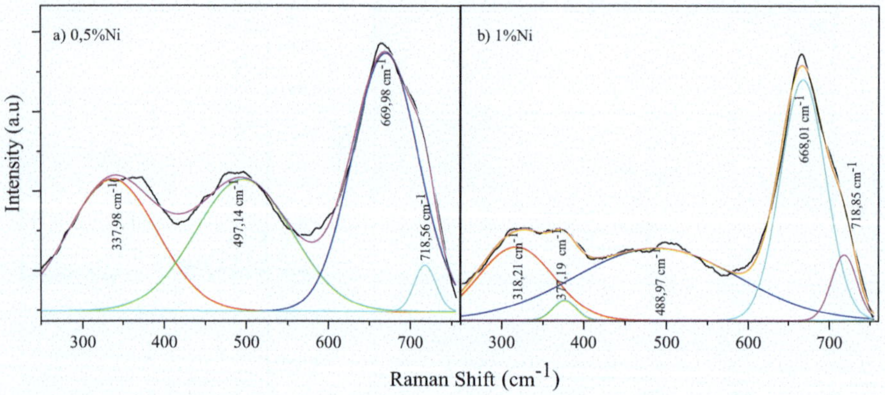

Fig. 4 Spectral deconvolution into Gaussian-shaped bands of Raman spectra of nickel-doped magnetite

obtained spectra are usable and comparable to each other, with no significant differences observed. As a result, the analysis of the profiles reveals three distinct regions: the first region between 400 and 700 cm^{-1}, characterized by the peak at 542 cm^{-1}, attributed to the vibration of the Fe–O bond in Fe_3O_4; the second region between 700 and 3000 cm^{-1} is attributed to the stretching vibrations of the CH, C=O, and C=C groups of sodium oleates absorbed on the surface of the Fe_3O_4 nanoparticles; and the third region corresponds to the stretching vibrations of OH, which is attributed to OH absorbed by the Fe_3O_4 nanoparticles in the precursor [28] (Fig. 3).

In recent years, Raman measurements were performed on magnetite by several groups of researchers. Some of these measurements were directly derived from Verwey's model, whose validity has recently come under increased scrutiny, mostly after a number of neutron and electron diffraction studies, but also NMR studies, have suggested that magnetite exhibits a more complex crystalline structure at low temperatures than that in Verwey's model. Careful consideration of the sizeable literature on the Raman spectroscopy of magnetite (Fe_3O_4) shows variety not only in the number of the observed Raman modes, but also in their spectral positions and their attribution to underlying mechanisms. Because of this diversity, researchers turned to the spinel structure to perform a symmetry analysis and determine the phonon modes attributable to the observed Raman bands. It is now accepted that magnetite has an inverse spinel structure described by the cubic space group Fd3m (OFh). Symmetry considerations predict five Raman-active modes that can be summed up by the following equation:

$$G = \mathrm{A1g} + Eg + 3T2g \quad (1)$$

Nanometric nickel-doped magnetite powders were characterized using Raman spectroscopy. Figure 4 shows the spectral deconvolution into Gaussian bands of the Raman spectra of the powders recorded between 250 and 800 cm^{-1}. Analysis of the two spectra reveals the absence of the 220, 290 cm^{-1} doublet, characteristic of hematite, and thus the formation of hematite in our samples is excluded. On the spectrum in Fig. 4a, obtained from the powder doped with 0.5% nickel, three peaks at 337.98, 497.14, and 669.98 cm^{-1} and a shoulder at 718.56 cm^{-1} are observed [29–32]. Given the specificity of the spectrum on Fig. 4a, the attribution of the modes to the observed bands is done following the extant literature. Hence, the three peaks are attributed respectively to the Eg, F2g (2), and A1g. The shoulder at 718.56 cm^{-1} essentially proves the superficial oxidation of magnetite into maghemite. The spectrum in Fig. 4b, obtained from the powder doped with 1% nickel, confirms that superficial oxidation with the presence of a shoulder at 718.85 cm^{-1}. In addition to the three bands, Eg, F2g (2), and A1g, at 318.21, 488.97, and 668.01 cm^{-1}, respectively [29–32], a fourth, weak-intensity band is observed at 377.19 cm^{-1}. We think, as shown by Monika Hanesch [33], that it is the sign of the onset of the transformation of magnetite into maghemite. This is due to the fact that the divalent iron renders magnetite easily prone to oxidation.

Moreover, the spectra show that the observed Raman bands are wide, a phenomenon usually explained by the conduction mechanism in magnetite, but also by the electronic disorder associated with the random arrangement of the Fe^{2+} and Fe^{3+} ions on the octahedral sites [30].

The iron oxide Fe_3O_4, a magnetized substance known for an extended period, has been a subject of extensive study in the field. The multitude of research efforts carried out on the magnetism of transition metal oxides, particularly magnetite, highlights the growing interest in this material [34–36]. In fact, the magnetism of Fe_3O_4 has long been a topic of discussion due to numerous controversies regarding the comprehension of some of its fundamental properties. Nowadays, this issue is addressed through purely quantum concepts based on two hypotheses that assimilate ions to magnetic dipoles:

The first, often overlooked, strictly treats the magnetic moment of the nuclei, which provides much information about nuclear magnetism.

The second that considers that the interactions are essentially due to only the spins of the Fe^{2+} and Fe^{3+} ions.

In the case of magnetite, its ferrimagnetic character is largely contingent upon the alignment of spin magnetic moments of Fe^{2+} and Fe^{3+} ions. It has been demonstrated that in octahedral sites, the moments are aligned in a parallel position, whereas in tetrahedral sites, ferric ions are distributed in an antiparallel alignment. Considering this configuration, the magnetic moments of ferric ions completely compensate each other. Thus, an overall magnetization of Fe_3O_4 resulting from the magnetic moments of ferrous ions is perceptible even in the absence of an external magnetic field [37–40].

The magnetic response of $Ni_xFe_{3-x}O_4$ powders with 0.5and 1% nickel was measured using a Vibrating Sample Magnetometer (VSM), allowing the extraction of general characteristics (saturation magnetization, coercivity, and remanence) for both powders [23]. Figure 5 compares the evolution of the magnetization as a function of the applied magnetic field of the two powders. The magnetization deduced from the curves and the mass of the powders is given in "emu/g". It is observed that both products exhibit superparamagnetic behaviour, featuring an open hysteresis loop

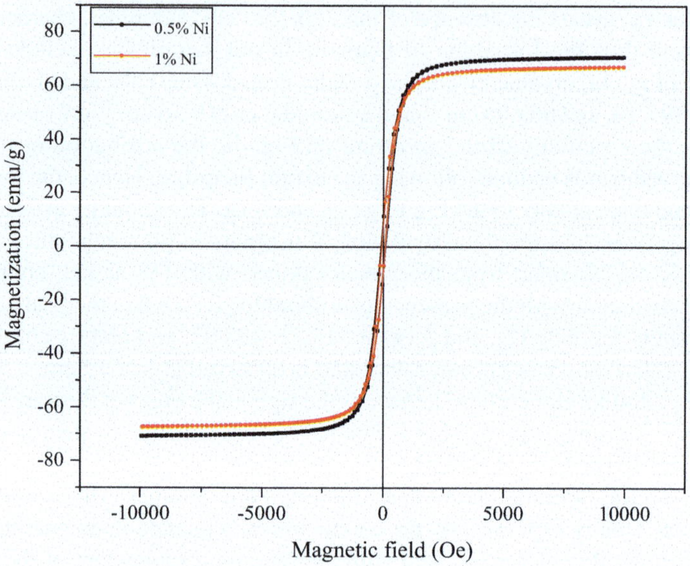

Fig. 5 Magnetic hysteresis loops of nickel-doped magnetite

Table 2 Magnetic characteristics of the magnetite

Characteristics	Fe_3O_4 with 0.5% Ni	Fe_3O_4 with 1% Ni
Saturation magnetization Ms (emu/g)	70.8	67.23
Coercivity Hc (emu/g)	9.8	8.75
Remanent magnetization (Oe)	90.77	72.99

with non-zero coercive field (Hc) and remanent magnetization (Mr). The saturation magnetization, coercivity, and remanence characteristics are summarized in Table 2.

A decrease in these characteristics is observed with increasing nickel doping rate. We attribute this decrease in the magnetic properties of nickel-doped magnetite to the difference in chemical composition of the two powders, by reason of their very close sizes. For both samples, saturation magnetization is achieved at an applied magnetic field of approximately 1.25 Oe.

4 Conclusion

The work presented in this paper aims to study nickel-doped magnetite nanoparticles synthesized by solvothermal technique using iron chloride and nickel acetate. The synthesis experimental parameters were optimized to obtain quasi-spherical nanoparticles with controlled size. Various characterization techniques were used to address the microstructural and magnetic properties of nickel-doped magnetite. The analysis of the results shows that the incorporation of nickel into the spinel structure of magnetite induces a slight evolution in the lattice parameter, as demonstrated by X-ray diffraction. In addition, variations in the magnetic properties have been recorded.

References

1. Nguyen MD, Tran H-V, Xu S, Randall Lee T (2021) Fe_3O_4 nanoparticles: structures, synthesis, magnetic properties, surface functionalization, and emerging applications. Appl Sci 11(23):1130
2. Lisjak D, Mertelj A (2018) Anisotropic magnetic nanoparticles. A review of their properties, syntheses and potential applications. Prog Mater Sci 95:286–328
3. Manh-Huong P, Javier A, Hafsa K, Paula L-K, Sayan C, Kristen SR, Zohreh N, Raja D, Óscar I, Hariharan S (2016) Exchange bias effects in iron oxide-based nanoparticle systems. Nanomaterials 6(11):221–251
4. Westphal WB, Von Hippel A (1957) Dielectric spectroscopy of ferromagnetic semiconductors. Rev Mod Phys 29(3):279–307
5. Evans BJ (1975) Experimental studies of the electrical conductivity and phase transition in Fe_3O_4. AIP Conf Proc 24(1):73–78
6. Laurent S, Forge D, Port M, Roch A, Robic C, Vander EL, Muller RN (2008) MagneticIron oxide nanoparticles: synthesis, stabilization, vectorization, physicochemical characterizations, and biological applications. Chem Rev 108(6):2064–2110
7. Ling D, Hyeon T (2013) Chemical design of biocompatible iron oxide nanoparticles for medical applications. Small 9(9–10), 1450–1466
8. Ulbrich W, Lamprecht A (2010) Targeted drug-delivery approaches by nanoparticulate carriers in the therapy of inflammatory diseases. J R Soc Interface 7(Suppl_1):S55–S66
9. Liu S, Yu B, Wang S, Shen Y, Cong H (2020) Preparation, surface functionalization and application of Fe_3O_4 magnetic nanoparticles. Adv Colloid Interface Sci 281:102165
10. Maggioni D, Arosio P, Orsini F, Ferretti AM, Orlando T, Manfredi A, Ranucci E, Ferruti P, D'Alfonsoa G, Lascialfari A (2014) Super para magnetic iron oxide nanoparticles stabilized by a poly(amidoamine)-rhenium complex as potential theranostic probe. Dalton Trans 43:1172–1183
11. Sapsford KE, Russ Algar W, Berti L, Gemmill KB, Casey BJ, Oh E, Stewart MH, Medintz IL (2013) Functionalizing nanoparticles with biological molecules: developing chemistries that facilitate nanotechnology. Chem Rev 113(3):1904–2074
12. Wulandari IO, Santjojo DJDH, Shobirin RA, Sabarudin A (2017) Characteristics and magnetic properties of chitosan-coated Fe_3O_4 nanoparticles prepared by ex-situ co-precipitation method. Rasayan J Chem 10(4):1348–1358
13. Zhou ZH, Wang J, Liu X, Chan HSO (2001) Synthesis of Fe_3O_4 nanoparticles from emulsions. J Mater Chem 11(6):1704–1709
14. Mirabello G, Lenders JJM, Sommerdijk NAJM (2016) Bioinspired: synthesis of magnetite nanoparticles. Chem Soc Rev 45:5085–5106
15. Liang Y, Jiang L, Xu S, Ju W, Tao Z, Yang Y, Peng X, Wei G (2023) Synthesis and characterization of Fe_3O_4 nanoparticles prepared by solvothermal method. J Mater Eng Perf. https://doi.org/10.1007/s11665-023-08431-1
16. Lian J, Duan X, Ma J, Peng P, Kim T, Zheng W (2009) Hematite (α-Fe_2O_3) with various morphologies: ionic liquid-assisted synthesis, formation mechanism, and properties. ACS Nano 3(11):3749–3761

17. Ge S, Shi X, Sun K, Li C, Uher C, Baker JR, Banaszak HMM, Orr BG (2009) Facile hydrothermal synthesis of iron oxide nanoparticles with tunable magnetic properties. J Phys Chem C 113(31):13593–13599
18. Häfeli UO, Riffle JS, Harris-Shekhawat L, Carmichael-Baranauskas A, Mark F, Dailey JP, Bardenstein D (2009) Cell uptake and in vitro toxicity of magnetic nanoparticles suitable for drug delivery. Mole Pharmaceut 6(5):1417–1428
19. Zhang J, Rana S, Srivastava RS, Misra RDK (2008) On the chemical synthesis and drug delivery response of folate receptor-activated, polyethylene glycol-functionalized magnetite nanoparticles. Acta Biomater 4:40–48
20. Geoffrey MC, Kenneth WA (2023) The cell: a molecular approach. Oxford University Press, Neuvièmeédition (2023)
21. Larumbe S, Gomez-Polo C, Pérez-Landazábal JI, García-Prieto A, Alonso J, Fdez-Gubieda ML, Cordero D, Gómez J (2012) Ni doped Fe_3O_4 magnetic nanoparticles. J Nanosci Nanotechnol 12(3):2652–2660
22. Hyeon T, Lee SS, Park J, Chung Y, Na HB (2001) Synthesis of highly crystalline and monodisperse maghemite nano crystallites without a size-selection process. J Am Chem Soc 123:12798–12801
23. Sorescu M, Diamandescu L, Tarabasanu-Mihaila D, Teodorescu V. Sequence of phases in the hydrothermal synthesis of zinc-dopemagnetite system 106(2–3):273–278 (2007)
24. Honami I, Shota A, Naoki N, Yoshio K, Yasuhiro Y. Wet chemical synthesis of zinc-iron oxide nanocomposite. Hyperfine Interact 238(1):79
25. Cullity BD, Graham CD (2008) Introduction to magnetic materials, 2nd ed. Wiley, IEEE Press
26. Smit J, Wijn HPJ (1959) Ferrites. Philips Technical Library, Eindhoven
27. Sing LR, Hussain MA (2020) Effect of doping concentration on the optical properties of nanocrystalline Zn doped PbS thin films deposited by CBD method. Chalcogenide Lett 17(11):583–591
28. Aztifa N, Darvina Y, Hidayat R (2023) Analysis of the optical properties of the synthesis of Fe_3o_4/Pani nanocomposites with the sol-gel method using spin coating. Pillar Phys 16(1):18–28
29. Atoui M, Benzeghda S, Touam T, Chelouche A, Djouadi D (2021) Ag doping effects on the microstructure, morphology, optical, and luminescence properties of sol–gel-deposited ZnO thin films. Semiconductors 55(12):922–976
30. Jiang W, Wu Y, He B, Zeng X, Lai K, Gu Z (2010) Effect of sodium oleate as a buffer on the synthesis of super para magnetic magnetite colloids. J Colloid Interf Sci 347:1–7
31. DIppolito V, Andreozzi GB, Lottici DPP (2015) Raman fingerprint of chromate, aluminate and ferrite spinels. J Raman Spectrosc 46:1255–1264
32. Larry Verble J (1974) Temperature-dependent light-scattering studies of the Verwey transition and electronic disorder in magnetite. Phys Rev B 9(12):5236–5248
33. Shebanova ON, Lazor P (2003) Raman spectroscopic study of magnetite ($FeFe_2O_4$): a new assignment for the vibrational spectrum. J Sol State Chem 174(2):424–430
34. Graves PR, Johnston C, Campaniello JJ (1988) Ramen scattering in spinel structure ferrites. Mat Res Bull 23:1651–1660
35. Hanesch M (2009) Raman spectroscopy of iron oxides and (oxy)hydroxides at low laser power and possible applications in environmental magnetic studies. Geophys J Int 177(3):941–948
36. Bragg WH (1915) The structure of magnetite and the spinels. Nature 95(2386):561–561
37. Marinca TF, Chicinas I, Isnard O, Pop V (2011) Structural and magnetic properties of nanocrystalline $ZnFe_2O_4$ powder synthesized by reactive ball milling. Optoelectron Adv Mater Rapid Commun 5(1):39–43
38. Mathew DS, Juang R-S (2007) An overview of the structure and magnetism of spinel ferrite nanoparticles and their synthesis in microemulsions. Chem Eng J 129(1–3):51–65
39. Maher BA, Taylor RM (1988) Formation of ultrafine-grained magnetite in soils. Nature 336(6197):368–370
40. Bercoff PG, Bertorello HR (1997) Exchange constants and transfer integrals of spinel ferrites. J Magn Magnetic Mater 169:314–322

41. Kechrakos D, Trohidou KN (1998) Magnetic properties of dipolar interacting single-domain particles. Phys Rev B 58(18):12169–12177
42. Ganapathe LS, Mohamed MA, Yunus RM, Berhanuddin DD (2020) Magnetite (Fe_3O_4) nanoparticles in biomedical application: from synthesis to surface functionalisation. Magnetochemistry 6(4):68

Effect of Ni-Doping on ZnO Nanostructures Properties

Sonia Attaf, Fouzia Abbas, Chourouk Namoune, and Rabeh Bensaha

Abstract In this work we have prepared and characterized nanoparticles of zinc oxides (ZnO) pure and doped with Nickel by the Sol Gel method. This non-costly, easy method makes it possible to have more homogeneous powders with a finer particle size. XRD analysis confirmed hexagonal würtzite-like structure of ZnO nanoparticles. Sizes of these nanoparticles are in an order of magnitude of a few tens of nanometers. The doping of ZnO nanopowders with 5% Ni improved the crystalline quality; this is revealed by the increase in the intensity of the peaks. The infrared absorption spectroscopy (FTIR) made it possible to identify the nature of the bonds present in the ZnO nanopowders. FTIR spectra confirmed two characteristic bands attributed to Zn–O stretching bands. The degradation rate of methylene blue (D) of pure ZnO reaches 99.6% after 90 min. Nickel doping greatly reduced the degradation (D) from 99.6 to 21%.

Keywords ZnO · Doping · Sol gel · Nanoparticules · Photocatalytic activity

S. Attaf (✉) · F. Abbas · C. Namoune · R. Bensaha
Laboratoire de Céramiques, Université Constantine 1, Route Ain El Bey, 25000 Constantine, Algeria
e-mail: sonia.attaf@umc.edu.dz

F. Abbas
e-mail: fouzia.abbas@umc.edu.dz

C. Namoune
e-mail: chouroukna@gmail.com

R. Bensaha
e-mail: bensaha@umc.edu.dz

S. Hamamda et al. (eds.), *Proceedings of the 2nd International Conference of Nanotechnology for Environmental Protection and Clean Energy Production*, Springer Proceedings in Materials 45, https://doi.org/10.1007/978-981-97-1916-7_20

1 Introduction

In recent years has promoted itself as an interesting metal oxide material because of its properties like high refractive index, binding energy, high thermal conductivity, antibacterial and UV protection. ZnO could be used in many materials and product like include medicine, cosmetics, rubber, solar cells and foods [1]. Zinc oxide has high biocompatibility and fast electric transfer kinetics, such phenomena encourage the use of this material as a biomimic membrane to immobilize and modify the biomolecules [2]. The addition of impurity atoms, or doping [3, 4] to the zinc oxide changes the optical and electrical properties of the material, making it suitable for new applications and devices.

Substitution of foreign impurities modifies; bandgap, optical, electrical and magnetic properties [5, 6]. Photoluminescence spectra defect level was reducing in the ZnO nanoparticules (NPs), by increasing of Fe concentration in the ZnO lattice sites [7]. In literature, the Cu doped ZnO NPs reduction in the energy gap from 3.49 to 3.43 eV revealed the substitution of Cu^{2+} ions in the ZnO lattice [8]. A variety of chemical techniques were used to synthesize ZnO NPs, such as solvothermal [9], hydrothermal [10], co-precipitation [11] and sol–gel methods [12], etc. Among these synthetic routes, sol Gel method has been popularly adopted to synthesize by using ZnO NPs due to its easy method, low cost and high yield of nanoparticles with uniform size. In the present work, Ni doped ZnO were synthesized by Sol Gel method. We have studied the structural and optical properties of the Ni doped ZnO NPs for using photocatalytic application.

2 Experimental Details

A first aqueous solution with a concentration of 0.1 mol/L of zinc acetate dihydrate ($Zn(CH_3COO)_2.2H_2O$) is dissolved in distilled water with continuous stirring at moderate speed at 80°C. In parallel, a second solution of citric acid $C_6H_8O_7$ dissolved in ethanol (CH_3CH_2OH) with magnetic stirring at room temperature for 15 min is then added drop by drop to the first solution with continuous magnetic stirring until a homogeneous gel. The gel is dried at 80 °C for 24 h to obtain Zinc oxide. After that, we crush the product manually in a mortar until the granules are then homogeneous and then we treat the powder obtained by annealing at a temperature of 400 °C for 4 h. Finally, we obtain the pure Zinc oxide nanoparticles.

3 Results and Discussion

3.1 X-Ray Diffractometer (XRD)

Figure 1 shows the X-ray diffraction (XRD) spectra of the pure ZnO sample. All the diffraction peaks observed are: (010), (002), (011), (012), (110), (013), (0220), (013) and (021) correspond to those of ZnO of hexagonal structure type würtzite (JCPDS 96-900-4180). All the peaks are intense, sharp and fine, affirm the good crystallinity of the elaborated nanostructures and their poly-crystalline natures and thus formed crystallites in different orientations. The three most intense and dominant peaks are located around the diffraction angle $2\theta = 31.737°$ and $36.225°$ correspond to the (010) and (011) planes respectively. The intensity of this last peak indicates that (011) is the preferred orientation. The variation in the intensity of the peaks implies a variation in the number of atoms which refract in each of the directions and the variation in the position of the peaks implies a variation in the structural parameters (cell parameters, cell volume, etc.). Indeed, the XRD spectra of the pure ZnO samples reveal an improvement in the intensity, thus indicating better crystallization. We also observe that there is no detected peak of the parasitic phase therefore all the lines observed belong to the hexagonal structure of ZnO.

The ZnO obtained was identified by comparison with the database (JCPDS card No: 96-900-4180). It has a hexagonal würtzite crystal system structure with space group P63 mc symmetry, with lattice parameters:

$a = b = 3.2530$ (Å) and $c = 5.2070$ (Å) and angles $\alpha = \beta = 90°$ and $\gamma = 120°$. This result is in good agreement with the results obtained in the literature [13, 14].

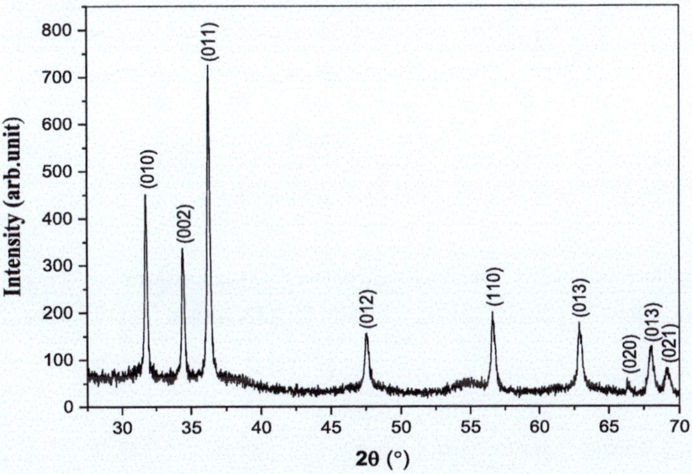

Fig. 1 XRD pattern of pure ZnO nanoparticles

Figure 2 shows the X-ray spectra (XRD) of undoped and doped with 5% Nikel ZnO samples. The diffraction peaks obtained with intensities correspond to the (110), (002), (011), (012), (110), (013), (020), (112) and (021) planes. Indeed the two diagrams represent the same dominant peak (011) which is located around the angle of 36.22° and which means that the two spectra present a single poly-crystalline phase which correspond to the hexagonal structure of the würtzite type (JCPDS 96-900-4182). The average particle size has been determined from full width at half maximum (FWHM) of the diffraction peaks using Scherrer's equation (Table 1). The average particle size of un-doped zinc oxide nanoparticles has been found to be 28.5 nm. Nickel doping reduced average particle size to 16.39 nm.

There is a slight shift of peaks towards high values. This leads to a decrease in parameter c (Table 2). Some authors attribute this shift to the incorporation of Ni ions in the substitution sites [15, 16]. We also observe a peak, located at 2θ = 44.47° which can be attributed to the parasitic phase NiO with low intensity

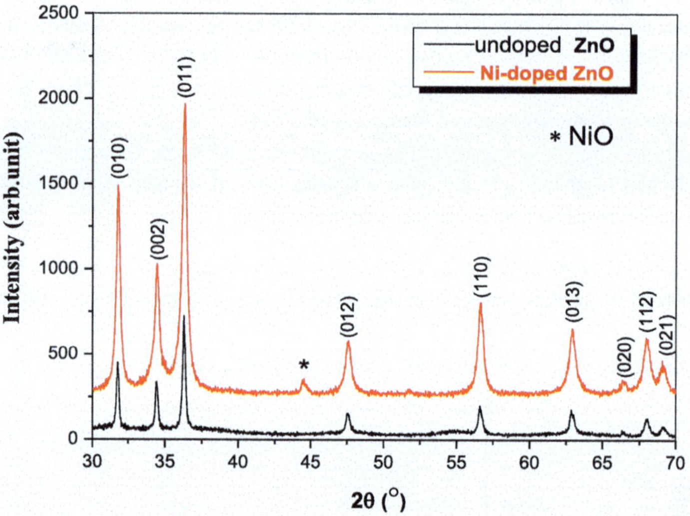

Fig. 2 XRD pattern of un-doped and doped ZnO nanoparticles

Table 1 Structural parameters of un-doped and doped ZnO nanoparticles

Doping rate (%)	Phase	2θ (°)	(hkl)	FWHM (°)	D_{aver} (nm)	δ (10^{14} lines/ m^2)	ε (10^{-4})
0	*Wurtzite*	*31.781*	*(010)*	*0.2323*	*28.5*	*12.6*	*34.1*
		36.274	(011)	0.2614			
5	*Wurtzite*	*31.809*	*(010)*	*0.398*	*16.39*	*23.28*	*16*
		36.287	(011)	0.496			

Strain (ε), dislocation density (δ), average crystallite size (D_{aver}), full width at half maximum (FWHM), the diffraction angle (2θ) and Miller indices of diffraction planes (hkl)

Table 2 The cell parameters and volume for un-doped and doped ZnO nanoparticles

Doping rate (%)	2θ (°)	(hkl)	d_{hkl}	Lattice parameters (Å)		*c/a*	Volume V (Å)
				a (Å)	*c* (Å)		
0	31.7806	(010)	0.2813	3.2486	5.2030	1.6016	54.9003
	34.4464	*(002)*	*0.2605*				
5	31.8096	(010)	0.2810	3.2457	5.2002	1.6021	54.7818
	34.4657	*(002)*	*0.2600*				

Spacing between consecutive parallel planes (d_{hkl}), the diffraction angle (2θ) and Miller indices of diffraction planes (hkl)

can be negligible. An increase in intensity is also clearly observed, indicating an improvement in crystallization. The addition of 5% Nickel therefore promotes the state of crystallization of the material [17, 18].

The values of the lattice parameters *a* and *c*, deduced from the X-ray spectra, are indicated in Table 2. The orders of magnitude of *a* and *c* are characteristic of the würtzite structure of ZnO.

3.2 Infrared Absorption (FTIR) of Zinc Oxide

The analysis of pure and doped ZnO powders by IR infrared spectroscopy allowed us to identify the specific vibrations as well as the characteristic peaks of the matrix of the samples produced. Figure 3 shows the transmission spectrum of the pure ZnO sample. The Zn–O bond was clearly observed at about 483.51 and 463.94 cm^{-1} [15] and the C=C bond was observed at about 1530.03 cm^{-1}. Two peaks of the absorption bands were observed at about 236.38 and 3740.44 cm^{-1} corresponding to O–H vibrations.

3.3 Photocatalytic Test of Zinc Oxide

In order to evaluate the photocatalytic activity of pure and Nikel doped ZnO nanoparticles, we used methylene blue (MB) as a target under irradiation from a lamp with a power of 120 W and a wavelength of 365 nm. Figure 4a shows the evolution of the UV–Vis spectrum of methylene blue as a function of time in the presence of pure ZnO nanoparticles. We clearly notice that the absorption peak at 663 nm characteristic of MB decreases sharply with time. Figure 5 shows the variation of the C/C0 ratio which represents the level of methylene blue in the presence of non-degraded zinc oxide nanoparticles over time. The results obtained show that the degradation

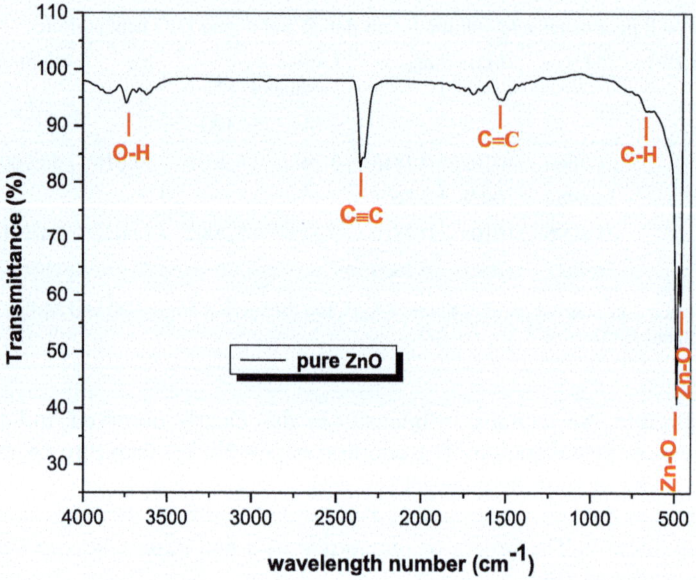

Fig. 3 FTIR of pure ZnO

rate (D) reaches 99.6% after 90 min. Figure 4b shows the evolution of the UV–Visible spectrum of methylene blue as a function of time in the presence of ZnO nanoparticles doped at 5% Nickel doping greatly reduced the degradation rate (D) from 99.6 to 21%. The highest photocatalytic activity is obtained for pure ZnO. The maximum absorption is located at 663.07 nm for both cases. The addition of Ni strongly decelerated the degradation of MB. The degradation rate (D) decreased from 97 to 73%.

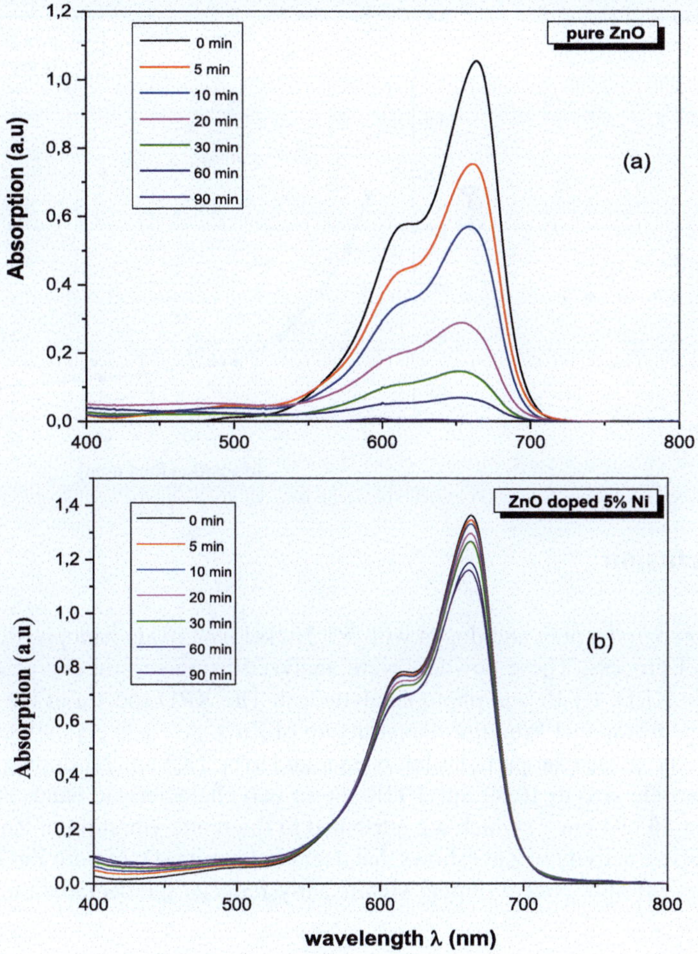

Fig. 4 Photocatalytic test for ZnO nanoparticles

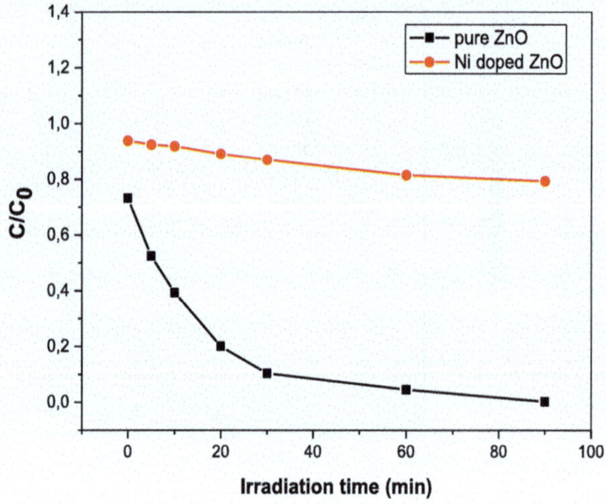

Fig. 5 Variation of c/c_0 as function of irradiation time

4 Conclusion

We have prepared a pure and doped with 5% Nickel zinc oxide nanopowders using the Sol–gel process. These powders were analyzed using several characterization techniques: XRD, FT-IR and photo-catalytic test. The XRD spectra of the samples revealed the hexagonal Wurtzite-like structure of ZnO. Average crystallites size of un-doped zinc oxide nanoparticles has been found to be 28.5 nm. Ni-doping reduced average particle size to 16.39 nm. FTIR shows two characteristic bands at around 483.51 and 463.94 cm^{-1}, which are attributed to the mode vibration of Zn–O. The photo-catalytic activity of ZnO shows that degradation rate (D) of pure ZnO reaches 99.6% after 90 min. Nickel doping significantly reduced the degradation rate (D) from 99.6 to 21%.

References

1. Klingshirn Z (2007) Material, Physics and application. Cheme Phy Chem 8:782–803
2. Chitra K, Annadurai G (2013) Antimicrobial activity of wet chemically engineered spherical Zno nano particles on food borne pathogen. Int Food Res J 20(1):59–64
3. Hsu LS, Yeh CS, Kuo CC, Huang BR, Dhar S (2005) J Optoelectron Adv Mater 7:3039–3046
4. Bae SY, Na CW, Kang JH, Park J (2005) J Phys Chem B 109:2526–2531
5. Lu JG, Fujita S, Kawaharamura T, Nishinaka H, Kamada Y, Ohshima T, Ye ZZ, Zeng YJ, Zhang YZ, Zhy LP, He HP, Zhai BH (2007) J Appl Phys 101:083705–083707
6. Leslie-Pelecky DL, Rieke RD (1996) Chem Mater 8:1770–1783
7. Pandiyarajan T, Udaya Bhaskar R, Karthikeyan B (2013) Spectrochim Acta A 103:173–178
8. Muthukumaran S, Gopalakrishnan R (2012) Opt Mater 34:1946–1953
9. Liu Z, Zhang Q, Shi G, Li Y, Wang H (2011) J Magn Magn Mater 323:1022–1026
10. Gao S, Li D, Lv X, Wang J, Li H, Yu Q, Guo F, Zhao L (2012) J Alloys Compd 539:200–204

11. Vijayaprasath G, Ravi G, HajaHameed AS, Mahalingam T (2014) J Phys Chem C 118:9715–9725
12. Arshad M, Azama A, Ahmed AS, Mollah S, Naqvi AH (2011) J Alloys Compd 509:8378–8381
13. Geetha Devi P, Sakthi Velu A (2016) Synthesis, structural and optical properties of pure ZnO and Co doped ZnO nanoparticles prepared by the co-precipitation method. J Theor Appl Phys
14. Geetha Devi P, Sakthi Velu A (2016) Structural, optical and photoluminescence properties of copper and iron doped nanoparticles prepared by co-precipitation method. J Mater Sci Mater Electron
15. Liu Y-Y, Jin H-J (2009) Analysis of photoluminescence for N-doped and undoped p-type ZnO thin films fabricated by RF magnetron sputtering method. J Trans Elect Elect 24–25
16. Zhong W-W, Liu F-M, Cai L-G, Peng-Ding (2011) Elaboration and characterization of Al doped ZnO nanorod thin films annealed in hydrogen. J Alloys Comp 3847–3851
17. Sahal M, Hartiti B, Mari B, Ridah A, Mollar M (2006) Etude des propriétés physiques des couches minces de ZnO dopées Al, préparées par la méthode de « sol - gel » associée au « spin coating ». J Afrique Sci 245–254
18. Kim K-K, Tampo H, Song J-O (2005) Effect of rapid thermal annealing on Al doped n-ZnO films grown by RF-magnetron sputtering. J Appl Phys 4776–4779

Synthesis, Characterization and Chemical-Sensor Applications of Zinc Oxide-Graphene Nanocomposite

Meriem Boutamine, Leila Grine, Abdelkadir Yessad, Hind Saidani, Hichem Hachemi, Samir Bellal, and Mourad Makhlouf

Abstract The proposed study focuses on the development of a gas sensor utilizing a quartz crystal microbalance (QCM) as the sensing platform. The sensor incorporates a nanocomposite of Zinc Oxide (ZnO) and Graphene as the sensitive layer. The synthesis of this nanocomposite involves a chemical method for thin film is deposition on QCM using spray pyrolysis technique. To evaluate the sensor's sensitivity, the frequency shift (Δf) of the coated QCM electrode is monitored when exposed to different gas concentrations such as ethanol, propane, and humidity. The frequency shift serves as an indicator of the gas concentration being detected. The isotherm response characteristics demonstrate good reproducibility and reversibility, suggesting the reliability of the sensor. Several techniques are employed to analyze various aspects of the coated QCM electrodes. Contact angle measurements (CA) are utilized to assess the surface wettability, providing information on how the nanocomposites interacts with the surrounding environment. Chemical compositions of the coated electrodes have been investigated by employing attenuated total reflectance-Fourier transform infrared (ATR- FTIR) spectroscopy. Scanning electron microscopy (SEM) is utilized to study the surface morphology of the coated QCM electrodes, providing detailed information on the structure and topography of the sensor's sensitive layer.

M. Boutamine (✉) · A. Yessad · H. Hachemi · S. Bellal
Electronics and Electrical Systems Engineering Research Laboratory, Military Academy DPHB-Cherchell, Cherchell, Algeria
e-mail: meriem.boutamine@gmail.com

L. Grine
Laboratoire d'Etude de Matériaux Electronique pour Applications Médicales, Université Mentouri Constantine 1, Constantine, Algeria

H. Saidani
Department of Mechanical Engineering, University of Bristol, Queens Building, University Walk, Bristol BS8 1TR, UK

M. Makhlouf
Combustion, Detonation and Ballistics Research Laboratory, Military Academy DPHB-Cherchell, Cherchell, Algeria

S. Hamamda et al. (eds.), *Proceedings of the 2nd International Conference of Nanotechnology for Environmental Protection and Clean Energy Production*, Springer Proceedings in Materials 45, https://doi.org/10.1007/978-981-97-1916-7_21

Keywords Spray pyrolysis deposition · Zinc oxide · Graphene · Gas sensor · Quartz crystal micro-balance (QCM) · Nanocomposites

1 Introduction

Over the past few decades, nanomaterials have gained significant attention and have found a wide range of applications across various industries particularly in electronic devices. due to their unique properties at the nanoscale. These materials are engineered to have specific properties based on their size, shape, and composition, making them versatile and valuable for numerous purposes [1–4].

Zinc Oxide (ZnO) nanoparticles are one of these metal oxyde nanomaterials and are highly used in solar cells, antibacterial applications, electronic optics, and chemical sensors. Chemical sensors play a crucial role in analyzers by converting chemical information into useful signals, enabling the detection and analysis of specific components or the overall composition of a sample [5–8]. Chemical sensors consist of a receptor part and a transducer part, and the ideal sensor is affordable, portable, and exhibits perfect selectivity and instantaneous response to the target analyte. The focus of current research lies in the development of Quartz Crystal Microbalance (QCM) gas sensors. QCM sensors are highly sensitive piezoelectric devices that can accurately detect mass changes due to the presence of molecules. The sensing material coated on the QCM's active electrode plays a crucial role in determining the sensor's characteristics. QCM sensors have been widely employed to selectively detect various gases (such as ammonia, alcohol, mercury, hydrocarbons, etc.) and humidity [9]. Strategies to enhance the selectivity of sensing materials on QCM platforms include heterostructure design, surface functionalization, and fabrication of nanocomposites. To improve the sensitivity of QCM sensors, researchers have incorporated sensing materials with high surface area, such as metal oxyde-graphene and its derivatives (graphene oxide and reduced graphene oxide) as nanocomposites [10, 11].

The proposed work aims 1—to investigate the synthesis of ZnO-graphene nanocomposites using solvothermal methods. 2—to elaborate a thin film using the as-prepared solution on commercial QCM electrodes using the spray pyrolysis technique. This technique allows for local deposition of thin films on the QCM electrode's surface in ambient air without causing damage to the piezoelectric quartz. 3—analyze various aspects of the coated QCM electrode by employing several techniques like Contact angle measurements (CA), Attenuated total reflectance-Fourier transform infrared (ATR-FTIR) spectroscopy, scanning electron microscopy (SEM) and energy-dispersive X-ray (EDX) spectroscopy.in order to provide detailed information on the structure and topography of the sensor's sensitive layer, aiding in the understanding of the gas sensing mechanism and potential optimization of the sensor's performance and 4—evaluate the sensor's sensitivity, by monitoring the frequency shift (Δf) of the coated QCM electrode when exposed to different gas concentrations such as ethanol, toluene, propane acetone and humidity.

2 Materials and Methods

2.1 Synthesis of Zinc Oxide-Graphene Nanocomposites

The pure zinc oxide solution was obtained using a simple solvothermal process by dissolving Zinc acetate dihydrate ($C_4H_6O_4Zn$, 99.99% purity, Sigma Aldrich) in the methanol. The molar mass of $C_4H_6O_4Zn$ is M = 219.51 g/mole. The methanol must completely dissolve the precursor (zinc acetate di-hydrate) in order to obtain a clear solution, calculate by chemical concentration relationship:

$$C = \frac{\mathrm{n}}{\mathrm{V}} = \frac{\mathrm{m/M}}{\mathrm{V}}$$

With

C concentration's solution in (mol/l);
n Mole number (mole);
V volume's solution in (ml).
m the dissolve's mass (g);
M molar mass (g/mol).

A solution with a concentration of 0.4 M have been prepared using stirrer by adding and heating 4.3902 g of $C_4H_6O_4Zn$ in 50 ml of methanol for 30 min.

In this study, graphite rods from electrical battery devices (as electrodes) and an electrolyte bath (electrochemical cell) composed of H_2SO_4 and H_2O were used for the electrochemical exfoliation process to obtain high-quality, large-area Graphene thin sheets, the process has been presented in previous works by Bakli et al. [12].

Different amounts of graphene powder were dispersed in a 50 ml of ZnO solution. Then, the suspension was mixed at room temperature for 1 h with ultrasonicating. Then, the well dispersed ZnO-Graphene nanocomposites solution was obtained. The ratio of ZnO and Graphene were controlled by the volume of each solution.

2.2 Coating of the Thin Films

In the described process, ZnO-graphene sensitive layers were prepared using the spray pyrolysis technique. Prior to deposition, all substrates were subjected to ultrasonic cleaning for 30 min in acetone, followed by rinsing with distilled water. The cleaned substrates were then placed on a hot plate maintained at a temperature of 350 °C with an accuracy of ± 2 °C. The as-prepared solution, containing the ZnO-Graphene nanocomposites, was loaded into a spray gun fixed at a distance of 20 cm above the substrate. To generate a fine mist of the precursor solution, compressed air was used as a carrier gas for spraying. The process parameters were set as follows: a spray time of 5 s, a break time of 1 s, and a total of 60 cycles. These parameters were determined to ensure a uniform deposition of the solution onto the substrates, resulting in the formation of a consistent and homogeneous sensitive layer.

2.3 *Sensor and Measurement System*

The principle of the QCM sensors is based on the change in the fundamental oscillation frequency Δf upon ad/absorption of molecules from the gas phase. To a first approximation, the frequency change Δf which results from an increase in the oscillating mass Δm and it was described by the Sauerbrey equation [13]. The piezoelectric crystals used were AT-Cut 6 MHz quartz crystal. The frequency shifts of the vibrating crystal were directly monitored measured using transistor Colpitts oscillator connected to the frequency counter TF930 TTi. The coated QCM-based sensor was evaluated for sensitivity by monitoring the frequency shifts of the quartz treated with different concentrations of VOC vapors with the aid of Labview interface software. A liquid of known volume and density was introduced in the testing cell by a Hamilton syringe and heated to induce free evaporation. Gas sensing measurement system is shown in Fig. 1.

The concentration of the introduced analyte was calculated in parts per million (ppm) according to:

$$C = \frac{\rho V_l}{V} = \frac{m_g}{V}$$

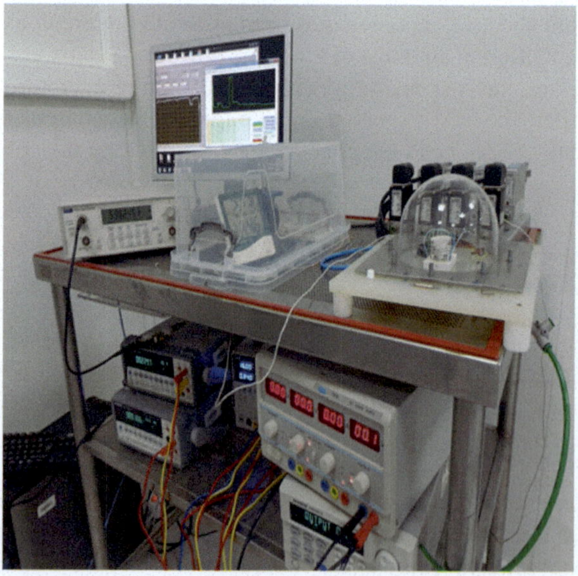

Fig. 1 Gas sensing measurement system

where

C is the concentration in ppm,
m_g is the weight of analyte in milligrams,
V is the volume of the testing cell in l,
ρ is the density of liquid sample in milligrams/l,
V_l is the volume of liquid sample in l.

After diffusion towards the electrode surface, the injected vapor was subsequently adsorbed on the functionalized QCM electrode which induced a frequency shift (Δf). After the adsorption process, the test chamber was purged with dry air until the vapor desorption process took place. The sensor sensitivity was evaluated against concentrations of propane, ethanol, and humidity vapours. Inside the chamber, the humidity and temperature evolution was monitored by THT22 temperature and humidity sensor with Arduino microcontroller.

3 Results and Discussion

3.1 Quartz Crystal Microbalance Responses

The kinetic responses of QCM-based sensor coated with thin layer of ZnO-Graphene nanocomposites are shown in Fig. 2a. The sensor responses in terms of frequency shift were plotted against time for concentrations of ethanol, propane, and humidity vapors ranging. For all vapors, the isotherms showed that the absolute value of the shift frequency increases gradually with time then reaches a steady state value. Between measurements, the sensor was exposed to dry air until full desorption was achieved. The frequency of the crystal back returning to its initial values indicates full desorption of analytes from the coated electrode surface. The elaborated sensor exhibited fairly rapid response, negligible baseline drift, and good reversibility for all vapors. From this figure, it is shown that the elaborated sensor presents a good affinity toward humidity vapors than other tested gases, indicating hydrophilic behaviour of the elaborated sensitive layer. In order to evaluate the response of sensor towards different types of gas. Further tests were carried out and responses have been recorded as shown in Fig. 2b. It is clear that, as a function of VOC concentrations, the saturated values of Δf increase proportionally with increasing analyte concentrations. As an example, the sensor response was determined to be 49, 82 and 100 Hz after contact with toluene (C_7H_8) vapors at concentrations of 47, 89 and 142 ppm, respectively.

The sensitivity which is an important performance parameter of a QCM sensor is usually defined as frequency change (Δf) divided by analyte vapors concentration change (ΔC). For this propose, the sensitivity was calculated by the slope of the regression curves. The sensitivity values are: 1.28, 0.4, 6.46 and 0.43 Hz/ppm for Ethanol, Propane, Acetone and Toluene, respectively. We can also observe that

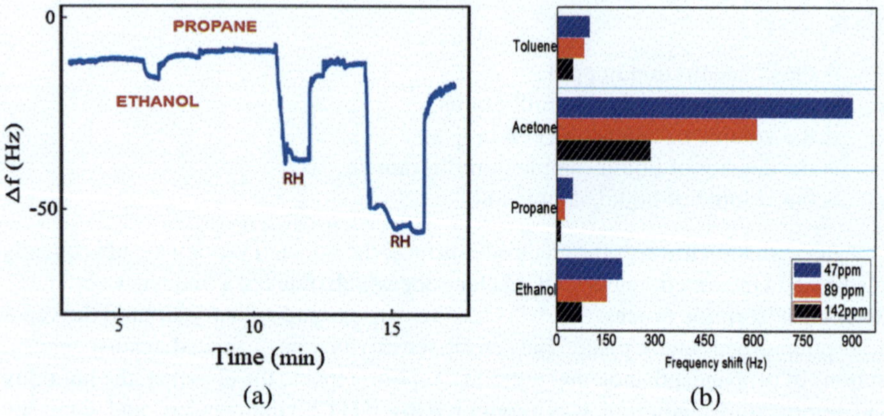

Fig. 2 **a** Gases kinetic responses of QCM-based sensor and **b** sensitivity toward different concentrations of VOCs vapors

Acetone (C_3H_6O) show the highest slope ratio compared to the other Vocs indicating the higher sensitivity toward acetone vapors which are polars molecules.

3.2 *Sensitive Layer Characterization*

In order to explain the sensing behaviour, it is necessary to perform a characterization of the ZnO-Graphene nanocomposites thin film. 1—Contact angle measurements were carried out at room temperature and atmospheric pressure to assess the wettability of the coated QCM surface. A 5 μl droplet of water was carefully deposited on the surface of the coated QCM using a microsyringe. Five measurements were carried out, and the average value of the contact angle was of the order of 30° indicating the hydrophilic bihaviour ot the elaborated film, this result is expected because zinc oxide and graphene have hydrophilic characters at it was reported by Srivastava et al. [14] and Accordino et al. [15]. 2—Chemical compositions of the coated electrodes have been investigated by employing attenuated total reflectance-Fourier transform infrared (ATR-FTIR) spectroscopy. The spectrum was recorded in the frequency range of 400–4000 cm^{-1} and was shown in Fig. 3. A broad and strong peak at 3600–3300 cm^{-1} is ascribed to the O–H stretching vibration of the hydroxyl group, the broadband observed around 2084–2350 cm^{-1} indicating vibrations related to the C–O bond. The spectrum also reveals the presence of oxygen functional groups, carbonyl (C=O), alkoxy (C–O), epoxy (C–O–C), and aromatic (C=C), around 1726, 1218, 1041, and 1622 cm^{-1} frequencies, respectively. On the other hand, a band around (400–800 cm^{-1}), indicate stretching modes of Zn–O bonds, corresponding to hexagonal ZnO's vibrating mode. The absorbance peaks at about 1561 and 450 cm^{-1} suggest the skeletal vibration of graphene sheets and the stretching vibration of Zn–O bonds, further supporting the concept of a composite structure involving both

ZnO and graphene [16, 17]. FTIR suggest the successful formation of a composite structure integrating ZnO and graphene. The identified vibrational modes and chemical groups provide clear evidence of the presence of these constituents within the composite material. FTIR analysis provides valuable information on the molecular composition and interactions within the composite, reinforcing the evidence for the existence of a ZnO-graphene matrix as well as hydrophilic behavior through the presence of hydroxyl groups, suitable for polar molecules detection which justifie good affinity toward acetone vapors.

SEM Zeiss EVO MA25 fitted with two Bruker xFlash 6I60 x-ray detectors for energy-dispersive X-ray spectroscopy «EDX» analysis was utilized to study the surface morphology and elemental analysis of the coated QCM electrodes, providing detailed information on the structure and topography of the sensor's sensitive layer (see Fig. 4a). The image suggest that the ZnO-Graphene nanocomposite has been successfully deposited on silicon substrates, forming a structure with decorated surfaces consisting of graphene nanoparticles and ZnO nanorods. The obtained EDX was shown in Fig. 4b, this analysis confirms that no contaminants are deposited and only carbon, zinc and oxygen are present in the deposited film on silicon substrate. All these results are in good agreement with the previous studies [18].

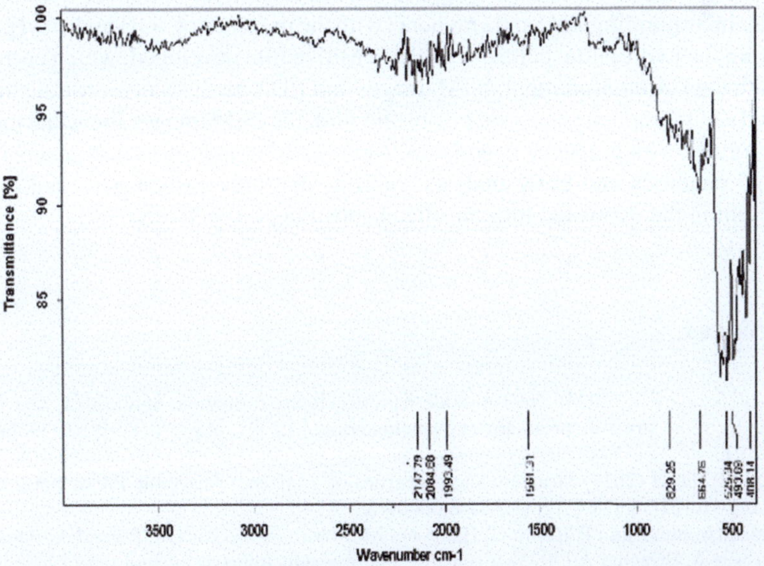

Fig. 3 ATR-FTIR spectrum of ZnO-graphene nanocomposite

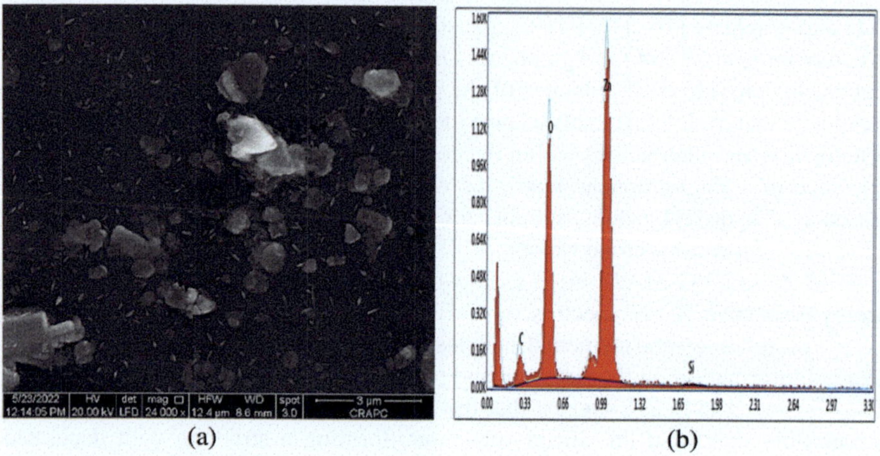

Fig. 4 **a** SEM image and **b** EDX analysis of ZnO-graphene nanocomposite

4 Conclusion

In this study, a gas sensor based on ZnO-Gaphene nanocomposite thin film coated on commercial quartz crystal microbalance as polar molecules sensor have been successfully fabricated using a spray pyrolysis method. The proprieties sensing in presence of gas vapors showed that the sensor was more sensitive to acetone vapors than others tested gases. Chemical and morphological properties' sensitive film have been studied, The FTIR results suggest the successful formation of a composite structure incorporating ZnO and graphene, with the presence of –OH hydroxyl groups indicating the hydrophilic behavior of the sensitive film developed, which correlates well with the CA measurement. SEM images and EDX analysis demonstrate that we obtained ZnO-Graphene nanocomposite thin film; the SEM images show that the film was decorated with graphene nanoflakes, Aggregated or agglomerated nanoparticles and ZnO nanorods and EDX analysis confirms that only carbon, zinc and oxygen are present in the deposited film on silicon substrate, respectively.

References

1. Liu P, Sivakov V (2023) Tin/Tin oxide nanostructures: formation, application, and atomic and electronic structure peculiarities. Nanomaterials 13:2391. https://doi.org/10.3390/nano13172391
2. Zhen Z, Zhu H (2018) Structure and properties of graphene. Academic Press, 1–12. https://doi.org/10.1016/B978-0-12-812651-6.00001-X
3. Madaci A, Suwannin P, Raffin G, Hangouet M, Martin M, Ferkous H, Bouzid A, Bausells J, Elaissari A, Errachid A (2023) A sensitive micro conductometric ethanol sensor based on an

alcohol dehydrogenase-gold nanoparticle chitosan composite. Nanomaterials 13:2316. https://doi.org/10.3390/nano13162316
4. Nadagouda M, Speth TF, Impellitteri C, Zhao Y (2011) Nanomaterials synthesis, applications, and toxicity. Publish Corp J Nanotechnol 2011:1–2. https://doi.org/10.1155/2011/218562
5. Huang K, Lu J, Li D, Chen X, Jin D, Jin H (2023) Au- or Ag-decorated ZnO-Rod/rGO nanocomposite with enhanced room-temperature NO_2-sensing performance. Nanomaterials 13:2370. https://doi.org/10.3390/nano13162370
6. Sharma DK, Shukla S, Sharma KK, Kumar V (2022) A review on ZnO: fundamental properties and applications. Mater Today Proc 49(Part 8):3028–3035. ISSN 2214-7853. https://doi.org/10.1016/j.matpr.2020.10.238
7. Barui AK, Kotcherlakota R, Patra CR (2018) Chapter 6—Biomedical applications of zinc oxide nanoparticles, inorganic frameworks as smart nanomedicines. William Andrew Publishing, pp 239–278. ISBN 9780128136614. https://doi.org/10.1016/B978-0-12-813661-4.00006-7
8. Moezzi A, McDonagh AM, Cortie MB (2012) Zinc oxide particles: synthesis, properties and applications. Chem Eng J 185–186:1–22. ISSN 1385-8947. https://doi.org/10.1016/j.cej.2012.01.076
9. Hydrogen sensing at room temperature using flame-synthesized palladium-decorated crumpled reduced graphene oxide nanocomposites
10. Mohammadi MM, Kumar A, Liu J, Liu Y, Thundat T, Swihart MT (2020) Hydrogen sensing at room temperature using flame-synthesized palladium-decorated crumpled reduced graphene oxide nanocomposites. ACS Sens. https://doi.org/10.1021/acssensors.0c01040
11. Wang Z, Gao S, Fei T, Liu S, Zhang T (2019) The construction of ZnO/SnO_2 heterostructure on reduced graphene oxide for enhanced nitrogen dioxide sensitive performances at room temperature. ACS Sens. https://doi.org/10.1021/acssensors.9b00648
12. Bakli H, Moualhi M, Makhlouf M (2022) High-sensitivity electrical properties measurement of graphene-based composites using interferometric near-field microwave technique. Meas Sci Technol 33:045012. https://doi.org/10.1088/1361-6501/ac3d09
13. Sauerbrey G, Angew Z (1959) Verwendung von Schwingquarzen zur Wägung dünner Schichten und zur Mikrowägung. Physics 155:206–222. https://doi.org/10.1007/bf01337937
14. Srivastava M, Basu BBJ, Rajam KS (2011) Improving the hydrophobicity of ZnO by PTFE incorporation. J Nanotechnol 2011:1–6. https://doi.org/10.1155/2011/392754
15. Accordino SR, de Oca M, Manuel J, Fris R, Ariel J, Appignanesi GA (2015) Hydrophilic behavior of graphene and graphene-based materials. J Chem Phys 143(15):154704. https://doi.org/10.1063/1.4933011
16. Li J, Jin Z, Chao Y, Wang A, Wang D, Chen S, Qian Q (2023) Synthesis of graphene-oxide-decorated porous ZnO nanosheet composites and their gas sensing properties. Chemosensors 11:65. https://doi.org/10.3390/chemosensors11010065
17. Salih E, Mekawy M, Hassan RYA, El-Sherbiny IM (2016) Synthesis, characterization and electrochemical-sensor applications of zinc oxide/graphene oxide nanocomposite. J Nanostruct Chem 6:137–144. https://doi.org/10.1007/s40097-016-0188-z
18. Venkidusamy V, Nallusamy S, Nammalvar G, Veerabahu R, Thirumurugan A, Natarajan C, Dhanabalan SS, Pabba DP, Abarzúa CV, Kamaraj SK (2023) ZnO/graphene composite from solvent-exfoliated few-layer graphene nanosheets for photocatalytic dye degradation under sunlight irradiation. Micromachines 14:189. https://doi.org/10.3390/mi14010189

HMDSO/ZnO-Graphene Heterostructure Coated Quartz Crystal Microbalance Transducer for VOCs Sensing

Leila Grine, Azzedine Bellel, Meriem Boutamine, and Salah Sahli

Abstract Accurate, reliable and fast, detection of different harmful volatile organic compounds (VOCs) is crucial criterion for medical, industrial, agricultural and environmental applications. To address the issue, in this study, porous HMDSO/ZnO-graphene heterostructure has been used as a sensitive layer. The thin HMDSO and ZnO-graphene nanocomposite layers have been elaborated on quartz crystal microbalance (QCM) transducer by plasma enhanced chemical vapor deposition (PECVD) and spray pyrolysis technique, respectively. The synthesized heterostructure was characterized by Fourier transform infrared spectroscopy (FTIR) and atomic force microscopy (AFM). The sensing performance of the coated QCM-based sensor was evaluated by expositing it to various concentrations of organic molecules. The coated QCM sensor demonstrated high sensitivity values of 1.26, 1.05, 1.58 and 0.90 Hz/ppm and low detection limit of 2.38, 2.58, 1.90 and 3.33 for toluene, ethanol, benzene and methanol, respectively. The sensing performance of the elaborated sensor correlates well its chemical and morphological properties. The observed results demonstrate the successful strategy of retaining the excellent property of graphene (high specific surface area) with plasma surface activation to improve the active sites for gas absorption.

Keywords QCM · VOCs sensor · Heterostructure · ZnO-graphene · HMDSO

L. Grine (✉) · A. Bellel
Laboratory for the Study of Electronic Materials for Medical Applications, Mentouri Constantine 1 University, Constantine, Algeria
e-mail: Leila.grine@doc.umc.edu.dz

M. Boutamine
Research Laboratory in Electronics and Electrical Systems Engineering, Scientific and Technological Research Directorate-Cherchell Military Academy DPHB, Cherchell, Algeria

S. Sahli
Laboratory of Microsystems and Instrumentation, Mentouri Constantine 1 University, Constantine, Algeria

S. Hamamda et al. (eds.), *Proceedings of the 2nd International Conference of Nanotechnology for Environmental Protection and Clean Energy Production*, Springer Proceedings in Materials 45, https://doi.org/10.1007/978-981-97-1916-7_22

1 Introduction

Volatile organic compounds (VOCs) are a chemical composites in liquid form which can evaporate at room temperature. These organic molecules differ in their chemical physical properties and safe limit of concentration [1]. Even at low concentration level, the exposition of VOC molecules causes different health effects [2, 3]. In this context, chemical sensor which can detect different types of VOCs reflects one of the primary goals of the actual research in the sensor area for medical and environmental applications [4–6]. Although highly sensitive and selective VOCs sensors have been recently reported in the literature [7], their ability to detect VOCs at low concentrations level remain insufficient. As an alternative, the use of nanostructured materials as sensitive layer makes gas monitoring practical at low concentration levels [8]. The commonly used methods to enhance the gas sensing properties are the elaboration of porous sensing layers (facilitating the diffusion of gas molecules) and the decrease in the particle size (increasing the surface to volume ratio) [9]. Due to its large surface to volume ratio, good electrical and chemical properties, graphene material has attracted great attention in a broad field of applications as good nanomaterial, during the last decade [10]. Despite these attractive properties, graphene-based gas sensors had low sensitivity due to the strong sp2 hybridization making pure graphene chemically inactive to various VOCs vapors [11, 12]. To overcome these drawbacks, many reported works have shown that, constructing heterostructure sensing layers have become a simple and effective way to improve the gas sensing properties [13].

The aim of this work is to investigate a high-performance VOCs sensor based on quartz crystal microbalance (QCM) coated with hexamethyldisiloxane (HMDSO)/ZnO-graphene heterostructure. The approach consists essentially on retaining all the excellent properties of graphene (high specific surface) with plasma surface activation. The morphological and structural characterization of ZnO-graphene nanocomposite and HMDSO/ZnO-graphene heterostructure were performed by atomic force microscopy (AFM) and Fourier transform infrared spectroscopy (FTIR), respectively. The VOCs sensing properties of the as-prepared HMDSO/ZnO-graphene heterostructure were evaluated towards various concentrations of toluene, benzene, methanol and ethanol ranging from 35 to 70 ppm.

2 Experimental

2.1 Deposition Method

Thin films of ZnO-graphene nanocomposite were firstly deposited on ultrasonically cleaned QCM surface using the pyrolysis spray deposition technique. For the preparation of the ZnO precursor solution, a 4.39 g of zinc acetate were dissolved in 100 ml of methanol and the solution was stirred vigorously. Then a 0.15 g of graphene was added to 25 ml of the ZnO precursor solution. The mixture was stirred with a thermal stirrer for a suitable time to ensure uniform dispersion of the graphene. During deposition, the temperature, the distance between spray and substrate and spray number were set to 300 °C, 7 cm and 40, respectively. After the deposition of the ZnO-graphene nanocomposite on QCM substrate, an ultra-thin layer of HMDSO was deposited on the synthetized ZnO-graphene by plasma enhanced chemical vapor deposition (PECVD) technique. During plasma deposition, the monomer partial pressure, the plasma power and the deposition time were set to 0.5 mbar, 200 W and 5 min, respectively.

2.2 *Structural and Morphological Characterizations*

The chemical structure of the ZnO-graphene and the HMDSO/ZnO-graphene heterostructure layers were characterized by FTIR spectroscopy. All spectra were acquired in absorbance mode in the 500–4000 cm^{-1} range using a 6300 type A spectrometer. The surface morphology of the elaborated layers was recorded in nanometer range by means of atomic force microscope. AFM images were acquired in tapping mode using Angstrom Advanced (AA2000) microscope.

2.3 *QCM-Based Sensor Evaluation*

Chemical sensing properties of the studied QCM-based VOCs sensor functionalized with HMDSO/ZnO-graphene heterostructure were evaluated towards methanol, ethanol, benzene and toluene molecules. The sensing properties of the fabricated sensor were evaluated in VOCs testing cell shown in Fig. 1. Organic liquid with specified volume and density are introduced in the measuring chambre by a micro syringe and allowed to evaporate freely on a hot plate. The chamber's temperature and atmosphere were meticulously controlled throughout all the experiments. The absorption of the analytes molecules induces a shift in the resonance frequency (Δf) of the quartz according to Sauerbrey equation in gas phase [14]. The responses of the elaborated sensor to the presence of different concentrations of VOCs vapors were recorded using LabVIEW software. The sensing properties of the heterostructure coated QCM sensor was monitored in presence of different concentrations of toluene, benzene, methanol and ethanol in the range of 35–70 ppm. After reaching the adsorption steady state, dry air has been introduced in the measurement chambre in order to obtain full desorption of VOCs molecules from the sensing layer.

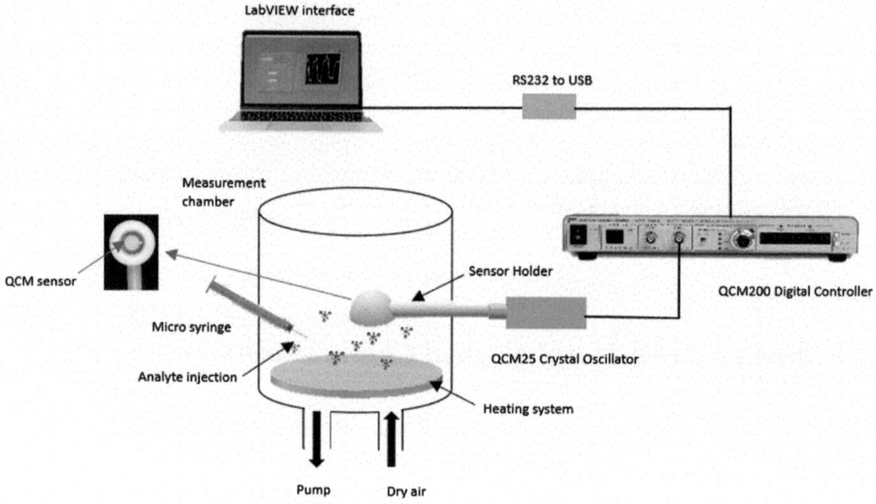

Fig. 1 Schematic representation of the testing system

3 Results and Discussion

3.1 Structural and Morphological Analysis

The FTIR spectra recorded on ZnO-graphene and HMDSO/ZnO-graphene heterostructure are presented in Fig. 2. The FTIR spectrum related to ZnO-graphene nanocomposite presents a vibration peak at 540 cm^{-1} corresponding to the stretching mode of the Zn–O. The spectrum also shows the presence of the C=C peak around 1600 cm^{-1} indicating the existence of graphene nanosheet [15–18]. The observed of Zn–O and C=C peaks in the FTIR spectrum indicates the formation of the nanocomposite in the synthesis [19]. For the HMDSO/ZnO-graphene heterostructure, the FTIR spectrum shows all the peaks related to the ZnO-graphene nanocomposite in addition to the presence of the Si–O–Si bond around 1020 cm^{-1} related to HMDSO layer and the CHx peak at 2924 cm^{-1}. The work of I. S. Bae et al. has shown that the structure of the film deposited by PECVD was more porous due to the existence of high proportion of CHx groups [20, 21]. The increase in the film porosity improves the gas sensing performance by facilitating the diffusion of gas molecules.

Figure 3a, b show AFM images (5 μm × 5 μm) obtained on the surface of the ZnO-graphene nanocomposite and the HMDSO/ZnO-graphene heterostructure, respectively. The elaboration heterostructure resulted in relatively rougher surface with the value of the Rms (root mean square roughness) of about 21.3 nm compared to 8.81 recorded on ZnO-graphene surface (Fig. 3). Increasing layer surface roughness causes the augmentation in the surface to volume ratio (specific surface area). In this case, the sensitive layer accommodates greater number of adsorption sites, leading to the enhancement of the sensor's adsorption properties.

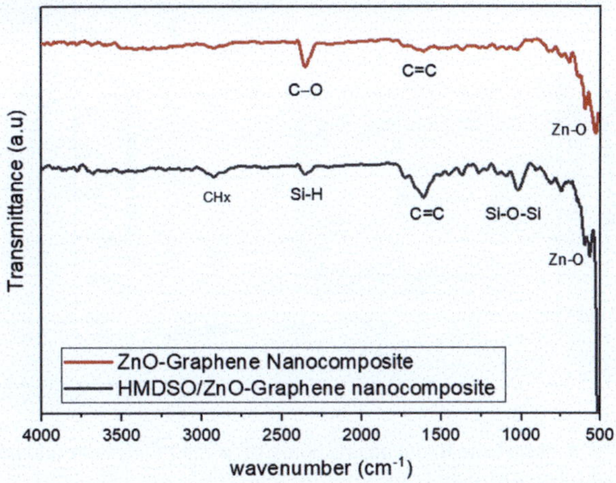

Fig. 2 FTIR spectra of ZnO-graphene and HMDSO/ZnO-graphene heterostructure

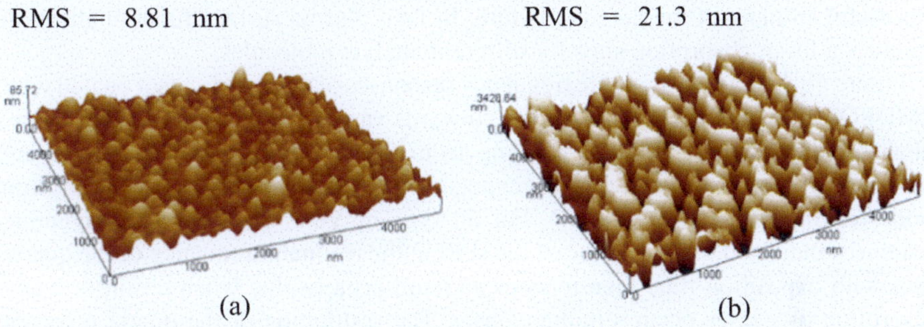

Fig. 3 Surface morphology of the: **a** ZnO-Graphene nanocomposite and **b** HMDSO/ZnO-graphene heterostructure

3.2 *Sensor Response Evaluation*

The sensing mechanism of the elaborated QCM-based sensors rely on the absorption of VOCs molecules accompanied by the shift in the resonant frequency of the coated quartz. Figure 4a shows the real time frequency response (in absorption and desorption mode) to 70 ppm concentration of toluene recorded on two different QCM sensors; one coated with ZnO-graphene nanocomposite and the other with HMDSO/ZnO-graphene heterostructure. When comparing the two responses, it is clearly noticed that the sensing behavior are drastically different for the two coating layers. The QCM sensor coated with ZnO-graphene nanocomposite was practically non sensitive to toluene molecules although the later has a high specific surface area. This behavior is probably due to the strong sp2 hybridization which makes graphene intrinsically inert to many gas molecules [12]. However, in case of heterostructure coating, the sensitivity to toluene vapor increases significantly confirming the

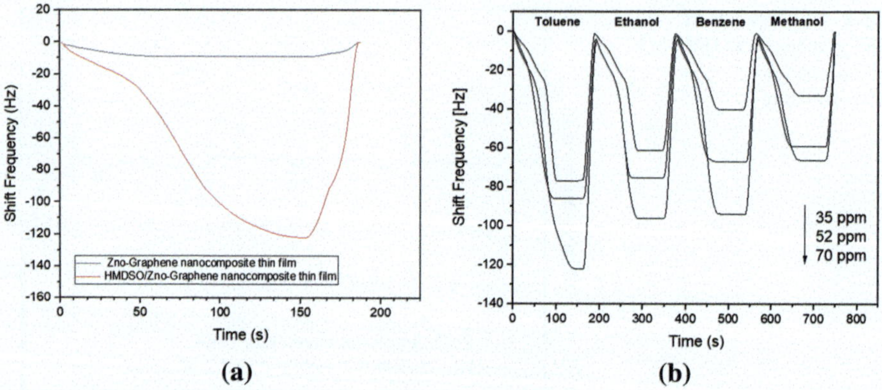

Fig. 4 Real time frequency response of: **a** ZnO-Graphene and heterostructure to 70 ppm of toluene and **b** heterostructure towards different VOCs concentrations

successful strategy of the heterostructure. In fact, plasma surface functionalization provides a lot of absorption sites for different analyte molecules.

Figure 4b depicts the dynamic frequency responses of the QCM sensor coated with HMDSO/ZnO-graphene heterostructure towards varying concentrations of toluene, ethanol, benzene and methanol ranging from 35 to 70 ppm. The responses are recorded in absorption mode (upon exposure to VOCs vapors) and in desorption mode (during purge with dry air). For all type of VOC molecules, the dynamic responses indicated that the value of the shift in the resonant frequency of the quartz rises with exposition time then reaches an equilibrium value when absorption and desorption processes occur simultaneously. The restoration of the quartz resonant frequency (the back returning of the sensor to its baseline value) was considered as an indication of complete desorption of analytes molecules from the coated QCM surface. The results also show that the studied sensor has good affinity to toluene molecules.

Figure 5 depicts the variations of the shift frequency (saturated values) as a function of analyte concentration. For all VOC molecules, the maximum response value increase with increasing VOC concentration and the sensing behavior is fairly linear. The sensors sensitivity S is defined as the ratio of the shift frequency (Δf) to the change in analyte concentration (Δppm) and the limit of detection (LOD) is calculated by the ration $3\sigma/S$ (where σ is the noise level estimated to 1 Hz). The calculated sensitivity was in the order of 1.26, 1.05, 1.58 and 0.90 Hz/ppm for toluene, ethanol, benzene and methanol, respectively. These values are very good compared to other QCM-based VOCs sensor [22, 23]. Concerning the LOD, the sensor coated with HMDSO/ZnO-graphene heterostructure presents a lower LOD values. It can detect up to 2.38, 2.58, 1.90 and 3.33 ppm of toluene, ethanol, benzene and methanol, respectively.

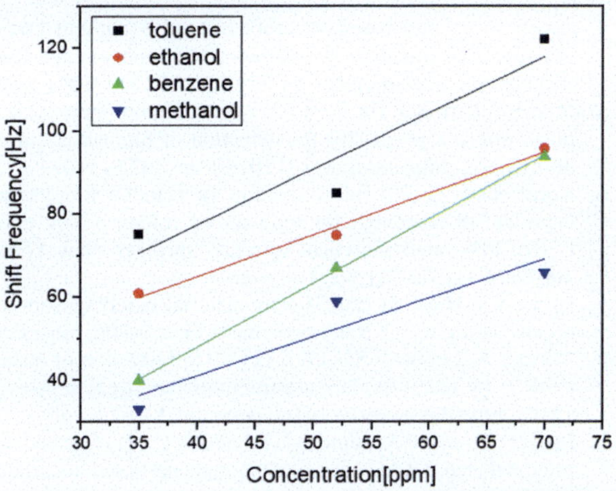

Fig. 5 Sensitivity as a function of analyte concentration recorded on the heterostructure coated QCM sensor

4 Conclusion

The aim of this study is the fabrication of a potential VOCs sensor using HMDSO/ZnO-graphene heterostructure as sensitive layer and QCM as transducer. In an attempt to improve the sensor sensitivity, an ultra-thin layer of HMDSO has been elaborated by PECVD over ZnO-graphene layer in order to create more active sites by plasma surface functionalization. The structural analysis deduced from FTIR results indicated the incorporation of groups related to the HMDSO material and the AFM data revealed the growth of rougher surface with high specific surface area, that can accommodate more absorption sites for the gas sensing. The fabricated QCM-based VOCs sensor showed good affinity to toluene molecules than other tested organic molecules. The performance of the fabricated sensors in terms of LOD and sensitivity, resulted from the measurements at various VOCs concentrations, indicates that the later showed improved detection properties in comparison with VOCs sensors reported in the literature. The fabricated sensor is therefore potentially useful in environmental and human health care applications.

References

1. Zeinali S, Homayoonnia S, Homayoonnia G (2019) Comparative investigation of interdigitated and parallel-plate capacitive gas sensors based on Cu-BTC nanoparticles for selective detection of polar and apolar VOCs indoors. Sens Act B Chem 278:153–164. https://doi.org/10.1016/j.snb.2018.07.006
2. Khan S, Le Calvé S, Newport D (2020) A review of optical interferometry techniques for VOC detection. Sens Act A Phys 302:111782. https://doi.org/10.1016/j.sna.2019.111782
3. Tsai W-T (2016) Toxic volatile organic compounds (VOCs) in the atmospheric environment: regulatory aspects and monitoring in Japan and Korea. Environments 3(3). https://doi.org/10.3390/environments3030023

4. An electronic nose based on carbon nanotube-titanium dioxide hybrid nanostructures for detection and discrimination of volatile organic compounds. ScienceDirect. https://www.sciencedirect.com/science/article/abs/pii/S0925400522000600 (consulté le 6 septembre 2023)
5. Jeon J-Y, Park S-J, Ha T-J (2022) Wearable nitrogen oxide gas sensors based on hydrophobic polymerized ionogels for the detection of biomarkers in exhaled breath. Sens Act B Chem 360:131672. https://doi.org/10.1016/j.snb.2022.131672
6. Korotcenkov G (2013) Introduction. In: Handbook of gas sensor materials: properties, advantages and shortcomings for applications volume 1: conventional approaches. In: Korotcenkov G (ed) Integrated analytical systems. Springer, New York, NY, pp 1–45. https://doi.org/10.1007/978-1-4614-7165-3_1
7. Broza YY, Haick H (2013) Nanomaterial-based sensors for detection of disease by volatile organic compounds. Nanomedicine 8(5):785–806. https://doi.org/10.2217/nnm.13.64
8. Mirzaei A, Leonardi SG, Neri G (2016) Detection of hazardous volatile organic compounds (VOCs) by metal oxide nanostructures-based gas sensors: a review. Cer Int 42(14):15119–15141. https://doi.org/10.1016/j.ceramint.2016.06.145
9. Fauzi F, Rianjanu A, Santoso I, Triyana K (2021) Gas and humidity sensing with quartz crystal microbalance (QCM) coated with graphene-based materials—a mini review. Sens Act A Phys 330:112837. https://doi.org/10.1016/j.sna.2021.112837
10. Zhang D, Chang H, Li P, Liu R, Xue Q (2016) Fabrication and characterization of an ultrasensitive humidity sensor based on metal oxide/graphene hybrid nanocomposite. Sens Act B Chem 225:233–240. https://doi.org/10.1016/j.snb.2015.11.024
11. Yuan W, Shi G (2013) Graphene-based gas sensors. J Mater Chem A 1(35):10078–10091. https://doi.org/10.1039/C3TA11774J
12. Zhou Y et al (2017) Study on gas sensing of reduced graphene oxide/ZnO thin film at room temperature. Sens Act B Chem 240:870–880. https://doi.org/10.1016/j.snb.2016.09.064
13. Jayawardena S, Kubono A, Rajapakse RMG, Shimomura M (2021) Effect of titanium precursors used in the preparation of graphene oxide/TiO_2 composite for gas sensing utilizing quartz crystal microbalance. Nano-Struct Nano-Obj 28:100780. https://doi.org/10.1016/j.nanoso.2021.100780
14. Sauerbrey G (1959) Z Phys 155:206
15. Abdel-Aal SK, Abdel-Rahman AS (2020) Graphene influence on the structure, magnetic, and optical properties of rare-earth perovskite. J Nanopart Res 22(9):267. https://doi.org/10.1007/s11051-020-05001-7
16. Abdel-Aal SK, Aly AE, Chanduví HHM, Gil Rebaza AV, Atteia E, Shankar A (2020) Magnetic and optical properties of perovskite-graphene nanocomposites $LaFeO_3$-rGO: experimental and DFT calculations. Chem Phys 538:110874. https://doi.org/10.1016/j.chemphys.2020.110874
17. Abdel-Aal SK, Attia SY, Mohamed SG (2019) Facile synthesis of Mn_3O_4-rGO Nanocomposite as an efficient electrode material for application in supercapacitors. J Electron Mater 48(8):4977–4986. https://doi.org/10.1007/s11664-019-07305-4
18. Abdel-Aal SK, Ionov A, Mozhchil RN, Naqvi AH (2018) Simple synthesis of graphene nanocomposites MgO–rGO and Fe_2O_3–rGO for multifunctional applications. Appl Phys A 124(5):365. https://doi.org/10.1007/s00339-018-1748-5
19. Ramadan R, Abdel-Aal SK. Facile synthesis of nanostructured ZnO-rGO based graphene and it's application in waste water treatment
20. Zhang C, Wyatt J, Russell SP, Weinkauf DH (2004) Methanol vapor sorption in plasma polymerized thin films. Polymer 45(22):7655–7663. https://doi.org/10.1016/j.polymer.2004.08.055
21. Bae I-S et al (2008) Characterization on polymerized thin films for low-k insulator using PECVD. Progr Org Coat 61(2):245–248. https://doi.org/10.1016/j.porgcoat.2007.09.031
22. Gupta M, Athirah N, Fahmi Hawari H (2020) Graphene derivative coated QCM-based gas sensor for volatile organic compound (VOC) detection at room temperature. IJEECS 18(3):1279. https://doi.org/10.11591/ijeecs.v18.i3.pp1279-1286
23. Detection of some volatile organic compounds with chitosan-coated quartz crystal microbalance. https://doi.org/10.1080/15685551.2014.918019?needAccess=true&role=button (consulté le 7 septembre 2023)

Correlation Between Structural and Optical Properties of CuO Nanoparticule Prepared by Sol–Gel Method: Degradation of Mythylene Blue

Fouzia Abbas, Sonia Attaf, Asma Zeibet, and Rabah Bensaha

Abstract Highly feature Copper Oxide CuO nanoparticles were prepared by sol–gel method at different temperature (400, 500 °C). The Copper chloride dehydrate, ethanol and citric acid were used as precursor, solvent and stabilizer, respectively. The influence of annealing temperature on the structural and optical properties of CuO nanoparticle was also investigated. In order to determine the information on our samples several characterizations were carried out structural characterization: X-ray diffraction (XRD); Optical properties: optical absorption and photodegradation of blue methylene of nanopowder. X-ray diffraction shows monoclinic structure with high crystallinity of all powder with preferential orientation and crystallite size is 22 nm. The intensity of peaks greatly increased with increasing of annealing temperature. An extreme absorption onset has been observed for all powder in the region between 380 and 650 nm. This attributed to the absorption of light by copper oxide (CuO) nanoparticle. The band gap was determined to be 1.47–1.39 eV (decreased with annealing temperature) that is significantly smaller than the band gap of bulk copper. These CuO nanoparticule show an excellent degradation of MB reach to 95% by advanced oxidation using H_2O_2 which enhanced the photocatalysis applications.

Keywords CuO nanoparticle · Structural properties · Photocatalysis

1 Introduction

Nanomaterials are the basis of a new scientific and technological revolution, and occupy a very important place in our lives. The interest of studying these nanomaterials in different forms lies mainly, on the one hand in their new properties which differ significantly from those at the macroscopic scale (massive materials), and on the other hand, their use in the industrial field [1]. Among these nanomaterials,

F. Abbas (✉) · S. Attaf · A. Zeibet · R. Bensaha
University Constantine 1, Laboratory Microstructure and Defects in Materials, 235 City Ain Elbey Constantine 25017, Constantine, Algeria
e-mail: fouzia.abbas@umc.edu.dz

S. Hamamda et al. (eds.), *Proceedings of the 2nd International Conference of Nanotechnology for Environmental Protection and Clean Energy Production*, Springer Proceedings in Materials 45, https://doi.org/10.1007/978-981-97-1916-7_23

nanoparticles of metals and metal oxides give rise to more and more interest. CuO is one of the most useful metal oxides for many applications in different fields such as solar cells and photovoltaic materials [2], electrochromic coatings [3], catalytic applications [4] and gas sensors [5].

Copper oxides (CuO) have attracted the attention of the scientific community because of their applications in various fields thanks to their chemical properties, their non-toxicity, and their low cost compared to other metals. These oxides are considered effective catalysts for the heterogeneous conversion of hydrocarbons into carbon dioxide and water and the degradation of organic molecules. The band gap (2.17 eV) of this material favors its use in applications oriented towards photoconduction, photothermic and solar cells [6, 7].

The development of a high quality nanoparticle of copper oxide (CuO) is a very important and delicate step; several techniques have been used for the production of CuO, among them the sol–gel technique [8]. They lead to products of high purity and make it possible to control the composition of the final material at the molecular level. They also allow a control of the doping of the coating particle; it can be very low thus offering a lower use of the product, a low toxicity emission and therefore a minimum of production cost.

The objective of this work is the elaboration of copper oxide (CuO) nanoparticles by sol–gel technique at different annealing temperature, then the structural and optical characterization of the samples obtained are investigated by different technic such as: XRD, UV-visible spectroscopy and degradation methylene blue for photocatalytic application.

2 Materials and Method

2.1 Materials

In our work, the CuO nanopwders was obtained by dissolving copper chloride dehydrate ($CuCl_2.2H_2O$) in solvent (Ethanol) with magnetic stirring at 50° C, after 30 min a volume of acid citric was added in a molar ratio (n (AC)\n (chloride) = 1), in order to dissolve motives of work hydroxide, the volume total of the solution is Vt = 15 ml. The solution is then brought to magnetic stirring at 80 °C for 2 h, the solution becomes yellow-green changed gradually to a gel stability, then dried at furnace for 24 h. Finally, the nanoparticles obtained and annealed at different temperature for 4 h (Fig. 1).

Fig. 1 Different step of preparation of CuO nanoparticles by sol–gel process

2.2 *Characterization*

The crystallography of the co-doped TiO2 was determined by Bruker D8 Advance X-ray diffractometer with a high-intensity Cu $K_{\alpha 1}$ irradiation (λ = 0.1540562 nm).The Photocatalytic activity studies were carried out using organic pollutant, Mythlyne blue. Moreover, the photocatalytic activities of CuO powder were estimated by using a 2.5 × 10^{-5} mol solution of MB under UV light source (λ = 360 nm, 2 W) placed about 10 cm above the solution under magnetic stirring. The test of duration without H_2O_2 and with H_2O_2 relevant the concentrations were calculated by a spectrophotometer. The maximum absorption peak (λmax) of MB at about 666 nm was considered to estimate the remaining concentration of MB [9]. The photodegradation of MB was usually determined by $\frac{C_0 - C_t}{C_0}$ × 100 where C_0 (initial absorbance) and C_t are the concentrations of MB at time (0) and (t), respectively. Subsequently, the photocatalytic activities of the photocatalyst were considered proportional to the relative degradation of MB.

3 Results

3.1 Structural Properties

4 XRD Analysis

X-ray diffraction patterns (XRD) shown in the Fig. 2 indicate the structural evolution of the CuO powders as a function of the annealing temperature from 400 to 500 °C for 4 h. They show that the powders crystallize greatly increases with increasing of annealing temperature. The XRD spectrum shows diffraction peaks located at 2θ(°): 32.49°, 34.99°, 35.51°, 38.72°, 43.72°, 48.81°, 53.46°, 58.25°, 61.56°, 66.12°, 68.018°, 72.40°, 75.13°. These peaks are assigned to reflection planes: (110), (002), (111−), (111), (200), (202−), (112), (202), (113−), (022), (220), (311), (004). All diffraction peaks correspond to the monoclinic phase of CuO (Tenorite) (JCPDS No. 48 1548), a space group C2/c with lattice constants a = 4.684 Å, b = 3.425 Å, c = 5.129 and β = 99.47 [10, 11]. We also see a quick increase in the intensity of the peaks as the annealing temperature increases. This result reflects an increase in the crystallinity rate of the material, which improved by an increase in grain size. Indeed, when the annealing temperature increases, the atoms have enough energy to diffuse and occupy a low-energy at normal position in the lattice, which results in an increase in rate of crystallinity and the crystallites size of the CuO. These results are in good agreement with the work [12–15]. The crystallite sizes D of CuO nanoparticle are calculated using a Debye–Scherrer equation and found around 15–22 nm approximately, this measured sizes of CuO nanopowders also found by Nithya et al. [16]. The sizes of crystallite are cited in Table 1

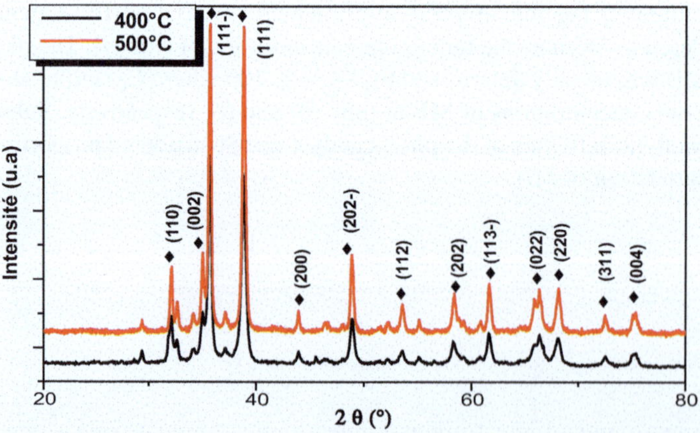

Fig. 2 XRD diffraction pattern of CuO nanopowders treated at different temperatures (400 and 500 °C) for 4 h. ♦: CuO

Table 1 The sizes of crystallite of CuO nanoparticles at different temperature

Powder	Temperature (C°) and annealing time (h)	Phase	(hkl)	Size of crystallite D (nm)
CuO	400 °C for 4 h	Tenorite	(110)	15,26
			(002)	20,84
			(111)	18,55
			(111−)	16,97
	500 °C for 4 h	Tenorite	(110)	16,76
			(002)	22,88
			(111)	19,30
			(111−)	17,27

4.1 Optical Properties

The UV-Visible absorption spectra of CuO nanoparticles measured in the range 200–800 nm (Fig. 3). The band gap energy calculated from $E = 1240/\lambda$ relation by plotting the absorption edge of CuO were found decrease from 1.47 to 1.39 eV by increasing of annealing temperature from 400 to 500 °C. CuO nanoparticle showed maximum absorption at wavelength 380 mm with red shift region which can be attributed to improved crystallinity of the CuO nanoparticles, which is confirmed by XRD results compared with bulk copper oxide. These results are in good agreement with this reported by other works [16].

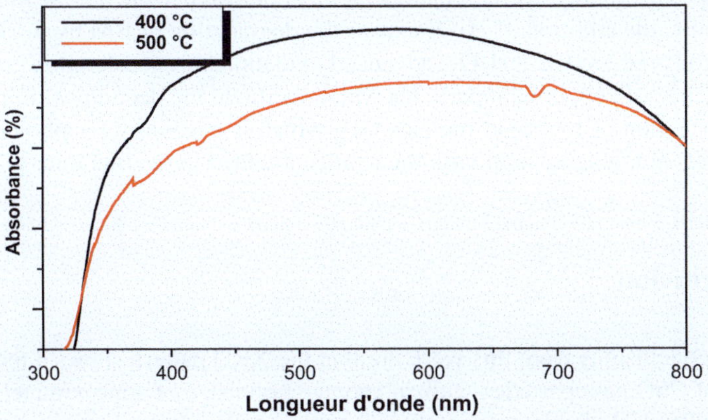

Fig. 3 UV-visible absorption spectra of CuO nanopowder at different annealing temperature

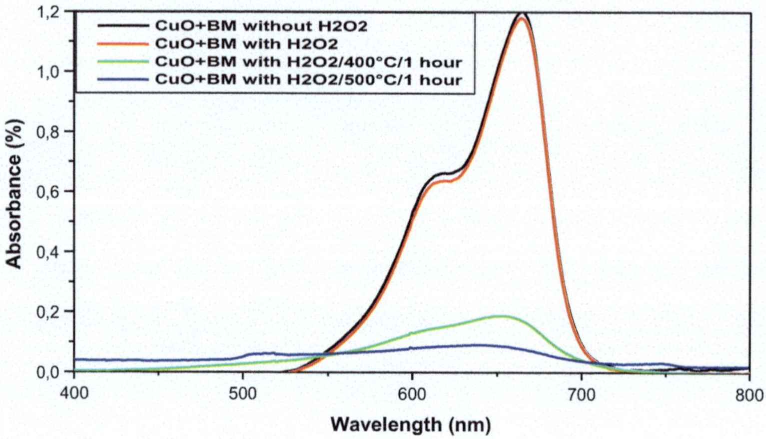

Fig. 4 Evolution of U.V-visible absorption spectra of methylene blue of CuO nanopowder annealed at different temperature (400, 500 °C) irradiation for 1 h without and with H_2O_2

4.2 *Photocatalysis Application*

The evolution of UV-Vis absorption spectra of BM as a function of annealing temperature in absence and in the presence of H_2O_2 for copper oxide (CuO) nanopowders is represented in Fig. 4. These spectra show a very small decrease in BM after the addition of copper oxide without H_2O_2, indicating that BM degradation is almost constant. On the other hand, the BM spectra of CuO in the presence of H_2O_2 obtained show that the absorbance decreased more rapidly compared to the state without H_2O_2. Consequently, the addition of H_2O_2 causes the degradation of BM by the heterogeneous Fenton process ($Cu + H_2O_2$) advanced oxidation, this acceleration may be due to the presence of copper nanoparticles which facilitate the decomposition of H_2O_2 and the formation of hydroxyl radicals OH, which is necessary for photocatalysis. These results are in agreement with the results obtained in the literature [17].

5 Conclusion

The main conclusions from this work are that the XRD pattern shows a monoclinic structure of CuO nanoparticles with an average grain size of approximately 22 nm. Moreover, The UV-visible pattern of CuO nanoparticles showed maximum absorption at 380 nm with red shift and the band gap energy decrease by increasing of annealing temperature to 1.39 eV. Finally, a test of photodegradation of methylene blue reaches to 95% of absorbance in the presence of H_2O_2 which enhance the photocatalysis application.

References

1. Jayaprakash J, Srinivasan N, Chandrasekaran P, Girija EK (2015) Synthesis and characterization of cluster of grapes like pure and zinc-doped CuO nanoparticles by sol-gel method. Spectrochimica Acta A Mol Biomol Spectrosc 136:1803–1806
2. Jundale DM, Joshi PB, Sen S, Patil VB (2012) Nanocrystalline CuO thin films: synthesis microstructural and optoelectronic properties. J Mater Sci Mater Electron: 1492–1499
3. Ristova M, Neskovska R, Mirceski V (2007) Chemically deposited electrochromic cuprous oxide films for solar light modulation. Sol Energy Mater Sol Cells 91(14):1361–1365
4. Santos A, Yustos P, Quintanilla A, Ruiz G, Garcia-Ochoa F (2005) Study of the copper leaching in the wet oxidation of phenol with CuO-based catalysts: causes and effects Study of the copper leaching in the wet oxidation of phenol with CuO-based catalysts: causes and effects. Appl Catal B Environ 61:323–333
5. Katti VR, Debnath KA, Muthe KP, Kaur M, Dua AK, Gadkari SC, Sahni VC (2003) Mechanism of drifts in H_2S sensing properties of SnO_2: CuO composite thin film sensors prepared by thermal evaporation. Sens Act B Chem 96(1–2):245–252. https://doi.org/10.1016/s0925-4005(03)00532-x
6. Dodoo-Arhin D, Leoni M, Scardi P (2012) Microemulsion synthesis of copper oxide nanorod-structures. Mol Cryst Liq Cryst 555(1):17–31
7. Sears WM, Fortin E (1984) Preparation and properties of Cu_2O/Cu photovoltaic solar cell. Solar Energy Mater 10(1):93–103
8. Ray SC (2001) Preparation of copper oxide thin film by the sol–gel-like dip technique and study of their structural and optical properties. Solar Energy Mater Solar Cells 68:307–312
9. Behzad KK, Rasoul SM (2012) Mater Res Bull 47:362
10. Zayyoun N, Jaber B, Laânab L, Ntsoenzok E, Bekkari R (2016) Effect of solvent on the morphological and optical properties of CuO nanoparticles prepared by simple sol-gel process. J Mater Environ Sci 7(5):1791–1797
11. Sert E, Sayrac M (2022) Investigation of optical and structural properties of tin-doped copper oxide thin films prepared by the drop cast method. J Aust Ceram Soc 58:93–100
12. He M, Wang Y, Wang H, Chen R (2016) A one-step sol–gel route derived Ag–CuO film as a novel solar selective absorber. Sol Energy Mater Sol Cells 144:264–272
13. Ha J, Oh J, Choi H, Ryu H, Lee W, Bae J (2017) Photoelectrochemical proper-ties of Ni-doped CuO nanorods grown using the modified chemical bath deposition method. J Ind Eng Chem 25:38–44
14. Yuan W, Qiu Z, Chen Y, Zhao B, Liu M, Tang Y (2018) A binder-free composite anode composed of CuO nanosheets and multiwall carbon nanotubes for high-performance lithium-ion batteries. Electrochim Acta 267:150–160. https://doi.org/10.1016/j.electacta.2018.02.081.
15. Kaur G, Saini K, Tripathi AK, Jain V, Deva D, Lahiri I (2017) Room temperature growth and field emission characteristics of CuO nanostructures. Vacuum 139:136–142. https://doi.org/10.1016/j.vacuum.2017.02.020
16. Nithya K, Yuvasree P, Neelakandeswari N, Rajasekaran N et al (2014) Preparation and characterization of copper oxide nanoparticles. Int J Chem Tech Res 6(3):2220–2222
17. Lanje AS, Sharma SJ, Pode RB, Ningthoujam RS (2010) Synthesis and optical characterization of copper oxide nanoparticles. Adv Appl Sci Res 1(2):36–40

Renewable Energy

Deposition of Copper Oxide Thin Films by Spray Polarisis Technique with Copper Chloride Solution for Different Spray Number

Ketita Nour Elhouda, Daira Radouane, Boudjema Bouzid, and Mohammedi Abdelkader

Abstract In this study, CuO thin films were produced using spray pyrolysis technique on glass substrates at a fixed temperature of 350 °C. The films had a 0.1 M deposit using ($CuCl_2$, $2H_2O$) as precursor, Using a spray number that ranges between 25 and 150 spray with a step of 25 spray to observe how this parameter affects the structural characteristics of our material. According to the plan (002), X-ray diffraction (XRD) analysis demonstrates the development of pure polycrystalline CuO with tenorite phase, which is a monoclinic structure. the characterizations utilized, such as X-ray diffraction (XRD), demonstrate the development of pure polycrystalline CuO with tenorite phase, which is a monoclinic structure. The parameter of the monoclinical cell a vary according to the spray number between 4.78 and 4.67 Å. also c vary according to the spray number between 5.12 and 5.3 Å. The variation in parameters was due to stresses. and Raman spectroscopy was used to perform the optical characterisation. What we discovered using XRD is confirmed by Raman spectroscopy.

Keywords Copper oxide · XRD · Spray pyrolysis · Spectroscopy Raman · Spray number · Copper chloride solution

1 Introduction

Properties of metal oxide thin films are what distinguish them most from other types of technological applications. Among these, extensive research has focused on cuprous oxide thin films as an encouraging substance with potential applications in

K. N. Elhouda (✉) · D. Radouane · B. Bouzid
LRPCSI, Université, 20 Août 1955-Skikda, B.P. 26, Route d'El-Hadaiek, 21000 Skikda, Algeria
e-mail: ketitanourelhouda@gmail.com

M. Abdelkader
Department of Physics, Higher School of Teachers of Bou Saada, 28001, Bou Saada, Algeria

S. Hamamda et al. (eds.), *Proceedings of the 2nd International Conference of Nanotechnology for Environmental Protection and Clean Energy Production*, Springer Proceedings in Materials 45, https://doi.org/10.1007/978-981-97-1916-7_24

various industries. There are three types of copper oxides (cupric oxide or tenorite (CuO), the cuprous oxide or cuprite (Cu_2O) and cu_3O_4, ,The most stable of them is cupric oxide (CuO) moreover, band gap ranging from 1.3 to 2 eV [1]. Cupric oxide which has a monoclinic polycrystalline structure, It is confirmed that the p-type conductivity of CuO thin films [2]. It has garnered significant attention over time due to its noteworthy electrical and optical characteristics [3]. Additionally, The material has attracted considerable interest from businesses due to its high optical absorption rate [4, 5], outstanding electrical conductivity [6], non-toxic nature [7]Furthermore, because of its strong solar absorbance and little thermal emittance [8]. Copper oxide (CuO) thin films found utility in various applications such as: gas sensor [9], solar cells [10], lithium battery [11]and material for the magnetic storage [12]. CuO can be acquired through both physical and chemical methods such as magnetron sputtering [13], molecular beam epitaxy [14], thermal evaporation [15], chemical precipitation [16], polyol process [17], spray pyrolysis [18], sol–gel method [19], solvo thermal [20], hydrothermal methods [21], etc. Chemical method are cheaper, faster, and more productive than physical method. This approach facilitates their straightforward adaptation for industrial purposes.

Among these methods, the growth of CuO thin films has conventionally been achieved through spray pyrolysis on various types of substrates. This is a straightforward and convenient method for depositing a range of metal oxide thin films. The two significant advantages of this approach are its operation at atmospheric pressure and its suitability for large-area deposition. The effect of deposition factors regarding the qualities of CuO thin films has been the subject of extensive research.

In this paper, we describe the creation of CuO thin films at the nano-scale using a cost-effective technique called discontinuous chemical spray pyrolysis. We explore the structural and optical properties of these films while altering the number of spray.

2 Experimental Methods

2.1 Elaboration of CuO Thin Film

By spraying an aqueous precursor solution of 0.1 M containing an acceptable amount of copper chloride dehydrate dissolved in an appropriate volume of distilled water, nano-sized CuO thin films have been produced. The precursor solution was discontinuously sprayed onto cleaned glass substrates with dimension (1 cm * 1 cm) that had been heated to 350 °C. Mechanically, spray was carried out utilising a perfume nozzle and air as the carrier gas. 25 cm separated the substrate from the nozzle. and the time between two successive 1 s. With a number of sprays ranging from 25 to 150 sprays with a step of 25 sprays.

where the formation of the CuO thin films begins following this reaction (Eq. 1)

$$CuCL_2 + H_2 + O \rightarrow CuO + 2HCL \tag{1}$$

We altered the deposition settings to generate good quality CuO films that were homogeneous, highly crystallised, and reproducible.

2.2 *Characterization of Materials*

The structural and optical characteristics of the deposition of CuO thin films against spray number were investigated using a variety of techniques. The symmetric—2 geometry was utilized to study the structural properties utilizing XRD on a Philips X'Pert diffractometer with a CuKα radiation source that has a wavelength of 0.154056 nm. The presence of the CuO phase was verified in this work using Raman spectroscopy. A Bruker Senterra Raman spectrometer with AlGaAs excitation at a 532 nm laser was used to record all Raman spectra over the range of 100–1400 cm^{-1} with an exposure time of 20s.

3 Results and Discussions

3.1 *XRD Pattern*

Figure 1 Show in XRD analysis was conducted to explore the structural characteristics of thin CuO nanostructured films formed on glass substrates using different numbers of spray.

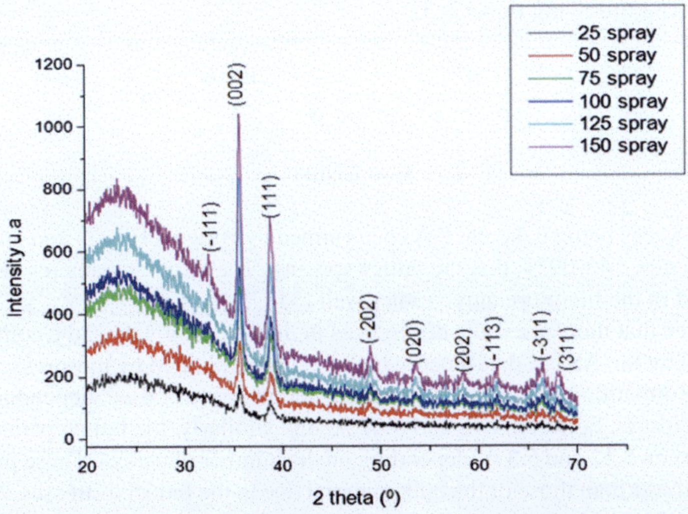

Fig. 1 CuO thin films with various spray numbers as seen in their XRD patterns

Figure 1 depicts the thin film's XRD pattern. It shows the crystalline nature of copper oxide and the diffraction peaks are in excellent agreement with the pdf file (JCPDS n = ° 45-0937). The enhancement of the CuO tenorite phase was verified by the single crystallographic planes. They have concluded the formation of CuO single phase. The crystallisation of two particular designs (002) and (111) in the CuO phase was good. According to JCPDS card numbers 85-1326 for copper and 78-2076 for copper dioxide, no peak of either of the other phases of copper or copper dioxide was seen in our investigation.

As can be observed, the XRD peak intensity increases with the number of sprays and noticeably decreases for the film created with 125 sprays. This signifies an enhancement in the crystalline quality of the deposited films, potentially attributed to the thermal annealing process that occurs in the cuo films due to the catalytic reaction The widening of the XRD peaks might be attributed to micro-strain, likely originating from the disparity in the thermal expansion coefficients between the substrate and the deposited films [22]. At the identical deposition temperature of 350 °C [23, 24].

As can be seen In the 1st layer spectrum of CuO with a spray count of 25 identified a matching polycrystalline CuO structure through spray pyrolysis. it is noticed that the film is composed of an amorphous fabric in which small crystallites are dispersed. In the 2nd to 6th layer spectra of CuO, the diagram exhibits the same spectrum. In all of the films, we did not observe any preferential orientation, but rather two directions: (111) and (002) with varying intensities, and minor peaks corresponding to (− 111), (− 202), (020), (202), (− 113), (− 311), (311). The CuO phase, where preferred orientation along (002) is seen, has also been described in other research [25–29]. The (− 111) plane is also a favourable orientation, according to Roy et al. [30].

Figure 2 show in the lattice characteristics of the CuO films, a, b, and c, are displayed as a function of the spray number. is used to calculate the lattice constants a, b, and c The following Eqs. (2, 3)

$$\frac{1}{d_{hkl}^2} = \frac{h^2}{\alpha^2 \sin^2\beta} + \frac{k^2}{c^2 \sin^2\beta} - \frac{2hl\cos}{\beta}\alpha c \sin^2\beta \tag{2}$$

$$\lambda = 2d\sin\theta \tag{3}$$

Miller indices being h, k, and l. As determined by Bragg's Eq. (2) and taken from the JCPDS n = ° 45-0937, d is the lattice spacing. The lattice parameter values and those found in the literature agree rather well [31].

We notice that there are some differences in the values of a, b, and c compared to those given by the ASTM datasheet, which may be explained by the stresses caused during the formation of our films. The lattice parameter 'a' varied, depending on the number of sprays, between 4.78 and 4.67 Å, and similarly, the lattice parameter 'c' varied between 5.12 and 5.3 Å, depending on the number of sprays. These parameter values are lower than those of the bulk material due to the induced stresses. Also, We can observe that the lattice parameters reach values that are close to those of the bulk

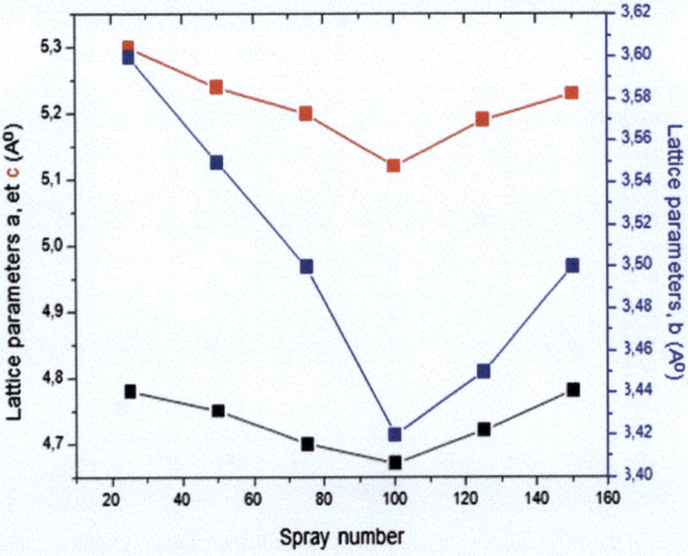

Fig. 2 The lattice parameters change depending on the spray number

material at 100 sprays. There are no differences in the lattice parameters whether using copper acetate precursor or copper chloride.

The following relationships were used used to determine the size of crystallites D for the films and the microstrain due to crystal imperfection and distortion in Fig. 3. The following [3, 4]:

$$D = \frac{K}{\lambda}\beta\cos\theta \tag{4}$$

$$\varepsilon = \frac{\beta}{4tg}\theta \tag{5}$$

The size of (002)-oriented grains is presented in Fig. 3 in relation to the quantity of sprays. In this figure, it is noticeable that the grain size expands. with two different rates until reaching 100 sprays, which have the largest grain dimension. Afterward, there is a decrease in grain size for 125 and 150 sprays. The growth in our curve indicates the enhancement in our films' crystalline structure (similar trend observed when using the copper acetate solution but with an improvement in grain size). At first to the quin relation indicates the improvement in our films crystalline structure. growth of grains during deposition (between 25 and 50 sprays) is faster than that observed between 75 and 100 sprays. Initially, our layers did not have a good crystalline structure, as shown in (see Fig. 1).

Figure 4 it can be observed a measure of the intensity of (002) orientation peak varies with quantity of sprays in a symmetrical manner, even when changing our

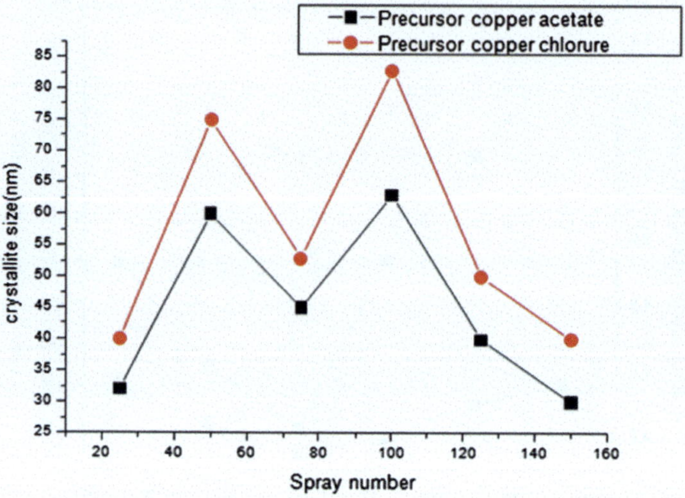

Fig. 3 Changes of the (002)-oriented crystallite size according on the spray number for both copper acetate (work already done in previous master's theses) and copper chloride solutions

precursor. The only difference is seen in the intensity, which is enhanced when using the copper chloride precursor compared to when using the copper acetate precursor. The maximum intensity is achieved at 125 sprays for both precursors.

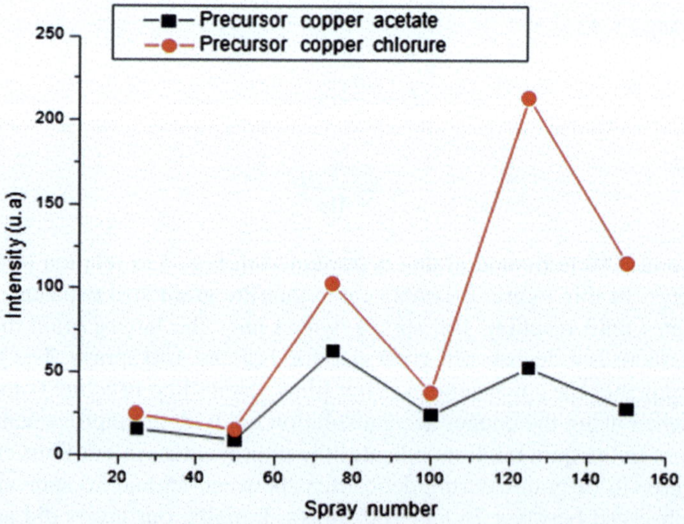

Fig. 4 Variation of expressing the peak intensity I(002) in relation to the spray number for two precursor

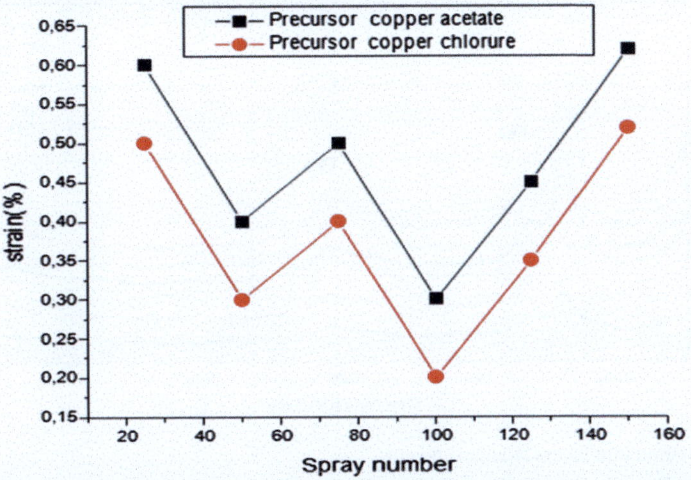

Fig. 5 Changes in the crystal strain with respect to the spray number for two precursor

Strain are calculated using Eq. (4) and represented in Fig. 5. It can be observed that the strain decreases at two different rates: the first from 25 to 50 sprays and the second from 75 to 100 sprays. The latter has the minimum strain value, allowing room for grain growth, which makes up our thin CuO layers, thus indicating an improvement in the crystalline state Fig. 5. After 100 sprays, the stresses increase until reaching a maximum value at 150 sprays. This trend is consistent for both copper acetate and copperchloride precursors, but the strain is lower for the copper chloride solution compared to the copper acetate solution.

3.2 Raman Spectra

Figure 6 displays three distinct peaks in the Raman spectra of the deposited films, situated at 277.2, 333.1, and 624.6 cm^{-1}. . The principal vibration modes of CuO, Ag(1), Bg(1), and Bg(2) are represented by these peaks, which have been documented in existing literature [22, 32].

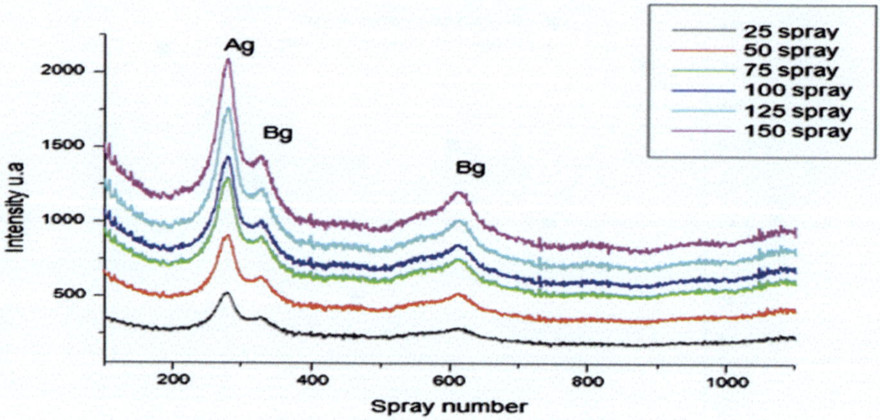

Fig. 6 Deposited thin layers' Raman spectra

4 Conclusion

Using the discontinuous chemical spray pyrolysis method, nanoscale CuO thin films were deposited on a glass substrate. We explored how the spray number affected the structural and optical characteristics of the deposited layers. XDR was used to characterise the thin layers' structural properties, while Raman spectroscopy was used to characterise their optical properties. The XDR has validated the crystalline condition of our thin layers and the favoured direction of CuO formation in accordance with the plan (002). We were able to verify the existence of a thin CuO layer using the characterisation and Raman spectroscopy.

References

1. Alkoy EM, Kelly P (2005) The structure and properties of copper oxide and copper aluminium oxide coatings prepared by pulsed magnetron sputtering of powder targets. Vacuum 79:221–230
2. Yeong YK, Choi GM (1996) Nonstoichiometry and electrical conduction of CuO. J Phys Chem Solid 57:81–84
3. Dhaouadi M, Jlassi M, Sta I, Miled IB, Mousdis G, Kompitsas M, Dimassi W (2018) Physical properties of copper oxide thin films prepared by sol–gel spin–coating method. Am J Phys Appl 6:43–50 [CrossRef]
4. Prabu RD, Valanarasu S, Ganesh V, Shkir M, AlFaify S, Kathalingam A, Srikumar S, Chandramohan R (2018) An effect of temperature on structural, optical, photoluminescence and electrical properties of copper oxide thin films deposited by nebulizer spray pyrolysis technique. Mater Sci Semicond Process 74:129–135 [CrossRef]
5. Aghayan M, Hussainova I, Kirakosyan K, Rodríguez MA (2017) The template-assisted wet-combustion synthesis of copper oxide nanoparticles on mesoporous network of alumina nanofibers. Mater Chem Phys 192:138–146
6. Özmenteş R (2019) Effect of thermal annealing time on the optical characteristics of CuO thin films. Yüzüncü Yıl Üniversitesi Fen Bilimleri Enstitüsü Derg 24:176–182

7. Mikami K, Kido Y, Akaishi Y, Quitain A, Kida T (2019) Synthesis of Cu_2O/CuO nanocrystals and their application to H_2S sensing. Sensors 19:211
8. Mazhir SN, Yasen H, Salih MM, Mohamed GH (2018) Structural and solar cell properties of CuO doped TiO_2 thin films prepared by laser induced plasma. J Eng Appl Sci 13:3555–3561
9. Chowdhuri A, Gupta V, Sreenivas K, Kumar R, Mozumdar S, Patanjali PK (2004) Appl Phys Lett 84:1180–1182
10. Han K, Tao M (2009) Sol Energy Mater Sol Cells 93:153
11. Gao XP, Bao JL, Pan GL, Zhu HY, Huang PX, Wu F, Song DY (2004) J Phys Chem B 108:5547–5551
12. Kumar RV, Diamant Y, Gedanken A (2000) Chem Mater 12:2301–2305
13. Dolai S, Dey R, Das S, Hussain S, Bhar R, Pal AK (2017) Cupric oxide (CuO) thin films prepared by reactive dc magnetron sputtering technique for photovoltaic application. J Alloys Compd 724:456–464 [CrossRef]
14. Yang KG, Hu P, Wu SX, Ren LZ, Yang M, Zhou WQ, Li SW (2016) Room-temperature ferromagnetic CuO thin film grown by plasma-assisted molecular beam epitaxy. Mater Lett 166:23–25 [CrossRef]
15. Sahu K, Choudhary S, Khan SA, Pandey A, Mohapatra S (2019) Thermal evolution of morphological, structural, optical and photocatalytic properties of CuO thin films. Nano-Struct Nano-Objects 17:92–102 [CrossRef]
16. Sagadevan S, Pal K, Chowdhury ZZ (2017) Fabrication of CuO nanoparticles for structural, optical and dielectric analysis using chemical precipitation method. J Mater Sci Mater Electron 28:12591–12597 [CrossRef]
17. Dobrozhan O, Vorobiov S, Kurbatov D, Baláž M, Kolesnyk M, Opanasyuk A (2020) Structural properties and chemical composition. Microstruct 140:106455 [CrossRef]
18. Moumen A, Hartiti B, Comini E, Arachchige HMM, Fadili S, Thevenin P (2019) Preparation and characterization of nanostructured CuO thin films using spray pyrolysis technique. Superlattices Microstruct 127:2–10 [CrossRef]
19. Lim YF, Chua CS, Lee CJJ, Chi D (2014) Sol–gel deposited Cu_2O and CuO thin films for photocatalytic water splitting. Phys Chem Chem Phys 16:25928–25934 [CrossRef] [PubMed]
20. Hong J, Li J, Ni Y (2009) Urchin-like CuO microspheres: synthesis, characterization, and properties. J Alloys Compd 481:610–615 [CrossRef]
21. Ozga M, Kaszewski J, Seweryn A, Sybilski P, Godlewski M, Witkowski BS (2020) Ultra-fast growth of copper oxide (II) thin films using hydrothermal method. Mater Sci Semicond Process 120:105279 [CrossRef]
22. Daira R, Kabir A, Boudjema B, Sedrati C (2020) Structural and optical transmittance analysis of CuO thin films deposited by the spray pyrolysis method
23. Singh I, Bedi RK (2011) Appl Surf Sci 257:7592
24. Gopalakrishna D, Vijayalakshmi K, Ravidhas C (2012) J Mater Sci Mater Electron 667:08
25. Maruyama T (1998) Copper oxide thin films prepared from copper dipivaloylmethanate and oxygen by chemical vapor deposition. Jpn J Appl Phys 37:4099–4102
26. Morales J, Sanchez L, Martin F, Ramos-Barrado JR, Sanchez M (2005) Use of low- temperature nanostructured CuO thin films deposited by spraypyrolysis in lithium cells. Thin Solid Films 474:133–140
27. Papadimitropoulos G, Vourdas N, Vamvakas VE, Davazoglou D (2005) Deposition and characterization of copper oxide thin films J Phys Conf Ser 10:182–185
28. Marabelli F, Parraviciny GB, Orioli FS (2005) Optical gap of CuO. Phys Rev B 52:1433–1436
29. Alkaya A, Kaplan R, Canbolat H, Hegedus SS (2009) A comparaison of fill factor and recombination losses in amorphous silicon solar cells on ZnO and SnO_2. Renew Energy 34:1595–1599
30. Roy S, Bhuiyan AH (2017) Properties of spray pyrolysied copper oxide thin films. Sens Transd 209(2):20–27
31. Kimura T, Sekio Y, Nakamura H, Siegrist T, Ramirez P (2008) Nat Mater 7:291
32. Volanti DP, Keyson D, Cavalcante LS, Simoes AZ, Joya MR, Longo E, Varela JA, Pizani PS, Souza AG (2008) J Alloys Compd 459:537

Stability Analysis of Gradient Zone of a Solar Pond

Ouzani Riadh, Alloui Imane, and Alloui Zineddine

Abstract Linear stability theory is applied to study the effect of feedback control on the onset of double diffusion convection in the non-convective layer of a solar pond. The system is stabilized using sensors and actuators that modify the intensity of the heat flow. The sensors are positioned at a given height of the layer. The influence of the application of active control on the appearance of convection in the case of an infinite layer is analyzed.

Keywords Active control · Solar pond · Double diffusion · Linear stability

1 Introduction

Solar energy is one of the most promising substitutes for traditional energy resources to produce clean and cheap energy. However, its intermittent and unstable nature is a major drawback, resulting in a mismatch between supply and demand [1]. A salinity gradient solar pond can provide an environmentally friendly, cost-effective and socially acceptable energy solution [2]. The characteristics that make it attractive are, on the one hand, a long-term storage capacity, which can provide enough heat for the year, and on the other hand, an annual capture efficiency of the order of 15–25% [3]. A review of the achievements and development of solar ponds over the decades is presented by [4]. A solar pond consists of three layers, an upper convective layer or zone of surface, a lower convective layer or zone of storage, and between these two layers there is the non-convective zone. The storage capacity of

O. Riadh (✉)
LESEI Laboratory, Mechanical Engineering Department, University of Batna 2, Batna, Algeria
e-mail: r.ouazani@univ-batna2.dz

A. Imane
Mechanical Engineering Department, Faculty of Applied Sciences, University of Kasdi Merbah Ouargla, Ouargla 30000, Algeria

A. Zineddine
LICEGS Laboratory, Mechanical Engineering Department, University of Batna 2, Batna, Algeria

S. Hamamda et al. (eds.), *Proceedings of the 2nd International Conference of Nanotechnology for Environmental Protection and Clean Energy Production*, Springer Proceedings in Materials 45, https://doi.org/10.1007/978-981-97-1916-7_25

a solar pond mainly rests on the area not convective that exhibits vertical stratification in salinity and temperature. The diffusivities of these two components being very different, instabilities may appear and lead to convective movements. The main problem in a solar pond is the maintenance of stability of this gradient zone. This shows why the researchers have spent as much time analyzing and study the thickness and structure of this zone, for optimal performance of the pond solar. Many of the physical processes important fundamentals for the dynamics of a solar pond have been described by [5], who also presented the stability criterion for systems with double diffusion with linear gradients. Among the first works concerning the study the linear stability of the repository zone [6, 7]. This area was considered as a layer confined with the taking into account of the absorption of the solar radiation in the layer. A weak formulation of the basic equations has been used as well as the Galerkin method for obtain approximate solutions. The dynamic processes that involve the stability and behavior limits between the gradient zone and convection zones has been studied by [8]. It has been established that the good operation of a solar pond requires three things: maintaining the temperature of the area of storage by heat extraction and injection of brine, surface washing to control the upper layer and an initial stratification of salt well designed to avoid the formation of instability.

A linear stability study of the area of gradient was performed by [9] using the mathematical formulation proposed by [6]. There linear stability was used by [10] with a surface boundary condition describing the transfer of heat at the surface of the pond in terms of mechanisms of radiation, convection and of evaporation. A theoretical study was carried out by [11] on the influence of a porous layer on the control of the onset of convection in the zone of storage.

Recently the quality of the area stability non-convective of a gradient solar pond salinity has been studied by [12]. Annual profiles surface temperature, salinity and extractors are built to different combinations of thickness of the area of upper convection and the temperature of the storage area.

The formation and maintenance of the gradient of salinity are essential elements for good functioning of a solar pond. In the case of pure water, studies have shown the possibility of the stabilization of Rayleigh- convection Bénard by active control [13]. The objective of present work consists in studying the active control of double diffusive convection in the layer layer of a solar pond, using a proportional feedback control.

2 Mathematical Modeling

In our study, we are mainly interested in the non-convective and insulating intermediate zone, height d (Fig. 1), because it is the stability of this area that depends on the proper functioning of the pond solar.

This area is modeled by a layer horizontal of infinite extension with surfaces upper and lower free, with constant concentrations C_1 and C_0 on both sides. We impose a heat flux by convection in top of the shell, and a conduction flux q_1 at low, taking

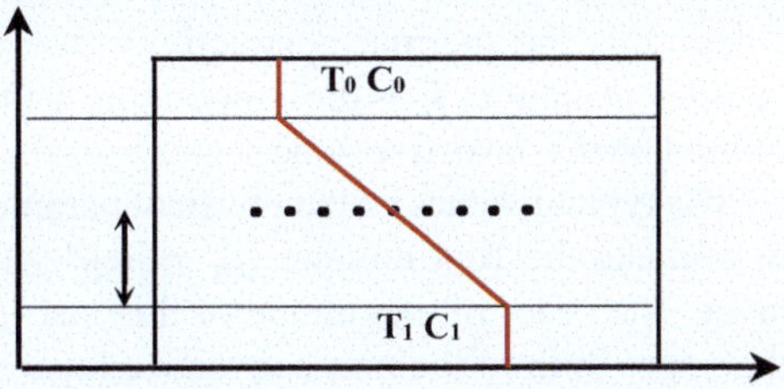

Fig. 1 Sketch of the solar pond structure

into account the heat extracted of the lower convective zone. In our work only the solar flux absorbed by the bottom of the pond and not that absorbed by the layer.

The equations governing the problem are the conservation of mass, quantity equations of motion, energy and species formulated together with the Oberbeck-Boussinesq approximation [9]. The basic equations are given by:

$$
\begin{aligned}
&\nabla.\mathbf{V} = 0\\
&\rho_0\left(\frac{\partial \mathbf{V}}{\partial t} + (\mathbf{V}.\nabla)\mathbf{V}\right) = -\nabla p + \mu\nabla^2\mathbf{V}\\
&\quad - \rho_0(1 - \beta_T(T - T_0) + \beta_C(C - C_0))g\mathbf{K}\\
&\frac{\partial T}{\partial t} + (\mathbf{V}.\nabla)T = \alpha\nabla^2 T\\
&\frac{\partial C}{\partial t} + (\mathbf{V}.\nabla)C = D\nabla^2 C
\end{aligned}
\tag{1}
$$

Here, $\mathbf{V} = (u, v, w)$ is the fluid velocity vector, t is the time, P is the pressure, μ is the dynamic viscosity of the water, D, α are the mass and thermal diffusivities, β_T, β_C are the coefficients of thermal and saline expansion respectively, C is the salt mass fraction, g is the acceleration due to gravity, $\boldsymbol{e_Z}$ is the unit vector along the z-axis.

The boundary conditions take into account the fact that the two horizontal free surfaces remain flat and impermeable; with continuity of velocities at each of the interfaces and that the fraction salt mass of the two free surfaces remains constant. This results in the following equations:

$$
\begin{aligned}
C - C_1 &= \lambda\frac{\partial T}{\partial y} + q_1 = 0\\
w &= \frac{\partial u}{\partial y} = 0 \quad \text{pour} \quad y = 0\\
C - C_0 &= \lambda\frac{\partial T}{\partial y} + h_d(T_d - T_\infty) = 0\\
w &= \frac{\partial u}{\partial y} = 0 \quad \text{pour} \quad y = d
\end{aligned}
\tag{2}
$$

where λ is the thermal conductivity of water, h_d is the convective heat transfer coefficient natural, T_d is the temperature at the limit upper layer T_∞, is the temperature of the upper convective zone.

The typical values of the physical parameters of a solar pond [6] make it possible to evaluate the Biot number ($\beta_i = h_d d/\lambda$) of the upper boundary of the stratified layer. The order of magnitude of this last ($\beta_i \approx 600$) makes it possible to consider the upper limit of the non-convective layer at constant temperature, equal to T_∞.

If the thermal losses between the bottom of the pond and the ground are neglected, the heat flux of the storage zone must be equal to the difference between the total flux q_0 absorbed in the storage zone and the flux q_{ext} extracted by the exchangers heat, $q_1 = q_0 - q_{ext}$. Since q_{ext} corresponds to a fraction f of the total heat flux absorbed: $q_{ext} = f q_0$, then we can deduce that:

$$q_1 = (1 - f)q_0 \tag{3}$$

The configuration of the non-convective layer is characterized by the fact that below the critical Rayleigh, the fluid is immobile and the heat transfer takes place by conduction and the mass transfer by diffusion. Above this threshold, a convection movement is generated, which changes the distribution of temperature and concentration. The idea of active control in this case is to adjust the heat flux q_l so as to delay the start of movement. To do this, a set of temperature sensors located at a height y_s detect temperature deviations from the conductive values T_{cond}. Then, heat source q_l is modulated proportionally to these deviations.

$$q(x, 0) = q_1 - G[T(x, y_s) - T_{cond}(x, y_s)] \tag{4}$$

where G is the gain of the proportional control.

The equations governing the problem are dimensionless using the length scale d, time scale d^2/α, velocity scale α/d, pressure scale $\rho(\alpha/d)^2$. We also introduce dimensionless temperature $\theta = (T - T_0)/\Delta T$ and dimensionless concentration $S = (C - C_0)/\Delta C$. Using these scales, Eq. (1) can be transformed into the following dimensionless form:

$$\begin{aligned}
&\nabla.\mathbf{V} = 0\\
&\rho_0\left(\frac{\partial \mathbf{V}}{\partial t} + (\mathbf{V}.\nabla)\mathbf{V}\right) = -\nabla P + \Pr \nabla^2\mathbf{V} + Pr(Ra_T\theta - Ra_s S)\mathbf{k}\\
&\frac{\partial \theta}{\partial t} + (\mathbf{V}.\nabla)\theta = \nabla^2\theta\\
&\frac{\partial S}{\partial t} + (\mathbf{V}.\nabla)S = \frac{1}{Le}\nabla^2 S
\end{aligned} \tag{5}$$

where the dimensionless parameters: $Ra_T = g\beta\Delta T^2 d^3/\alpha\nu$ is the thermal Rayleigh number, $Ra_S = g\beta\Delta C d^3/\alpha\nu$ is the solute Rayleigh number, $\Pr = \nu/\alpha$ is the Prandtl number, $Le = D/\alpha$ is the Lewis number,.

The dimensionless boundary conditions are given by the following equations:

$$\theta = (T - T_0)/\Delta T$$

The system (5) and (6) has a basic solution associated with a situation of pure double diffusion without active control given by:

$$\begin{aligned}
&S - 1 = \frac{\partial \theta}{\partial y} + (1 - f) - G(\theta_S - \theta_{CS}) = 0\\
&w = \frac{\partial u}{\partial y} = 0 \quad \text{pour} \quad y = 0\\
&S = \theta = 0\\
&w = \frac{\partial u}{\partial y} = 0 \quad \text{pour} \quad y = 1
\end{aligned} \tag{6}$$

The system (5) and (6) has a basic solution associated with a situation of pure double diffusion without active control given by:

$$\begin{aligned}
&\mathbf{V}_B = 0\\
&\theta_B = (1 - f)(1 - y)\\
&S_B = (1 - y)
\end{aligned} \tag{7}$$

3 Linear Stability Study

We carried out a study of the linear stability of the previous basic solution relatively to infinitesimal disturbances, in the case of an infinite horizontal extension layer. The disturbances are developed in normal modes and introduced into the disturbance equations. These equations are discretized using finite differences. For N points of calculation we have a matrix system of dimension 3N. The analysis of linear stability is thus reduced to an eigenvalue problem, which can be found using the standard subroutine DGVCCG of the IMSL library of FORTRAN. There value of the Rayleigh number, for which the rate of growth of the amplitude is zero, is calculated by the iterative method of Newton. For details on the method and the solution procedure as well as its validation see [14].

4 Results and Discussions

The problem studied is governed by seven parameters, namely Ra_T, Ra_s, Pr, Le, G, y_s and f. In the present study, the Prandtl number and the number are fixed at Pr $= 7$ and $Le = 100$ [6].

Figure 2 presents the marginal stability curves, in terms of the Rayleigh number, *Ra*, as a function of the wave number, a. The results were obtained for different values of the gain of control, *G*, when the sensors are located at position $y_s = 0.725$. For a given value of the gain, the minimum of the curve gives the critical Rayleigh number Ra_{Tc} and the corresponding critical wavenumber a_c. For an uncontrolled system ($G = 0$), we find that, $Ra_{Tc} = 419$ to $a_c = 1.77$. A global view on the graph shows that the feedback control ($G > 0$) has a stabilizing effect on the system, so that Ra_{Tc} increases with increasing *G*. It should be noted that for all results presented in this work, it was found that the onset of convection occurs through an oscillatory instability (Hopf bifurcation).

The effect of sensor position, y_s, on the marginal stability curves is shown in Fig. 3 for $G = 5, f = 0$ and $Ra_S = 2000$. The results indicate that the position of the sensors affects the value of the critical Rayleigh number for the onset of convection. This point has been demonstrated in the past by [13] in the case of a classical Rayleigh-Bénard problem, i.e. a horizontal layer of pure water heated from below and cooled from above. The optimal position of the sensors is at the center of the layer height, i.e. at $y_s = 0.5$. This position gives the control the maximum efficiency in terms of stabilization of the layer, this value will thus be adopted for all the results which follow.

Figure 4 shows the variation of the critical Rayleigh number as a function of the control gain for different values of the fraction extracted from the heat flux. The curve for $f = 0$ corresponds to the case where there is no heat extraction from the solar pond. $G \approx 13$, which means that beyond this value the increase in control gain leads to the destabilization of the system instead of stabilizing it. All the curves show

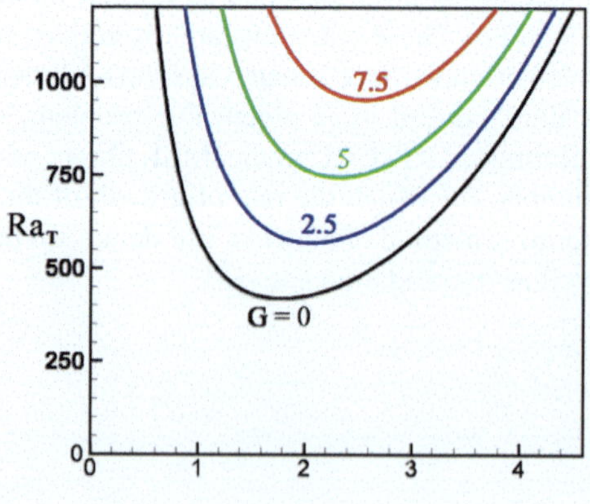

Fig. 2 Effect of control gain G on the marginal stability curve for $Ra_s = 2000$, $y_s = 0.725, f = 0$

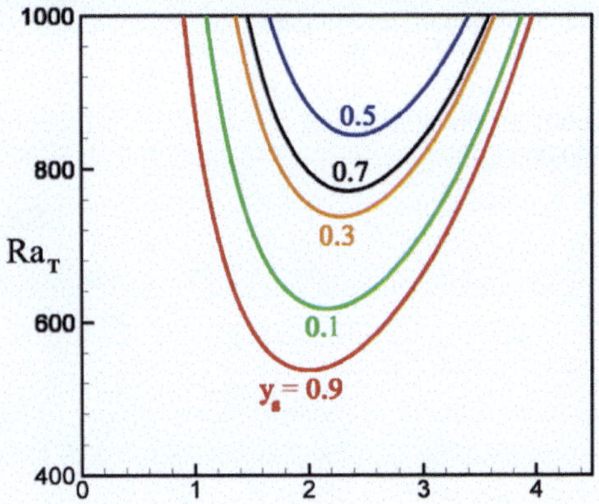

Fig. 3 Effect of sensor height ys on the marginal stability curve for $\mathrm{Ra}_s = 2000$, $f = 0$, $G = 5$

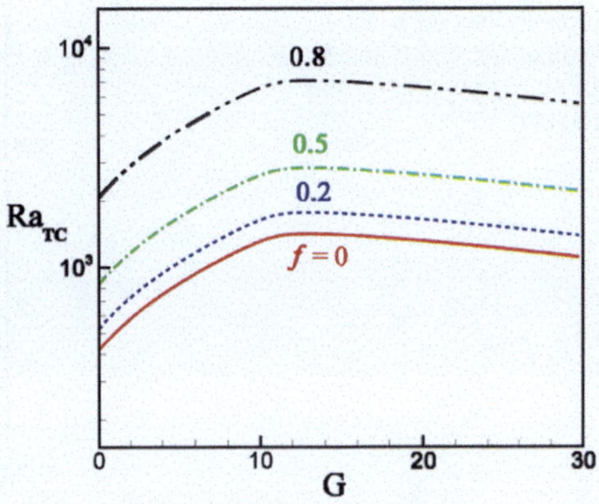

Fig. 4 Critical Rayleigh number as a function of control gain for different values of the fraction of the flux extracted f and $Ra_s = 2000$

a maximum at around G. We see that convection starts later for larger values of f, the system is more stable when more heat is extracted from the layer.

The effect of the control gain G on the critical Rayleigh number Ra_{TC} is shown in Fig. 5 for several values of the Rayleigh number solute Ra_s. For an uncontrolled system ($G = 0$), the critical Rayleigh number increases with increasing Ra_s (stabilizing effect). The shape of the curve changes with the increase in the value of the solutal Rayleigh number. For $Ra_s = 105$ the proportional active control destabilizes the layer instead of stabilizing it.

The critical wave number a_c associated with the critical Rayleigh (Fig. 5) is shown in Fig. 6. We see that a_c varies with the gain of imposed control, the curves also have a maximum value which changes with the value of the Rayleigh solutal Ra_s imposed. When the critical wave number increases it means that the structure of the disturbances at the beginning of the instability is more confined, as shown in Fig. 7. The obtained result can be merged with some techniques in control theory [14–17].

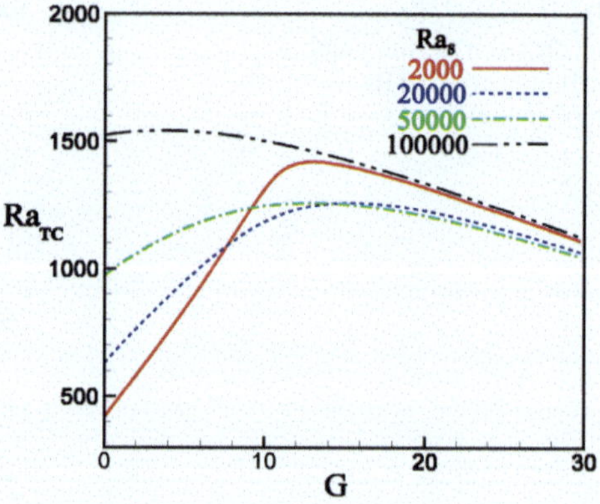

Fig. 5 Effect of control gain G on critical Rayleigh number for: $f = 0$ and $Ra_s =$ 2000, 20,000, 50,000 and 100,000

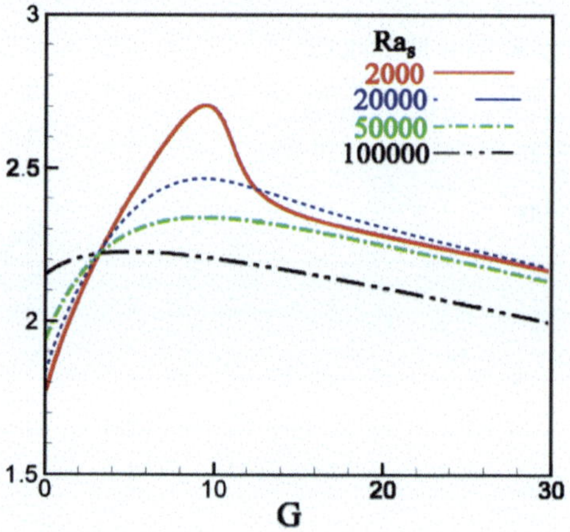

Fig. 6 Effect of control gain G on the critical wave number for: $f = 0$ and $Ra_s =$ 2000, 20,000, 50,000 and 100,000

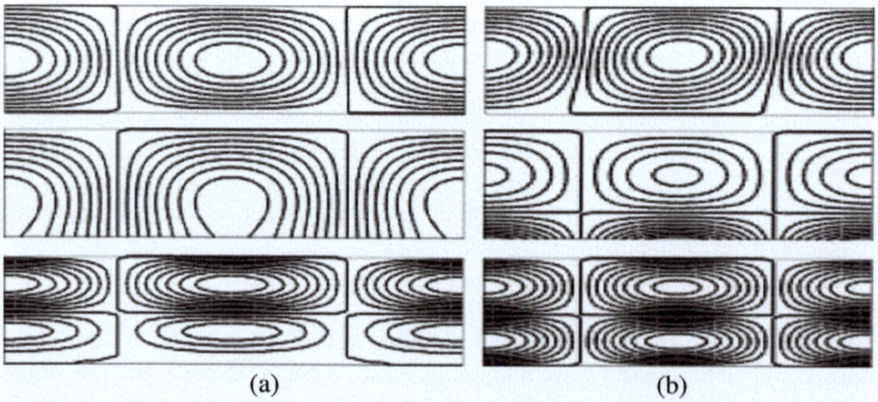

Fig. 7 Streamlines (top), isotherms (middle) and isoconcentrations (bottom) at the onset of convection for $f = 0$ and RaS = 2000: **a** $G = 0$, ac = 1.77, **b** $G = 10$, ac = 2.69

5 Conclusion

A linear stability analysis was performed to investigate the active control of the layer with vertical temperature and concentration stratification, which represents the non-convective zone of a solar pond. It has been shown that the onset of diffusive double convection in this layer can be controlled. However, depending on the imposed solutal Rayleigh number, the control can become a destabilizing factor. The results show that the position optimal vertical position of the collectors is located in the center of the non-convective layer.

The optimal vertical position of the sensors is at the center of the non-convective layer. In the range of parameters considered in this study, the linear stability shows that the triggering of convection is via an oscillatory instability.

References

1. Valderrama C, Luis Cortina J, Akbarzadeh A (2015) Solar ponds, in: storing energy: with special reference to renewable energy sources. Chem Int 37(5–6):273–289
2. Ito M, Azam S (2019) Feasibility of saline gradient solar ponds as thermal energy sources in Saskatchewan, Canada. J Environ Inform Lett 1(2):72–80
3. Alcaraz A, Montalà M, Cortina JL, Akbarzadeh A, Aladjem C, Farran A, Valderrama C (2018) Design, construction, and operation of the first industrial salinity-gradient solar pond in Europe: an efficiency analysis perspective. Sol Energy 164:316–326
4. Kasaeiana A, Sharifia S, Yanb WM (2018) Novel achievements in the development of solar ponds: a review. Sol Energy 174:189–206
5. Weinberger H (1964) The physics of the solar pond. J Solar Energy 8:45–56
6. Giestas M, Pina H, Joyce A (1996) The influence of radiation absorption on solar pond stability. Int J Heat Mass Transfer 39(18):3873–3885
7. Giestas M, Pina H, Joyce A (1997) The influence of nonconstant diffusivities on solar ponds stability. Int J Heat Mass Transfer 40(18):4379–4391
8. Ouni M, Guizani A, Lu H, Belghith A (2003) Simulation of the control of a salt gradient solar pond in the south of Tunisia. Sol Energy 75(2):95–101
9. Rebaï LK, Mojtabi AK, Safi MJ, Mohamad AA (2006) A linear stability study of the gradient zone of a solar pond. J Sol Energy Eng 128(3):383–393
10. Boudhiaf R, Moussa AB, Baccar M (2013) A two-dimensional numerical study of hydrodynamic, heat and mass transfer and stability. Adv Water Resour 52:1–6
11. Hill AA, Carr M (2013) The influence of a fluid-porous interface on solar pond stability. Adv Water Resour 52:1–6
12. Abdullah AA, Lindsay KA (2017) Assessing the maximum stability of the nonconvective zone in a salinity-gradient solar pond. J Solar Energy Eng 139(4):041010 (12)
13. Tang J, Bau HH (1995) Stabilization of the no-motion state in of a horizontal fluid layer heated from below with Joule heating. J Heat Transfer 117:329–333
14. Bounemeur A, Chemachema M, Essounbouli N (2018) Indirect adaptive fuzzy fault-tolerant tracking control for MIMO nonlinear systems with actuator and sensor failures. ISA Trans 1(79):45–61
15. Bounemeur A, Chemachema M (2021) Adaptive fuzzy fault-tolerant control using Nussbaum-type function with state-dependent actuator failures. Neural Comput Appl 33:191–208
16. Zahaf A, Bououden S, Chadli M, Chemachema M (2022) Robust fault tolerant optimal predictive control of hybrid actuators with time-varying delay for industrial robot arm. Asian J Control 24(1):1–5

17. Zahaf A, Bououden S, Chadli M, Boulkaibet I, Neji B, Khezami N (2023) Dynamic sensorless control approach for markovian switching systems applied to PWM DC–DC converters with time-delay and partial input saturation. Sensors 23(15):6936

Heat Transfer Enhancement in a Parabolic Trough Solar Collector Using Inner Wall Fins and Nanofluids

Halla Diafi, Sihem Djouimaa, and Djemaa Guerraiche

Abstract In this study, two different absorber tube configurations were examined: a smooth absorber tube and an absorber tube with inner wall fins at the bottom. The study primarily focused on the effects of nanoparticle additives (Al_2O_3 and SiO_2) to the base fluid, Syltherm 800. The nanoparticles were introduced at concentrations of 1, 2, 3, and 4% within a smooth tube. The study also explored their impact when introduced in a tube with inner wall fins at the bottom. The investigation centered on 2D heat transfers under turbulent flow conditions, with Reynolds numbers ranging from 6.10^4 to 15.10^4. The research uncovered several key observations. The incorporation of nanofluids resulted in a 4.90% increase in the Nusselt number for the smooth tube. Additionally, the friction factor experienced a 2.64% increase, and the Performance Evaluation Criteria (PEC) improved by 1.04. Interestingly, a combination of SiO_2 with Syltherm 800 and the addition of fins exhibited even better performance, showcasing, 88.76% enhancement compared to individual nanofluids. This configuration led to a Nusselt number and PEC increase of 1.80, while the friction factor initially decreased by 9.10% at Reynolds numbers below 10^5, and subsequently increased by 16.73%.

Keywords Parabolic trough collector · Nanofluid · Receiver tube · Forced convection · Nusselt number

H. Diafi (✉) · S. Djouimaa · D. Guerraiche
Applied Physic Energetic Laboratory (LPEA), Department of Physics, Faculty of Matter Sciences, University of Batna 1, Batna, Algeria
e-mail: halla.diafi@univ-batna.dz

S. Djouimaa
e-mail: sihem.djouimaa@univ-batna.dz

D. Guerraiche
e-mail: djemaa.guerraiche@univ-batna.dz

S. Hamamda et al. (eds.), *Proceedings of the 2nd International Conference of Nanotechnology for Environmental Protection and Clean Energy Production*, Springer Proceedings in Materials 45, https://doi.org/10.1007/978-981-97-1916-7_26

1 Introduction

Solar energy, especially from parabolic trough collectors (PTC), has proven to be an efficient source of power and high-temperature heat production [1] and heat at very high temperatures [2]. However, optimizing PTC performance remains a challenge. To address this, researchers have explored the use of nanofluids in heat transfer, which can enhance PTC performance. Various studies refer to the use of nanoparticles, among them, Vahidinia et al. [3] conducted an analysis study on PTC operating with different nanofluids. The results point to that the Al_2O_3/SiO_2-Syltherm 800 hybrid nanofluid is the best choice with the greatest PEC of 1.76 with highest energy and exergy efficient. The study conducted by Guerraiche et al. [4] investigated the impact of incorporating nanoparticles into the operational Heat Transfer Fluid, using Al_2O_3/water based nanofluid at diverse volume fractions. Their results show significant enhancement in thermal efficiency, with an increase ranging from 3 to 14% for volume fractions of nanoparticles varying from 2 to 6%. The latest research revolves around altering the geometry of the PTC and introducing nanoparticles into the heat transfer fluid. Khan et al. [5] compared the smooth tube, the tube with longitudinal fins and the tube with twisted tape insert with Al_2O_3/water, when combining the nanofluids insertion with the twisted tape, the best thermal performance is attempts by 72.26%, but the results in a severe pressure drop that requires energy to pump. Amina et al. [6] presented the use of nanofluids of 1% fraction volume and the presence of two longitudinal fins in the tube side of the PTC, they noted that the introduction of fins increases the Nusselt number, and nanofluids have been shown to improve heat transfer.

The objective of this work is to improve heat exchange between the fluid and the absorber tube wall. The enhancement is done primarily by inserting nanoparticles into the fluid flow. Furthermore, the study requires an analysis and comparison of the performance of two distinct absorber tube configurations: one with a smooth absorber tube and the other with inner wall fins positioned in areas of elevated heat concentration along the lower wall.

2 Physical Model

In our study, we consider a two-dimensional absorber tube of PTC shown in Fig. 1 with length 1 m, internal diameter $D_{ri} = 64$ mm and 22 fins are placed in the lower region with 1 mm height. Figure 1c show the position of the fins used for the study.

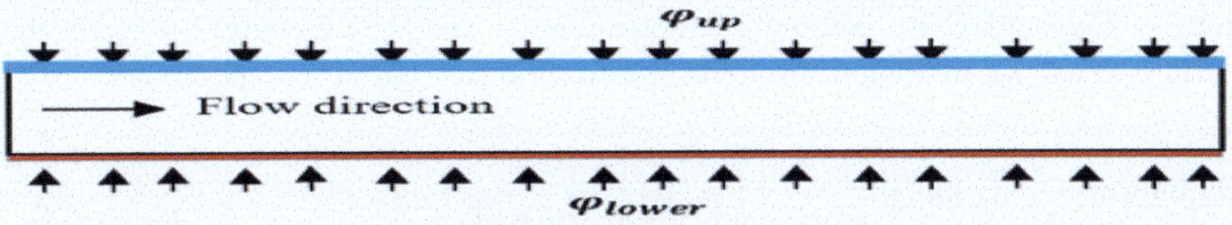

(a) Smooth absorber tube

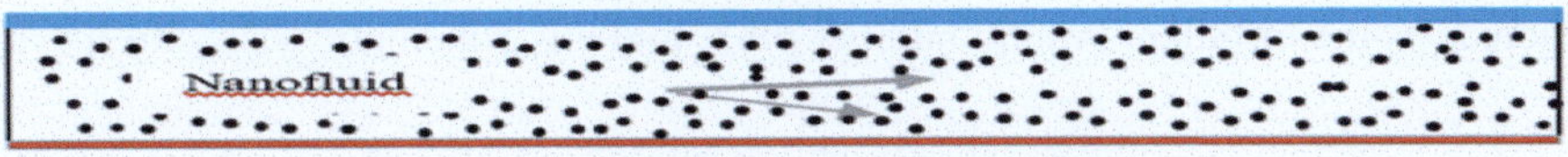

(b) Receiver tube with nanofluid.

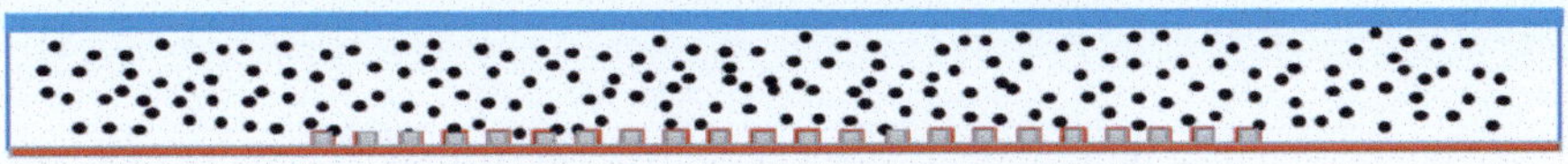

(c) Receiver tube with nanofluid and fins

Fig. 1 2D receiver tube of PTC

3 Numerical Study

3.1 Governing Equations

The governing equations in the case of a monophasic, turbulent, incompressible and steady fluid flow in the receiver tube are given as [7]:

$$\text{Continuity: } \frac{\partial(\rho\overline{u_i})}{\partial x_i} = 0 \tag{1}$$

$$\text{Momentum: } \frac{\partial\left(\rho\overline{u_i u_j}\right)}{\partial x_i} = -\frac{\partial\overline{P}}{\partial x_i} + \frac{\partial}{\partial x_j}\left[\mu\left(\frac{\partial\overline{u_i}}{\partial x_j} + \frac{\partial\overline{u_j}}{\partial x_i}\right) - \frac{2}{3}\mu\frac{\partial\overline{u_i}}{\partial x_i}\delta_{ij} - \rho\overline{u'_i u'_j}\right] \tag{2}$$

where $-\rho\overline{u_i u_j}$ and δ_{ij} are Reynolds stress and Kronecker delta respectively

$$\text{Energy: } \frac{\partial}{\partial x_j}\left(\rho\overline{u_i}C_P\overline{T}\right) = \frac{\partial}{\partial x_i}\left(\lambda\frac{\partial\overline{T}}{\partial x_i} + \frac{\mu_T}{Pr_i}\frac{\partial\left(C_P\overline{T}\right)}{\partial x_i}\right) + S_T \tag{3}$$

where μ_T and Pr_i are Eddy Viscosity (Pa.s) and the turbulent Prandtl number for energy

$$\text{k equation: } \frac{\partial}{\partial x_j}\left(\rho \mathrm{K}\overline{u_j}\right) = \frac{\partial}{\partial x_j}\left[\left(\mu + \frac{\mu_T}{\delta_k}\right)\frac{\partial \mathrm{K}}{\partial x_j}\right] + G_K - \rho\varepsilon \tag{4}$$

where k, G_K and ε are turbulent kinetic energy (m^2/s^2), Generation of turbulent kinetic energy (kg/ms^3) and turbulent dissipation rate (m^2/s^3) respectively.

$$\varepsilon\text{ equation: } \frac{\partial}{\partial x_j}(\rho\varepsilon\overline{u_j}) = \frac{\partial}{\partial x_j}\left[\left(\mu + \frac{\mu_T}{\delta_\varepsilon}\right)\frac{\partial\varepsilon}{\partial x_j}\right] + \rho C_{1\varepsilon}S_\varepsilon - \rho C_2\frac{\varepsilon^2}{K+\sqrt{\nu\varepsilon}} \quad (5)$$

where δ_ε is the turbulent Prandtl number for ε, and ν is Kinematic Viscosity (m^2/s)

$$G_k = \mu_T S^2 \quad (6)$$

where S is Rate of strain tensor (1/s)

$$\text{The turbulent viscosity equation: } \mu_T = \rho C_\mu \frac{K^2}{\varepsilon} \quad (7)$$

The Eqs. (4) and (5) presents the transport equations of the realizable k–ε turbulent model. The constants for the realizable k–ε turbulent model are expressed bellow: $C_2 = 1.9, \sigma_k = 1, \sigma_\varepsilon = 1, 2$, and S_T = Source Term (concentrated flux).

3.2 *Mathematical Modeling*

$$\text{The average Nusselt number: } Nu = \frac{h.D_{ri}}{\lambda} \quad (8)$$

$$\text{The heat transfer coefficient: } h = \frac{\varphi_u}{(\pi.D_{ri}.L)(T_r - T_{fm})} \quad (9)$$

When the average working fluid temperature T_{fm} is equal to $T_{fm} = \frac{T_{inlet}+T_{outlet}}{2}$ and T_r is temperature in the inner of an absorber tube.

$$\text{The Reynolds number is given as: Re} = \frac{4.\dot{m}}{\pi.D_{ri}.\mu} \quad (10)$$

where $\dot{m}$ is the mass flow rate (kg/s).

$$\text{The Prandtl number is given us: } Pr = \frac{\mu C_P}{k} \quad (11)$$

Table 1 Correlations of liquid Syltherm 800

Density	$\rho = -4.153495 \times T + 1.1057 \times 10^3$
Specific heat capacity	$C_P = 0.001708 \times T + 1.107798$
Thermal conductivity	$\lambda = -5.753496 \times 10^{-10} \times T^2 - 1.875 \times 10^{-4}T + 1.9 \times 10^{-1}$
Viscosity	$\mu = 6.672 \times 10^{-7} \times T^4 - 1.566 \times 10^{-3}T^3 + 1.388 \times T^2 - 5.54 \times 10^2 \times T$

Table 2 Thermophysical properties of nanoparticles

Nanoparticles	ρ (kg/m^3)	K (W/m k)	C_P (J/kg k)
Al_2O_3	3970	40	765
SiO_2	2200	1.2	703

3.3 *Thermophysical Properties of Nanofluid:*

The base fluid employed is Syltherm 800, and its thermophysical properties, which change with temperature, can be found in Table 1 [8]. These physical properties of the base fluid undergo modification when solid nanoparticles are introduced (see table 2).

The following formulas are semi-empirical correlations were used to calculate the thermophysical properties of nanofluid for Density [9], Heat capacity [10], Dynamic viscosity [11] and Thermal conductivity [12].

$$\rho_{nf} = (1-\varphi)\rho_f + \varphi\rho_s \tag{12}$$

$$(\rho C_p)_{nf} = (1-\varphi)(\rho C_p)_f + \varphi(\rho C_p)_s \tag{13}$$

$$\mu_{nf} = \frac{\mu_f}{(1-\varphi)^{2.5}} \tag{14}$$

$$\frac{k_{nf}}{k_f} = \frac{k_s + 2k_f + 2(k_s - k_f)\varphi}{k_s + 2k_f - (k_s - k_f)\varphi} \tag{15}$$

3.4 *Boundary Conditions*

The boundary conditions for resolving differential equations are listed as follows.

At the inlet, the uniform and constant velocity and temperature conditions are imposed, $u_x = u_{in}$, $u_y = 0$, $T = T_{in} = 500$ K.

At the outlet, the outflow (conservation of mass flow and energy) is used.

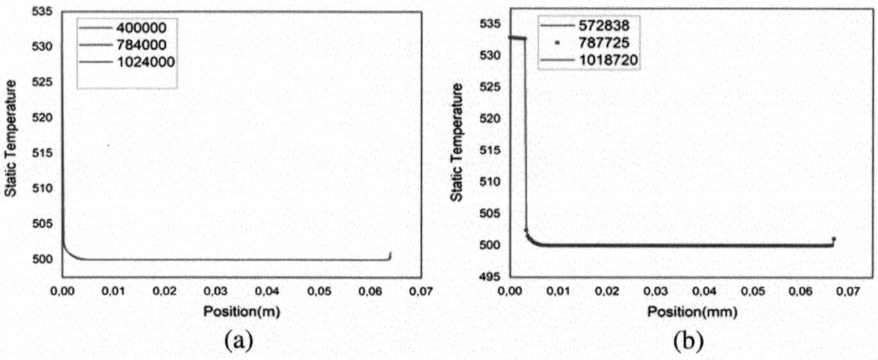

Fig. 2 Grid independence test for different cell numbers: **a** smooth tube, **b** tube with fins

The heat flux displays irregularity on both sides of the absorber tube, with a greater concentration observed at the tube's lower portion. This phenomenon occurs due to the concentration of solar radiation at the top of the collector, which is then reflected onto the lower section of the tube, as illustrated below:

The flux at the upper half is given by: $q_{up} = \varphi_{solaire} \times \tau_{verre} \times \alpha_{absorbeur} = 912$ W/m^2 The flux at the lower half is defined as: $q_{lower} = \varphi_{solaire} \times \tau_{verre} \times \alpha_{absorbeur} \times C_R$ = 26320 W/m^2

Where (τ_{verre}) Transmissivity for the cover, ($\alpha_{absorbeur}$) Absorptivity of the absorber tube, ($\varphi_{solaire}$) Incident solar radiation (w/m^2) and (C_R) concentration ratio equal to 0.96, 0.95, 1000 and 28.86 respectively.

3.5 Grid Independency

To create a mesh that is independent of numerical solution, we use a well-structured quadratic mesh for both tube shapes, after several refinements, it can be seen in Fig. 2 that the element of the mesh does not affect the results of the solution for different cell numbers for the smooth tube and for the tube with fins.

4 Results and Discussion

The effect of a volume fraction on the average Nusselt number for two nanoparticles types Al_2O_3 and SiO_2 across different volume fractions, for both a smooth absorber tube and an absorber tube with fins, for Re = 6.10^4 can be observed in Fig. 3a the average Nusselt number increases with increase of volume fraction. The heat transfer improvement achieved 2.51% and 8.20%, respectively, when Al_2O_3 and SiO_2 nanoparticles were introduced compared to the smooth tube using a basic heat

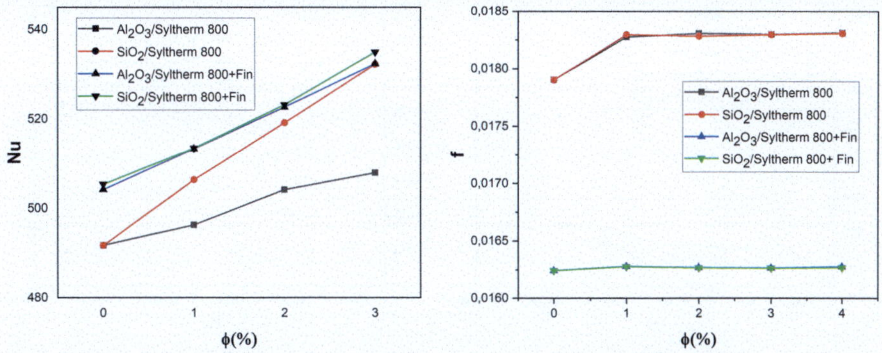

(a) (b)

Fig. 3 **a** The Nusselt number versus volume fraction for different nanofluid with absence and presence of fins. **b** Friction factor versus volume fraction for different nanofluid with absences and presence of fins

transfer fluid. Moreover, the Nusselt number increases for the absorber tube with fins, and the enhancement in Nusselt values is most pronounced in the case of SiO_2/Syltherm 800 with fins, showing an improvement of 8.76%. Adding nanoparticles led to improved thermal conductivity due to their high thermal efficiency. The fins implementation is also made to increase the thermal conductivity, and to provide heat homogenization.

This can be observed in Fig. 3b, which illustrates the variation in the friction factor for the two nanofluid types at various volume fractions, in both a smooth absorber tube and an absorber tube with fins. It becomes evident that the friction factor increases as the volume fraction rises. The increased concentration of nanoparticles results in a higher fluid density, leading to a corresponding rise in pressure drop. The inclusion of fins, on the other hand, leads to a reduction in friction factor values. Interestingly, in the area exposed to concentrated heat flux, the fluid's viscosity is reduced due to the expanded exchange surface. In the case of the tube with fins, the friction factor decreases by 12.46% when the nanofluid concentration is 4%, as compared to a smooth tube using nanofluid. Figure 4a presents the Nusselt number at different Reynolds number for various types of nanofluids studied, for a fixed volume fraction of $\varphi = 3\%$ for both geometries. The results are compared with those of a smooth tube containing Syltherm 800 only. It's clear that the average Nusselt number increases as the Reynolds number rises. We observed that introducing nanofluid into a smooth tube yields better results with SiO_2 nanoparticles, with a thermal improvement of 4.90%. This improvement can be attributed to the distinctive properties of SiO_2 nanofluid compared to Al_2O_3. Notably, the highest Nusselt number is observed in the case of a tube containing fins, where the thermal transmission was enhanced by the incorporation of nanofluids (SiO_2/Syltherm 800) by 88.76%. In Fig. 4b, the relationship between the friction factor and Reynolds number is presented for a volume fraction of nanoparticles $\varphi = 3\%$. It is evident that as the Reynolds

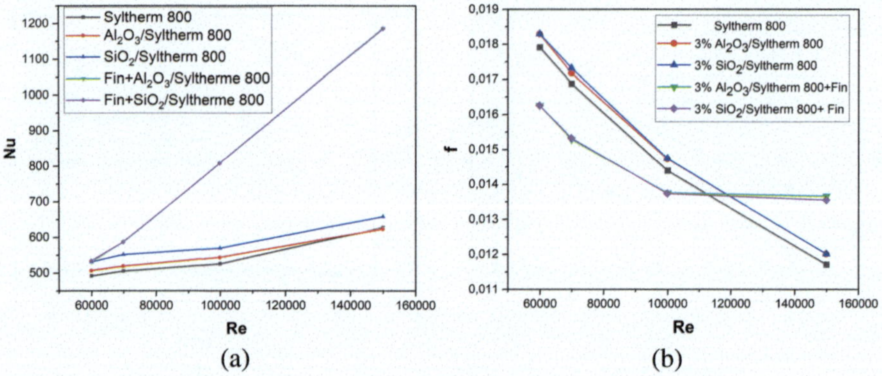

Fig. 4 **a** Nusselt number versus Reynolds number for different nanofluid with absence and presence of fin. **b** Friction factor versus volume fraction for different nanofluid with absence and presence of fins

number increases, the friction factor decreases. This phenomenon occurs due to the correlation between the Reynolds number and velocity. With an increase in Reynolds number, the velocity also increases, leading to a decrease in the friction factor. The presence of nanoparticles causes the friction factor to be higher in comparison to the base fluid, where we observed an increase in the friction factor by 2.64% at SiO_2/Syltherm 800, this discrepancy is attributed to the effects of nanofluids on density and viscosity, both of which contribute to the higher friction factor. The presence of fin results in a reduced friction factor when compared to Syltherm800 by 9.10%.

Observed an increase in the friction factor by 2.64% at SiO_2/Syltherm 800, This discrepancy is attributed to the effects of nanofluids on density and viscosity, both of which contribute to the higher friction factor. The presence of fins results in a reduced friction factor when compared to Syltherm800 by 9.10%. This outcome is due to the increased surface area provided by the fins, leading to a higher pressure drop. Consequently, this rise in the pressure drop lowers the friction factor. However, this trend holds true only for Reynolds numbers below 10^5. Beyond this threshold, the friction factor begins to rise again by 16.73%.

To assess the improvement in heat transfer, we employed the Performance Evaluation.

Criteria (PEC), which is determined using the following relationship [13]:

$$PEC = \frac{Nu/Nu_0}{(f/f_0)^{\frac{1}{3}}} \tag{16}$$

The PEC of PTC receiver tube using nanofluid in smooth tube and tube with fins for different Reynolds number is illustrated in Fig. 5. As can be seen for these cases, the PEC values are greater than 1, the best performance evaluation criteria are when $Re = 15.10^4$ and $\varphi = 3\%$ for SiO_2/Syltherm 800 in tube with fins. We can also

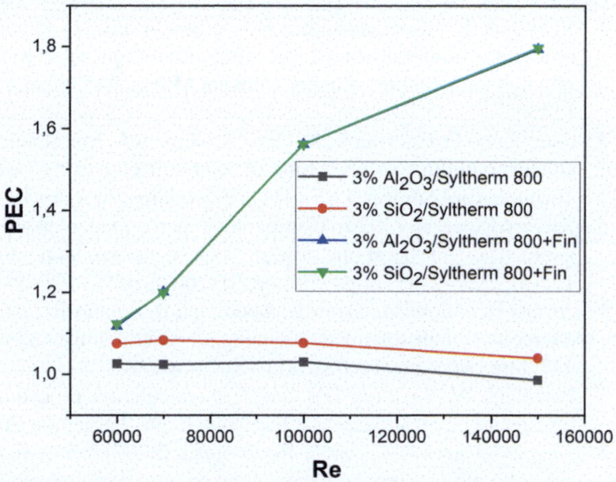

Fig. 5 Performance evaluation criteria versus Reynolds number

note that for the tube with nanofluids and without fins, the PEC does not make much difference when the Reynolds number increases.

5 Conclusions

A 2D steady state turbulent flow numerical simulation of an absorber PTC was performed to study the effects of two types of nanofluids (Al_2O_3/Syltherm 800 and SiO_2/Syltherm 800) with and without adding fins with 1 mm high, to improve heat transfer and decrease the temperature concentration along the lower wall of the absorber tube. The outcomes drawn from our study are as follows:

The study found that introducing nanofluids and fins into the absorber tube of a PTC could enhance heat transfer. The choice of nanoparticles, particularly SiO_2, resulted in significant improvements in heat transfer, with increased Nusselt numbers. The study also found that friction factor increased with nanoparticle volume fraction, but decreased with the addition of fins, especially at low Reynolds numbers. Overall, the study provided valuable insights into improving heat transfer in PTCs using nanofluids and modified tube geometries.

References

1. Fernández-García A, Zarza E, Valenzuela L, Pérez M (2010) Parabolic-trough solar collectors and their applications. Renew Sustain Energy Rev 14(7):1695–1721. https://doi.org/10.1016/j.rser.2010.03.012
2. Kalogirou SA (2004) Solar thermal collectors and applications. Prog Energy Combust Sci 30(3):231–295. https://doi.org/10.1016/j.pecs.2004.02.001

3. Vahidinia F, Khorasanizadeh H, Aghaei A (2021) Comparative energy, exergy and CO_2 emission evaluations of a LS-2 parabolic trough solar collector using Al_2O_3/SiO_2-Syltherm 800 hybrid nanofluid. Energy Convers Manag 245. https://doi.org/10.1016/j.enconman.2021.114596
4. Guerraiche D, Guerraiche K, Driss Z, Chibani A, Merouani S, Bougriou C (2022) Heat transfer enhancement in a receiver tube of solar collector using various materials and nanofluids. Eng Technol Appl Sci Res 12(5):9282–9294. https://doi.org/10.48084/etasr.5214
5. Khan MS et al (2020) Comparative performance assessment of different absorber tube geometries for parabolic trough solar collector using nanofluid. J Therm Anal Calorim 142(6):2227–2241. https://doi.org/10.1007/s10973-020-09590-2
6. Amina B, Miloud A, Samir L, Abdelylah B, Solano JP (2016) Heat transfer enhancement in a parabolic trough solar receiver using longitudinal fins and nanofluids. J Therm Sci 25(5):410–417. https://doi.org/10.1007/s11630-016-0878-3
7. Patankar SV, Spalding DB (1972) A calculation procedure for heat, mass and momentum transfer in three-dimensional parabolic flows. Pergamon Press
8. Delgado-Torres AM, García-Rodríguez L (2007) Comparison of solar technologies for driving a desalination system by means of an organic Rankine cycle. Desalination 216(1–3):276–291. https://doi.org/10.1016/j.desal.2006.12.013
9. Pak BC, Cho YI (1998) Hydrodynamic and heat transfer study of dispersed fluids with submicron metallic oxide particles. Exper Heat Transfer 11(2):151–170. https://doi.org/10.1080/08916159808946559
10. Xuan Y, Roetzel W (2000) Conceptions for heat transfer correlation of nanofluids. Int J Heat Mass Transf [online]. Available www.elsevier.com/locate/ijhmt
11. Brinkman HC (1952) The viscosity of concentrated suspensions and solutions. J Chem Phys 20(4):571. https://doi.org/10.1063/1.1700493
12. Hamilton RL, Crosser OK (1962) Thermal conductivity of heteogeneous two-component systems. Ind Eng Chem Fundam
13. Webb RL (1981) Performance evaluation criteria for use design. Int J Heat Transf 24:715–726

Fe_2HfZ (Z = Si, Ge, Sn), Promising New Candidate Materials for Electronic and Thermoelectric Applications

Djennane Khaoula, Dehbaoui Mourad, Dehimi Nour Elhouda, and Meliani Kawther

Abstract Full-Heusler materials have attracted intensive attention of scientific research in recent years due to their physical properties, the materials are considered potential candidates for high performance thermoelectric applications. Structural, elastic, and electronic properties of full-Heusler Fe_2HfZ (Z = Si, Ge, Sn) compounds have been studied using a pseudopotential plane-wave method within the framework of density functional theory, the exchange correlation effects have been treated using the generalized gradient approximation GGA-PBEsol as implemented in the CASTEP code. The calculated results for lattice constant, bulk modulus (B) and its pressure derivative $B_0\prime$ are in good agreement with the theoretical data. We have calculated the elastic constants and checked the mechanical stability of the Fe_2HfZ (Z = Si, Ge, Sn) materials. Fe_2HfZ (Z = Si, Ge, Sn) are found to be mechanically stable. Independent single-crystal elastic properties like bulk modulus, shear modulus, Young's modulus and universal index for the corresponding polycrystalline phase indicate that Fe_2HfZ (Z = Si, Ge) are brittle and Fe_2HfSn is ductile. The values of Young's modulus demonstrate that, among the investigated materials, Fe_2HfSi is the stiffer one. The universal anisotropy index differs significantly from zero, indicating that our materials are elastically anisotropic. We also calculated electronic band structures, total and partial electronic density of states. According to our calculations, Fe_2HfZ (Z = Si, Ge, Sn) compounds are not spin polarized. While Fe_2HfSn have a metallic behavior, Fe_2HfSi and Fe_2HfGe exhibit a semiconductor behavior with indirect band gaps of 0.325 eV and 0.097 eV respectively. The partial densities of states show that the valence bands due to Fe-d orbitals and the conduction band due to Fe-d and Hf-d orbitals.

Keywords Full-Heusler compounds · Density functional theory · GGA-PBEsol · Structural properties · Elastic properties · Electronic properties

D. Khaoula (✉) · D. Mourad · D. N. Elhouda · M. Kawther
Laboratoire de Physique des Techniques Experimentales et ses Applications, University of Médéa, 26000 Médéa, Algeria
e-mail: djennane.khaoula@univ-medea.dz

D. Mourad
e-mail: mourad.dehbaoui@gmail.com

S. Hamamda et al. (eds.), *Proceedings of the 2nd International Conference of Nanotechnology for Environmental Protection and Clean Energy Production*, Springer Proceedings in Materials 45, https://doi.org/10.1007/978-981-97-1916-7_27

1 Introduction

The increasing importance of sustainable and clean energy sources can be attributed to the serious energy crisis and environmental pollution. The science and industry communities have directed their attention towards thermoelectric materials for their capability to convert waste heat into electrical energy. The high stability and reliability of thermoelectric materials have made them a popular choice [1]. Full-Heusler materials have become a significant area of research due to their physical properties, which makes them potential candidates for high-performance thermoelectric materials [2, 3], this class of materials offers low band gaps, high electrical conductivity, low thermal conductivity and large thermopower [4]. The full-Heusler materials with a narrow band gap are promising for a variety of applications, including thermoelectric devices [5]. Particularly Fe-based alloys and many others [4, 6]. Full-Heusler materials typically consisting of X_2YZ formula, with X and Y being transition metals and Z being a p-block element (space group Fm3m; No: 225) [5]. Within the full-Heusler family investigating the electronic, elastic and structural properties of Fe_2HfZ (Z = Si, Ge, Sn) full-Heusler compounds is the main objective of this paper.

2 Computational Details

The calculations were conducted using the Cambridge Serial Total Energy (CASTEP) Package based on the density functional theory (DFT) [7, 8]. The electron–electron interaction considered within the generalized gradient approximation GGA-PBE [9]. A 12 × 12 × 12 Monkhorst and Pack grid of k-points is used for integration in the irreducible Brillouin zone [10]. The plane wave basis set cut-off is set as 500 eV. where the convergence criteria on the total energy was set to 5×10^{-6} eV/atom. The maximum force and displacement were set to be less than 0.01 eV/Å and 5×10^{-4} Å respectively, and a maximum stress within 0.002 GPa. The stress–strain method approach [11] is used to calculate the single-crystal elastic constant Cij from the first principles calculation. The Voigt–Reuss–Hill (V–R–H) approximations were used to estimate the elastic moduli.

3 Results and Discussion

3.1 Structural Properties

The Fe_2HfZ (Z = Si, Ge, Sn) full-Heusler material crystallize in the space group Fm3m with Fe, Hf and Z atoms at 4c (1/4, 1/4, 1/4), 4b (1/2, 1/2, 1/2) and 4a (0, 0, 0) Wyckoff positions [5].

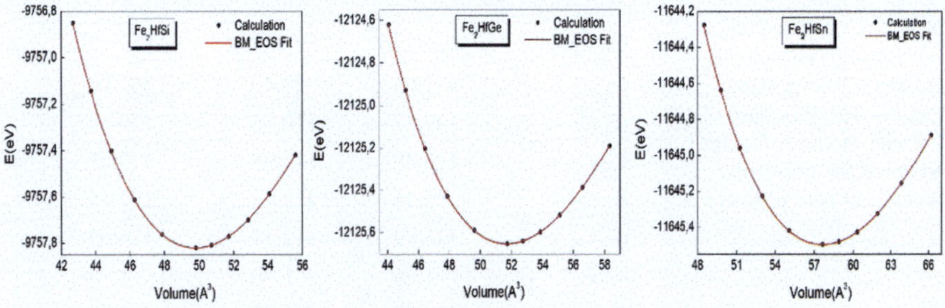

Fig. 1 Energy–volume curve for Fe_2HfZ (Si, Ge, Sn) full-Heusler compounds

Table 1 Calculated equilibrium lattice parameter (a), bulk modulus (B) and its pressure derivative for Fe_2HfZ (Z = Si, Ge, Sn) full-Heusler compounds

Compound	*a (Å)*	B_0 *(GPa)*	B_0'	References
Fe_2HfSi	5.84	230.95	4.36	This work
	5.92	213.5	4.18	[13]
Fe_2HfGe	5.1	217.72	4.47	This work
Fe_2HfSn	6.13	196.92	4.46	This work

A common approach to determine structural properties such as lattice constant (a), bulk modulus (B) and the pressure derivative of bulk modulus B_0’ near equilibrium is to use the GGA-PBE scheme to calculate the total energy volume at different values and simulate the results combined into a semi-empirical equation of state.

The E(V) curve shown in Fig. 1 was fitted using the following Birch Murnaghan equation of state [12]:

$$E(V) = E_0 + \frac{9B_0V}{16}\left(\left[\left(\frac{V_0}{V}\right)^{2/3} - 1\right]^2\left(6 - B_0^{'}\left[\left(\frac{V_0}{V}\right)^{2/3} - 1\right]\right) - 4\left(\frac{V_0}{V}\right)^{2/3}\right) \tag{1}$$

Table 1 summarizes the structural parameter results of the Fe_2HfZ (Z = Si, Ge, Sn) full-Heusler compounds. These results are in good agreement with the theoretical data [13].

3.2 *Elastic Properties*

We calculated the elastic constant Cij using the stress strain method [14], to determine the mechanical behavior of the Fe_2HfZ (Z = Si, Ge, Sn). The cubic structure has 3 independent elastic constants, namely C_{11}, C_{12}, and C_{44} [15]. To ensure the

Table 2 Calculated elastic constants *Cij* (GPa), bulk modulus *B* (GPa), shear modulus *G* (GPa), ratio of *G/B*, Young modulus *E* (GPa) and universal anisotropy index A^U at zero pressure

Compound	Fe_2HfSi	Fe_2HfGe	Fe_2HfSn
C_{11}	437.65	388.40	352.09
C_{12}	149.37	110.86	119.68
C_{44}	112.39	143.95	117.41
B	245.46	225.43	197.15
G	125.09	115.41	116.93
B/G	1.96	1.90	1.68
E	320.77	295.75	292.88
A^U	0.07	0.01144	0.00013

mechanical stability of cubic crystals, the elastic constants must satisfy the Born stability criteria [16, 17]:

$$C_{11} > 0; C_{44} > 0; C_{11} > |C_{12}|; C_{11} + 2C_{12} > 0. \quad (2)$$

The calculated elastic constants satisfy these constraints as shows in Table 2, which means that the studied materials are mechanically stable. Since C_{11} and C_{12} are larger than C_{44} and C_{11} is larger than C_{12} the hardness in the direction of the *x* axis is larger than that in the direction of the *y*- and *z*-axes. Independent single-crystal elastic properties like bulk modulus, shear modulus, Young's modulus and universal index for the corresponding polycrystalline phase have been calculated from the independent elastic constants Cij, the results are summarized in Table 2. The calculated bulk modulus values from the elastic constants Cij are very close to the estimation from Murnaghan's equation of state for these compounds.

We predicted the brittle or ductile nature of this compound based on the Pugh criteria (*B*/*G* ratio). That is, when *B*/*G* ratio is higher than 1.75, then a material shows ductile behavior otherwise the material is brittle [18]. The Pugh's ratio of Fe_2HfZ (Si, Ge, Sn) indicate that Fe_2HfSi and Fe_2HfSn are brittle and Fe_2HfSn is ductile. The Young's modulus related to the stiffness of materials [19]. The values of Young's modulus demonstrate that Fe_2HfSi is the stiffer material.

Another important parameter to estimate elastic anisotropy is the so-called universal anisotropy index [20]. For isotropic crystals, A^U equals zero, deviations from zero characterize the degree of elastic anisotropy. From the calculated A^U, it can be inferred that the anisotropy factor of the studied compounds differs significantly from zero, indicating the presence of significant elastic anisotropy in these materials.

3.3 Electronic Properties

In order to understand the electronic conduction of Fe_2HfZ (Z = Si, Ge, Sn), we calculate the energy band structures, total and partial electronic densities of states.

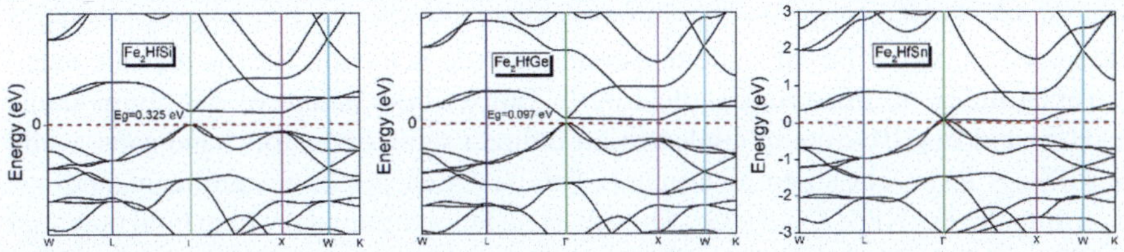

Fig. 2 Electronic band structure of Fe_2HfZ (Si, Ge, Sn) full-Heusler compounds

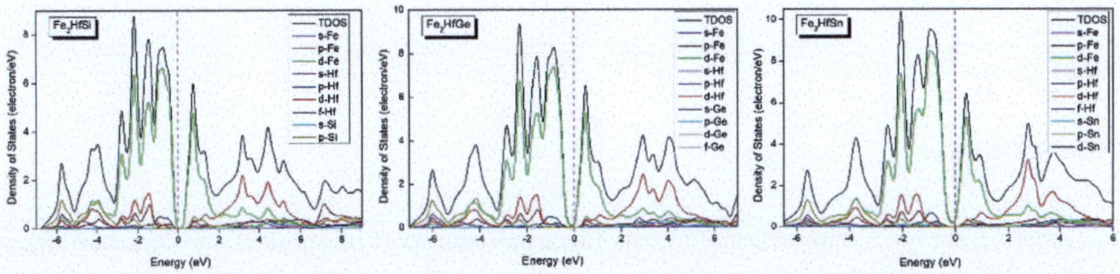

Fig. 3 Total and partial density of state of Fe_2HfZ (Si, Ge, Sn) full-Heusler compounds

The band structures of the studied compounds are shown in Fig. 2. According to our calculations, Fe_2HfZ (Z = Si, Ge, Sn) compounds are not spin polarized. Fe_2HfSi and Fe_2HfGe exhibit a semiconductor behavior, in which the valence band maximum is at Γ point and the conduction band minimum is at X point, with indirect band gaps of 0.325 eV, 0.097 eV for Fe_2HfSi and Fe_2HfGe, respectively. While Fe_2HfSn have metallic behavior, where the valance band and the conduction band are overlapping. It is important to note that our results show reasonable agreement with other theoretical data. A. Azam et al., for example, have found an indirect energy band gap of 0.317 eV for Fe_2HfSi through the use GGA-PBE [13].

Figure 3 shows the total and partial densities of states of the Fe_2HfZ (Z = Si, Ge, Sn) full-Heusler compounds. We can note that the valence bands are mainly due to Fe-d orbitals. In the conduction band above the Fermi energy level, the primary contribution comes from Fe-d and Hf-d orbitals. The obtained result can be merged with some techniques in control theory [21–25].

4 Conclusion

Based on DFT calculations, the plane-wave pseudopotential method was utilized as implemented in the CASTEP code to investigate the structural, electronic, and elastic properties of Fe_2HfZ (Z = Si, Ge, Sn) full-Heusler materials. The structural properties calculations successfully determined the optimized lattice constant and bulk modulus, which showed excellent correlation with existing theoretical data. Notably, the calculated elastic constants confirmed the mechanical stability of our compounds. Furthermore, the Paugh ratio values revealed the brittleness of Fe_2HfSi and Fe_2HfSn, while Fe_2HfSn demonstrated ductile behavior. Lastly, the calculated universal index of anisotropy indicating the anisotropic nature of our compounds. Fe_2HfSi and Fe_2HfGe compounds exhibit indirect semiconductors band gaps, while Fe_2HfSn is to be classified as a metal. This research presents significant findings regarding the Fe_2HfZ (Z = Si, Ge, Sn) full-Heusler compounds, which hold potential for thermoelectric applications. The narrow band gaps of Fe_2HfSi and Fe_2HfGe make them particularly attractive choices for such applications. The Fe_2HfSn full-Heusler compound, with their metallic behavior, can be utilized as electrical contact in thermoelectric devices.

References

1. Cutler M, Leavy J, Fitzpatrick R (1964) Electronic transport in semimetallic cerium sulfide. Phys Rev 133(4A):1143
2. Cha P-R et al (2019) The n-and p-type thermoelectric response of a semiconducting co-based quaternary Heusler alloy: a density functional approach. J Mater Chem C 7(25):7664–7671
3. Berche A, Noutack MT, Doublet M-L, Jund P (2020) Unexpected band gap increase in the Fe_2VAl Heusler compound. Mater Today Phys 13:100203
4. Bilc DI, Hautier G, Waroquiers D, Rignanese G-M, Ghosez P (2015) Low-dimensional transport and large thermoelectric power factors in bulk semiconductors by band engineering of highly directional electronic states. Phys Rev Lett 114(13):136601
5. Graf T, Felser C, Parkin SS (2011) Simple rules for the understanding of Heusler compounds. Prog Solid State Chem 39(1):1–50
6. Jong J-Y, Zhu J, Jon M-G, Zhou Y, Kim J, Yan J (2017) Theoretical investigation of stabilities and physical properties of low cost Fe-based full-Heusler materials. J Alloy Compd 693:462–467
7. Clark S, Segall M, Pickard C, Hasnip P, Probert M, Refson K (2005) Mc payne zeitschrift fur kristallographie-cryst. Mater 220:567
8. Hernandez-Haro N, Ortega-Castro J, Martynov YB, Nazmitdinov RG, Frontera A (2019) DFT prediction of band gap in organic-inorganic metal halide perovskites: an exchange-correlation functional benchmark study. Chem Phys 516:225–231
9. Perdew JP, Burke K, Ernzerhof M (1996) Generalized gradient approximation made simple. Phys Rev Lett 77(18):3865
10. Monkhorst HJ, Pack JD (1976) Special points for Brillouin-zone integrations. Phys Rev B 13(12):5188
11. Milman V, Warren M (2001) Elasticity of hexagonal BeO. J Phys Condens Matter 13(2):241
12. Sofi SA, Yousuf S, Gupta DC (2019) Prediction of robustness of electronic, magnetic and thermoelectric properties under pressure and temperature variation in Co_2MnAs alloy. Comput Condensed Matter 19:e00375
13. Azam A, Sharma R, Behera D, Raza HH, Ali HS, Abdelmohsen SA, Abdelbacki AM, Mukherjee SK (2023) Insight into the structural, optoelectronic, and thermoelectric properties of Fe_2HfSi Heusler by DFT investigation. RSC Adv 13(23):15437–15447
14. Mehl MJ (1993) Pressure dependence of the elastic moduli in aluminum rich Al-Li compounds. Phys Rev B 47(5):2493
15. Kittel C (2005) Introduction to solid state physics. Wiley
16. SinKo G, Smirnov N (2002) Ab initio calculations of elastic constants and thermodynamic properties of bcc, fcc, and hcp al crystals under pressure. J Phys Condens Matter 14(29):6989

17. Wu S-C, Fecher GH, Shahab Naghavi S, Felser C (2019) Elastic properties and stability of Heusler compounds: Cubic Co_2YZ compounds with $L2_1$ structure. J Appl Phys 125(8):082523
18. Pugh S (1954) XCII. Relations between the elastic moduli and the plastic properties of polycrystalline pure metals. Lond Edinburgh Dublin Philos Mag J Sci 45(367):823–843
19. Balluffi RW (2016) Introduction to elasticity theory for crystal defects. World Scientific Publishing Company
20. Ranganathan SI, Ostoja-Starzewski M (2008) Universal elastic anisotropy index. Phys Rev Lett 101(5):055504
21. Zahaf A, Bououden S, Chemachema M. Contribution to robust fault-tolerant predictive control with constraints for hybrid systems (Doctoral dissertation, Université Frères Mentouri-Constantine 1)
22. Zahaf A, Bououden S, Chadli M, Boulkaibet I, Neji B, Khezami N (2023) Dynamic sensorless control approach for Markovian switching systems applied to PWM DC–DC converters with time-delay and partial input saturation. Sensors 23(15):6936
23. Abdelmalek Z, Bououden S, Chadli M, Zelinka I, Boulkaibet I (2019) Observer based model predictive control of hybrid systems. In: Advanced control engineering methods in electrical engineering systems 2019. Springer International Publishing, pp 198–207
24. Abdelhamid B, Chemachema M, Bouzina S (2023) Fuzzy fault-tolerant control applied on two inverted pendulums with nonaffine nonlinear actuator failures. Int J Robot Control Syst 3(2)
25. Bounemeur A, Chemachema M (2023) General fuzzy adaptive fault-tolerant control based on Nussbaum-type function with additive and multiplicative sensor and state-dependent actuator faults. Fuzzy Sets Syst 7:108616

First-Principles Calculations of Structural, Elastic and Electronic Properties of Cubic $CsBeI_3$ Compound

Amel Meziani

Abstract The structural, elastic, and electronic properties of the cubic $CsBeI_3$ compound have been calculated using the full potential linearized augmented plane wave (FP-LAPW) method based on density functional theory (DFT). The exchange–correlation potential for structural and elastic properties has been evaluated using the generalized gradient approximation (GGA-PBE), while the modified Becke-Johnson potential (mBJ-GGA) has been employed in the calculation of the band structure. The structural parameters (lattice constant and bulk modulus) were found to be in reasonable agreement with the available data. Additionally, the elastic constants (C_{11}, C_{12}, and C_{44}), elastic moduli and Debye temperature have been studied for the first time. We have predicted the electronic band structure and density of states, and $CsBeI_3$ exhibits semiconductor behavior with an indirect band gap of 0.4628 eV. The upper valence bands are essentially dominated by I-p states. The estimated elastic parameters of this material indicate that it is anisotropic, ductile in nature, and mechanically stable.

Keywords DFT · FP-LAPW · Electronic properties · Elastic properties · $CsBeI_3$

1 Introduction

In recent years, halide-based peroveskites with the general formula ABX_3 (where A represents alkali, B represents metal, and X represents halide ions) have been the focus of many studies due to their physical properties and potential applications in optoelectronics and photovoltaics. The ideal perovskite structure is simple cubic and belongs to space group $Pm\overline{3}m$ (No 221), with A located at the corners of the cube, B at the body center, and X at the centers of the faces. Only a few studies have been conducted on the electronic band structure of ABI_3 type perovskite. However,

A. Meziani (✉)
Département des Sciences de la Matière, Faculté des Sciences Exactes et des Sciences de la Nature et de la Vie, Université d'Oum El Bouaghi, Oum El Bouaghi, Algeria
e-mail: mezianiphys@gmail.com

S. Hamamda et al. (eds.), *Proceedings of the 2nd International Conference of Nanotechnology for Environmental Protection and Clean Energy Production*, Springer Proceedings in Materials 45, https://doi.org/10.1007/978-981-97-1916-7_28

there are no reports on the electronic and elastic properties from experimental or theoretical studies on the electronic band structure and elastic properties of cubic $CsBeI_3$. Therefore, the objective of the present work is to investigate the structural, elastic and electronic properties of this compound using the FP-LAPW approach.

2 Method of Calculation

In the cubic halide perovskite $CsBeI_3$, the atomic positions of the atoms are as follows: Cs: (0,0,0)a, Be: (1/2,1/2,1/2)a and I: (0,1/2,1/2)a, (1/2,0,1/2)a, and (1/2,1/2,0)a. The crystal structure of the $CsBeI_3$ is depicted in Fig. 1.

The calculations were performed using the FP-LAPW method [1] implemented in WIEN2k code [2]. The exchange–correlation potential was evaluated using the generalized gradient approximation (GGA-PBE) [3]. The modified Becke–Johnson potential (mBJ-GGA) potential [4] was used in the calculation of the band structure. In our work, the number of plane waves was limited to $R_{mt} \times K_{max} = 9$ (where R represents the smallest of the muffin-tin sphere radii and $K_{\max}$ is the maximum value of the wave vector K), and the expansion of the wave functions was set to $l_{max} = 10$ inside the muffin-tin spheres. The sphere radii values are 2.5 Bohr for Cs and I and 2.08 Bohr for Be. The K integration over the full Brillouin zone was performed using a $7 \times 7 \times 7$ k-points mesh.

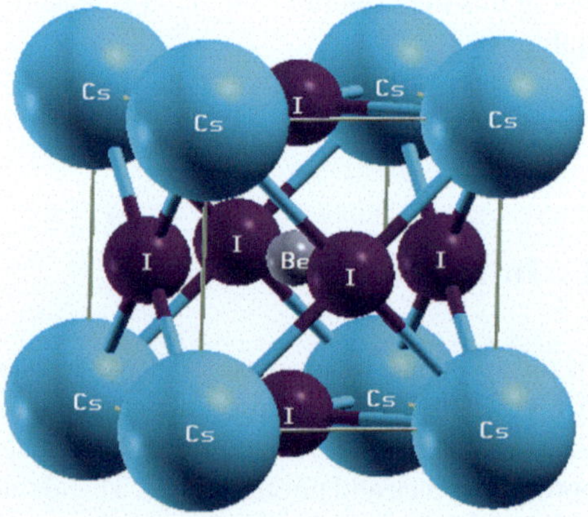

Fig. 1 Crystal structure of $CsBeI_3$ obtained with XcrysDen

3 Results

3.1 *Structural and Elastic Properties*

The equilibrium structural parameters of this compound were determined by fitting the total energy as a function of volume to Murnaghan's equation of state [5]. Table 1 presents the calculated structural parameters (lattice constant a_0, bulk modulus B and its pressure derivative B') obtained from the GGA-PBE, along with the corresponding theoretical values.

The lattice constant a_0 and the bulk modulus B were estimated to be 5.6103 Å and 23.1877 GPa, respectively. We found that our results are in good agreement with the available theoretical results. To our knowledge, there are no experimental results for the structural parameters of this compound.

In this work, we have also determined elastic properties such as the elastic constants C_{ij} and other elastic moduli using the IRelast package [7]. The computed elastic constants C_{ij} have been presented in Table 2. It is noteworthy that all the constants are positive and satisfy the mechanical stability conditions for cubic crystals [8]: $(C_{11} - C_{12}) > 0$; $(C_{11} + 2C_{12}) > 0$; $C_{11} > 0$; $C_{44} > 0$, and the bulk modulus B also should satisfy a criterion: $C_{12} < B < C_{11}$.

Furthermore, Table 2 includes the estimation of other elastic parameters, such as the bulk (B) and shear modulus (G), Poisson's ratio (ν), Young's modulus (E) and Zener anisotropy factor (A). These parameters have been calculated using Voigt-Reuss-Hill approximations [9–11]:

$$B = \frac{1}{3}(C_{11} + 2C_{12}) \tag{1}$$

$$G = \frac{1}{2}(G_V + G_R) \tag{2}$$

Table 1 Calculated structural equilibrium lattice constant a_0 (in Å), bulk modulus B (in GPa) and its pressure derivative B' for $CsBeI_3$

	a_0	B	B'
This work GGA-PBE	5.6103	23.1877	4.8473
Theo [6]	5.555	18.844	–

Table 2 Elastic constants (C_{11}, C_{12} and C_{44}), bulk modulus B, shear modulus G (in GPa), Poisson's ratio ν, Young's modulus E (in GPa), Zener anisotropy factor A and B/G ratio for $CsBeI_3$

C_{11}	C_{12}	C_{44}	B	G	ν	E	A	B/G
24.15	20.39	18.28	21.64	7.89	0.33	21.11	9.72	2.74

$$G_V = \frac{1}{5}(C_{11} - C_{12} + 3C_{44}) \tag{3}$$

$$G_R = \frac{5C_{44}(C_{11} - C_{12})}{4C_{44} + 3(C_{11} - C_{12})} \tag{4}$$

$$\nu = \frac{3B - 2G}{2(3B + G)} \tag{5}$$

$$E = \frac{9BG}{3B + G} \tag{6}$$

$$A = \frac{2C_{44}}{C_{11} - C_{12}} \tag{7}$$

The elastic anisotropy factor is equal to one for an isotropic material, while any value smaller or larger than one indicates anisotropy. From Table 2, we can see that the value of the anisotropy factor A is 9.72. This indicates that our compound is anisotropic. The value of the *B/G* ratio is 2.74, which is larger than the critical value (1.75) separating ductile and brittle materials [12]. This suggests that $CsBeI_3$ belongs to ductile materials.

The Debye temperature θ_D can be derived from the calculated elastic constants using the following equation [13]:

$$\theta_D = \frac{h}{k}\left[\frac{3n}{4\pi}\left(\frac{N_A\rho}{M}\right)\right]^{1/3} \upsilon_m \tag{8}$$

where h is Plank's constant, k is Boltzmann's constant, N_A is Avogadro's number, ρ is the density, M is the molecular weight, n is the number of atoms in the molecule, and υ_m is the average sound velocity, which is approximately given by:

$$\upsilon_m = \left[\frac{1}{3}\left(\frac{2}{\upsilon_t^3} + \frac{1}{\upsilon_l^3}\right)\right]^{-(1/3)} \tag{9}$$

where υ_l and υ_t are the longitudinal and transverse sound velocities, which are obtained by the following expressions [14]:

$$\upsilon_l = \left(\frac{3B + 4G}{3\rho}\right)^{1/2} \tag{10}$$

and

$$\upsilon_t = \left(\frac{G}{\rho}\right)^{1/2} \tag{11}$$

Table 3 The transverse, longitudinal and average sound velocities (v_t; v_l; v_m in m/s), Debye temperature (θ_D in K) and melting temperature (Tm in K) for $CsBeI_3$

	v_l	v_t	v_m	θ_D	Tm ± 300
This work	2558.57	1267.45	1422.61	129.092	695.7406

Table 3 gives the calculated sound velocities, Debye temperature and melting temperature for cubic $CsBeI_3$. The Debye temperature is predicted to be 129.092 K.

To our knowledge, there are no experimental or theoretical results available for the anisotropy factor, shear modulus, Young's modulus, Poisson's ratio, Debye temperature and melting temperature of $CsBeI_3$. We can consider the present results as predictions for this compound.

3.2 Electronic Properties

Figure 2 shows the calculated band structure for $CsBeI_3$ along the high symmetry lines using the mBJ-GGA approximation.

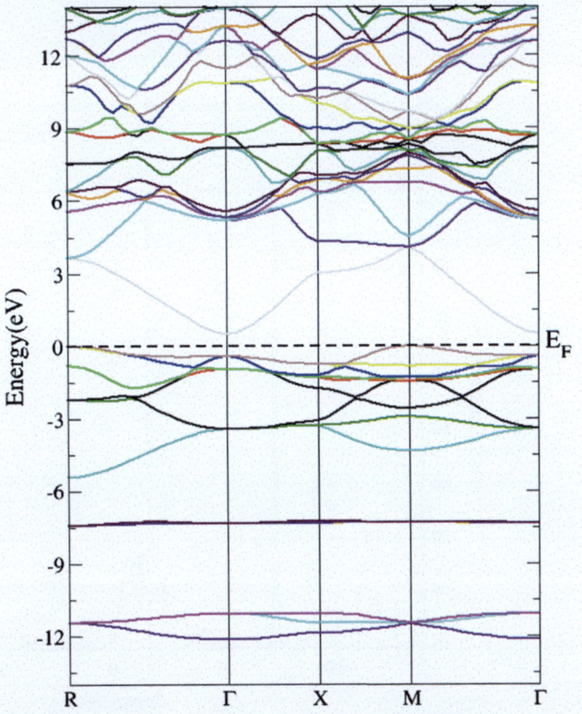

Fig. 2 The calculated band structure of $CsBeI_3$ using mBJ-GGA approximation

The valence band maximum occurs at the R point, whereas the conduction band minimum occurs at the Γ point. Therefore, $CsBeI_3$ has an indirect band gap (R-Γ) of 0.4628 eV. Unfortunately, to the best of our knowledge, there are no experimental or theoretical band gap studies to compare our results with. The obtained result can be merged with some techniques in control theory [15–19].

The calculated total and partial densities of states (DOS) are presented in Figs. 3 and 4 using the mBJ-GGA approximation. At lower energy, the structure situated at around − 11.40 eV, below the Fermi level, reflects the I-s states. The peak centered at about − 7.29 eV originates from the Cs-p states. The upper valence bands (between − 5.5 and 0 eV) are essentially formed by I-p states with a small contribution of Be-s states. The lowest conduction bands are composed of Be-s and Be-p states. The conduction bands from 5 to 14 eV are essentially dominated by Cs-d states, with a small contribution of I-p and Be-p states.

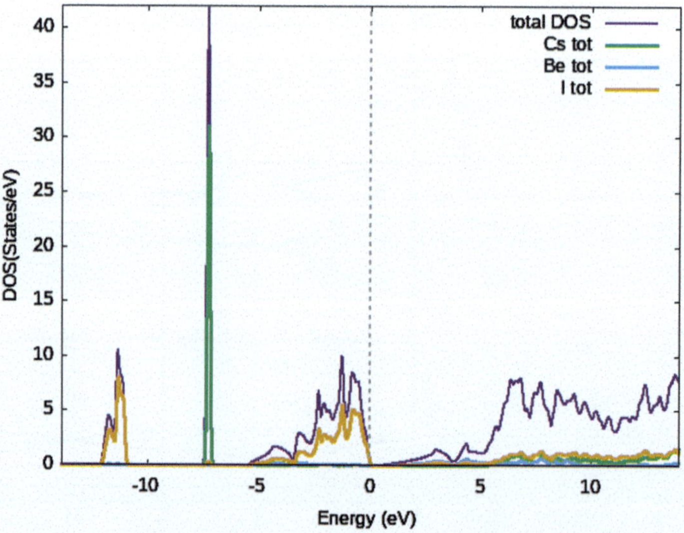

Fig. 3 The total density of states (DOS) of $CsBeI_3$ using mBJ-GGA approximation

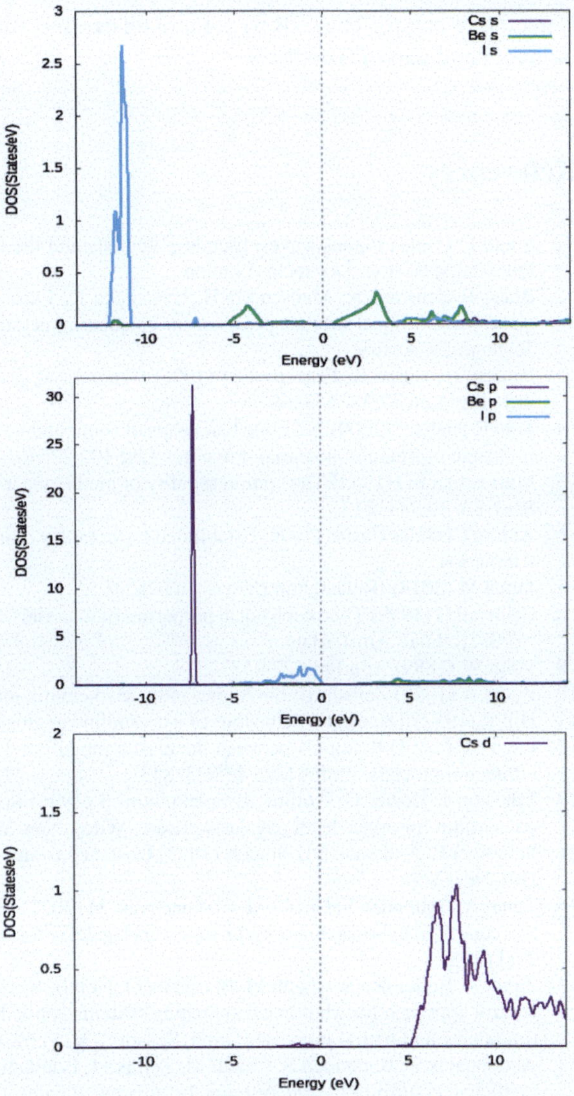

Fig. 4 The partial densities of states of $CsBeI_3$ using mBJ-GGA approximation

4 Conclusion

We have examined the structural, elastic and electronic properties of the $CsBeI_3$ compound using the FP-LAPW method. The structural parameters are in good agreement with the theoretical results. The electronic and elastic properties have been calculated for the first time in this study. The elastic constants were found to meet the criteria for the stability of the cubic crystal. The B/G ratio indicates that $CsBeI_3$

is ductile in nature. The $CsBeI_3$ compound exhibits semiconductor behavior with an indirect band gap of 0.4628 eV.

References

1. Singh D (1994) Planes waves, pseudo-potentials and the LAPW method. Kluwer Academic Publishers, Boston, Dortrecht, London
2. Blaha P, Schwarz K, Madsen GKH, Hvasnicka D, Luitz J (2001) WIEN2k, an augmented plane wave + local orbitals program for calculating crystal properties. Vienna University of Technology, Austria
3. Perdew JP, Burke K, Ernzerhof M (1996) Generalized gradient approximation made simple. Phys Rev Lett 77:3865–3868
4. Tran F, Blaha P (2009) Accurate band gaps of semiconductors and insulators with a semi local exchange-correlation potential. Phys Rev Lett 102:226401
5. Murnaghan FD (1944) The compressibility of media under extreme pressures. Proc Natl Acad Sci USA 30:244–247
6. Aurora Costales Castro (1996) Estabilidad estructural y quimica de halurocompuestos binarios y ternarios
7. Jamal M (2014) IRelast, http://www.wien2k.at/
8. Grimvall G (1999) Thermophysical properties of materials. Enlarged and revised edn. Elsevier/North-Holland, Amsterdam
9. Voigt W (1889) Ann Phys 274:573–587
10. Reuss A (1929) Zeitschrift fur Angewandte Mathematik und Mechanik. 9:49–58
11. Hill R (1952) The elastic behaviour of a crystalline aggregate. Proc Phys Soc A 65:349–354
12. Pugh SF (1954) Relations between the elastic moduli and the plastic properties of polycrystalline pure metals. Philos Mag 45:823–843
13. Johnston I, Keeler G, Rollins R, Spicklemire S (1996) Solid state physics simulations. The consortium for upper-level physics software. Wiley, New York
14. Schreiber E, Anderson OL, Soga N (1973) Elastic constants and their measurements. McGraw-Hill, New York
15. Zahaf A, Bououden S, Chadli M, Chemachema M (2022) Robust fault tolerant optimal predictive control of hybrid actuators with time-varying delay for industrial robot arm. Asian J Control 24(1):1–5
16. Zahaf A, Bououden S, Chadli M, Boulkaibet I, Neji B, Khezami N (2023) Dynamic sensorless control approach for Markovian switching systems applied to PWM DC–DC converters with time-delay and partial input saturation. Sensors 23(15):6936
17. Abdelmalek Z, Bououden S, Chadli M, Zelinka I, Boulkaibet I (2019) Observer based model predictive control of hybrid systems. In: Advanced control engineering methods in electrical engineering systems 2019. Springer International Publishing, pp 198–207
18. Abdelhamid B, Chemachema M, Bouzina S (2023) Fuzzy fault-tolerant control applied on two inverted pendulums with nonaffine nonlinear actuator failures. Int J Robot Control Syst 3(2)
19. Bounemeur A, Chemachema M (2023) General fuzzy adaptive fault-tolerant control based on Nussbaum-type function with additive and multiplicative sensor and state-dependent actuator faults. Fuzzy Sets Syst 7:108616

Impact of High Temperature and Germanium Rate on the Electrical Performance of the SiGe Heterojunction Bipolar Transistor with DPSA-SEG Architecture

Chems El Ghizlane Lachkhab, Maya Lakhdara, Abdelaaziz Boulgheb, and Saida Latreche

Abstract The aim of this work is to model and analyze the electrical characteristics of a SiGe Heterojunction Bipolar Transistor HBT integrated in BiCMOS055 technology with DPSA-SEG (Double Poly silicon, Self-Aligned, Selective Epitaxial Growth) architecture, using COMSOL Multiphysic software. The goal is to analyze the impact of high temperatures on component characteristics in the range from 300 to 450 K. At high temperatures, electrical performances decrease considerably, the maximum oscillation frequency/transition frequency drops from 397/334 to 343/268 GHz when the temperature ranges from 300 to 450 K. Secondly, we optimized the impact of the Germanium fraction rate introduced into the SiGe base of the HBT on electrical properties. For Germanium concentration less than 20%, static and dynamic performances improved but these performances decreased when germanium concentration is over 20%.

Keywords SiGe · HBT · BiCMOS55nm · DPSA-SEG · COMSOL · High temperature · Germanium fraction rate

1 Introduction

The bipolar transistor was the first active semiconductor component and it had a significant impact on the electronics industry. Despite this, the homojunction bipolar transistor has been supplanted by a new type of transistor the Heterojunction Bipolar Transistor HBT due to the technological limitations of its frequency increase. Today, these components provide fast responses at low cost, a reduction in energy consumption and a reduction in noise compared to standard Silicon technology [1, 2].

C. E. G. Lachkhab (✉) · M. Lakhdara · A. Boulgheb · S. Latreche
Laboratoire Hyperfréquences et Semi-Conducteurs LHS, Université Mentouri Constantine 1, Constantine, Algeria
e-mail: chemselghizlane.lachkhab@doc.umc.edu.dz

S. Hamamda et al. (eds.), *Proceedings of the 2nd International Conference of Nanotechnology for Environmental Protection and Clean Energy Production*, Springer Proceedings in Materials 45, https://doi.org/10.1007/978-981-97-1916-7_29

The decrease in component size has resulted in frequencies of 300 THz for terahertz active imaging applications using innovations for high performance technology (IHP) [3, 4]. This scenario permits us to attain the terahertz frequencies while also achieving associated applications that are able to be mass-produced at a moderate cost.

The Heterojuction Bipolar Transistor (HBT) has the advantage of being implemented in a BiCMOS technology, since it employs both bipolar and MOS technology [5, 6].

This technology combines the benefits of both types of components, high operating frequencies and low noise for the bipolar technology, high integration density and low power consumption for MOS technology. Indeed, BiCMOS technology satisfies the present demands for speed and high integration density in circuit fabrication while enabling greater transition frequencies of up to 300 GHz [7, 8].

The aim of this work is to simulate the impact of temperature on the electrical performance of a SiGe Heterojunction Bipolar Transistor HBT integrated in BiCMOS55 nm technology. In addition, we will optimize the germanium profile in the SiGe alloy, which is the main element in the study of the HBT's electrical properties. As the base is the active part of the HBT it largely determines the static and dynamic component's performance to highlight the choices to make to achieve transistors with ultimate performance.

2 The Considered HBT Structure

The studied structure is a SiGe Heterojunction Bipolar Transistor HBT integrated in a BiCMOS55nm, with a DPSA-SEG (Double Poly-silicon, Self-Aligned, Selective Epitaxial Growth) architecture. This structure includes a collector module consisting of a Selectively Implanted Collector (SIC), a highly doped buried layer and a collector well. The emitter–base module contains of an intrinsic monocrystalline SiGe base with a concentration of germanium of 20% and an extrinsic Polysilicon base. The polycrystalline emitter and internal spacers reduce the width of the emitter window and isolate the extrinsic base from the emitter [9–12]. Deep trench isolation DTI is used in this architecture to isolate the transistor from the rest of the plate environment, while shallow trench isolation STI separates the device core from the collector contact. This minimizes device size as well as collector/substrate capacitance [11, 13].

The drift–diffusion model (DDM), which based on the Poisson equation and continuity equations for electrons and holes is used in COMSOL software to model and analyze the semiconductor device.

Poison equation:

$$\nabla * (-e_r * V * \nabla) = q\left(P - N + N_D^+ + N_A^-\right) \tag{1}$$

Continuity equations for electrons and holes:

$$\frac{\partial \mathrm{N}}{\partial \mathrm{t}} = +\frac{1}{q}(\nabla * J_n) - U_n \tag{2}$$

$$\frac{\partial \mathrm{P}}{\partial \mathrm{t}} = -\frac{1}{q}(\nabla * J_n) - U_p \tag{3}$$

e_r: relative permittivity, N_A^- and N_D^+: concentration of dopant, P and N: carrier concentrations, q: Electric charge.

Jn and Jp are the current densities of electrons and holes which are affected by concentrations N and P, carrier mobility μ_n, μ_p and electron and quasi-Fermi hole potentials ϕn, ϕp [14, 15].

$$J_n = +qN\mu_n \frac{d\phi_n}{d_x} \tag{4}$$

$$J_p = -qP\mu_p \frac{d\phi_p}{d_x} \tag{5}$$

The energies of the valence band and the conduction band (E_v, E_c) are presented in the Eqs. (6) and (7), Where ΔE_g is the gap energy, χ_0 electronic affinity, E_g energy band discontinuity [15, 16].

$$E_c = -(V + \chi_0) \tag{6}$$

$$E_v = -\left(V + \chi_0 + E_{g0}\right) + \Delta E_g \tag{7}$$

Figure 1 illustrates the final structure of BiCMOS055, while Fig. 1a shows an image of a BiCMOS55 with DPSA-SEG structure and Fig. 1b shows the final structure with COMSOL software.

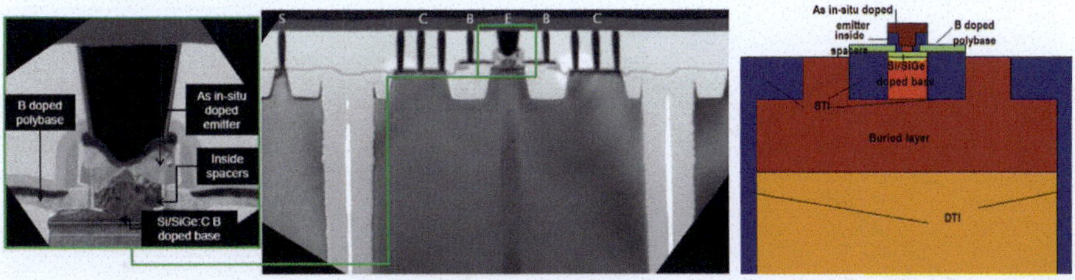

(a) (b)

Fig. 1 Final structure of BiCMOS55nm **a** in STMicroelectronics [17], **b** under COMSOL [18]

3 Results and Discussion

3.1 Effect of High Temperature on the Electrical Characteristic of the HBT

The effect of temperature especially high temperature is a fundamental aspect of accurately characterizing the electrical properties of SiGe Heterojunction Bipolar Transistor. The use of these components in microwave applications that may be exposed to different temperatures and operate at high current densities accentuate is very common.

Static characteristics at high temperatures. The evolution of base and collector currents with increasing temperature (300, 350, 400 and 450 K) is shown in Fig. 2 where curve (a) illustrates the variation of collector current I_c and curve (b) shows the variation base current I_b. An increase in collector current is distinguished when the temperature rises, this is attributed to the decrease in mobility and band gap with the increase in temperature. However I_b has a coefficient of ideality η close to 1, this allows an increase of the base current more than the collector current.

In the case of high polarization of V_{BE}, the rate of increase of the current with the temperature is reduced because of the influence of the emitter and the base's access resistors.

Figure 3a illustrates the way that current gain changes with temperature. The current gain β changes inversely with the increase of the temperature, we observe that when the temperature increases (more than 300 K), the maximum value of gain decreases and shifts towards the low polarizations of V_{BE}. The simulated I_c as a function of V_{CE} at high temperatures is shown in Fig. 3b, the output characteristic was simulated for a fixed base current of 1 μA, it can be seen that the output characteristics I_C (V_{CE}) change when the temperature rises, the maximum collector current density increases at high temperatures the heating can become very important.

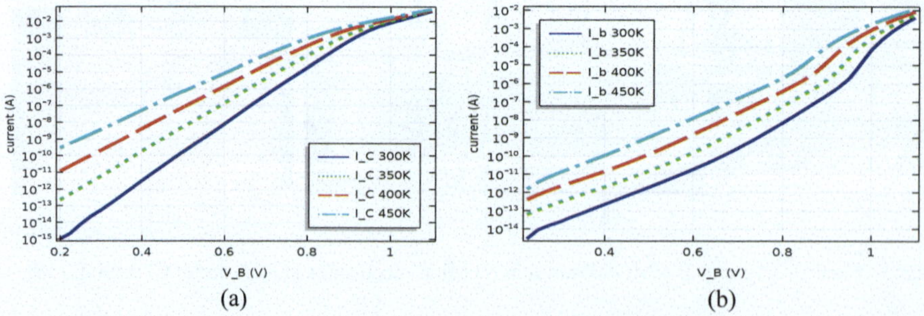

Fig. 2 Current variation against V_{BE} for T = 300, 350, 400, 450 K **a** collector current I_C, **b** base current I_B

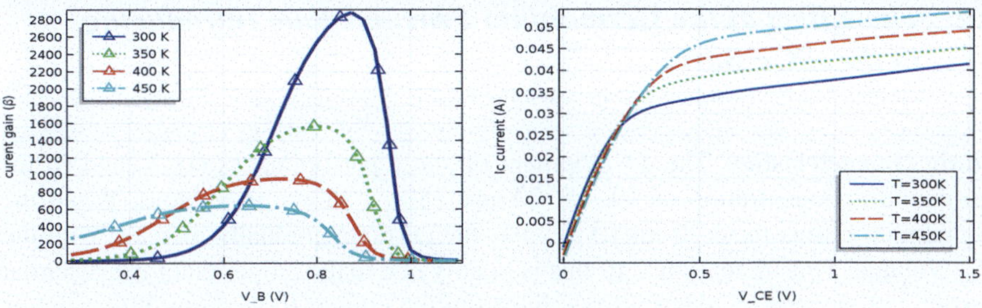

(a) (b)

Fig. 3 **a** Current gain, **b** output characteristic as a function of high temperature

Dynamic characteristics at high temperatures. The decrease at high temperatures of the static characteristics of the transistor (transconductance, access resistances) due to the decrease of the carrier's mobility and the band gap, this in turn generates a degradation of the HBT's dynamic performances. The transition frequency f_T decreases when the temperature rises by 150 K (Fig. 4a) due to a decrease in transconductance g_m that allows an increase in the loading time of the capacities C_{BE} and C_{BC}, as well as a decrease in mobility that leads to an increase in transit time τ_T, which is inversely proportional to the transition frequency.

The maximum oscillation frequency f_{max} decreases as the transition frequency f_T reduces for high temperature, as shown in (Fig. 4b). Table 1 gives a summary of the dynamic results.

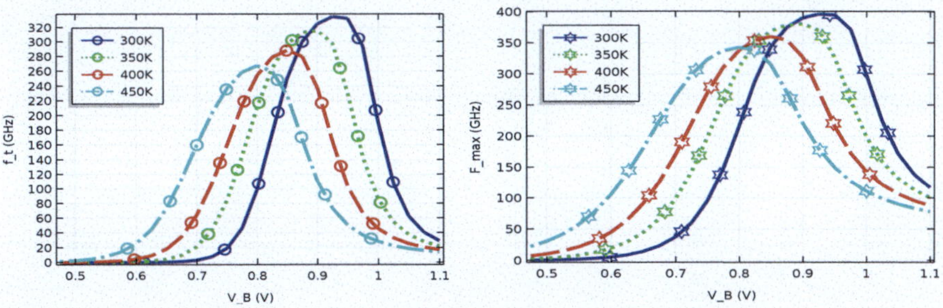

(a) (b)

Fig. 4 Cut-off frequencies as function as temperature, **a** transition frequency, **b** maximum oscillation frequency

Table 1 Frequency performances of BiCMOS55nm against the high temperature

Temperature	300 K	350 K	400 K	450 K
f_T (GHz)	334	316	288	268
F_{max} (GHz)	397	382	360	343

3.2 *The Effect of the Germanium Fraction Rate Introducing in the SiGe Base*

Static characteristics. The development of static performances (I_c, I_b and β) for germanium concentration of 10, 15, 20 and 25% is shown in Table 2. When the germanium rate is raised from 10 to 25%, the currents gradually increase with the V_{BE} voltage. This increase in currents is mainly due to the reduction of the potential barrier experienced by electrons passing from the emitter to the base, this explains the increase in the electron injection efficiency. For low concentrations (10 and 20%) the current gain increases until it reaches the maximum value (2800), this is mainly due to the increase of I_C with the germanium rate. For concentrations more than 20%, the maximum gain is smaller (2327), which is due to a considerable rise in the base current to the collector current.

Dynamic characteristic. The variation in cut-off frequencies (f_t and f_{max}) for different germanium concentration in the $Si_{1\text{-}x}Ge_x$ alloy is shown in Fig. 5a, b There is a slight increase in cut-off frequency when the germanium rate below 20% (for x_{Ge} = 20%, f_t/f_{max} = 334/397 GHz), this is mainly attributable to a small rise in collector current. In the case where the rate of germanium becomes more than 20%, there is a decrease in the cut-off frequencies (ft/f_{max} = 325/395 GHz). It is interesting to observe that when the germanium rate rises in the alloy, the physical properties of the material change and approach to those of germanium rather than silicon, thus modifying the electrical characteristics of the component studied (see Table 3). The obtained result can be merged with some techniques in control theory [19–22].

Table 2 Static performance according to the germanium rate introduced into the SiGe base

Germanium %	10%	15%	20%	25%
I_C (A)	6.23×10^{-5}	3.81×10^{-5}	1.3×10^{-5}	9×10^{-4}
I_B (A)	3.4×10^{-8}	1.77×10^{-8}	0.5×10^{-8}	4×10^{-7}
β	1833	2164	2600	2327

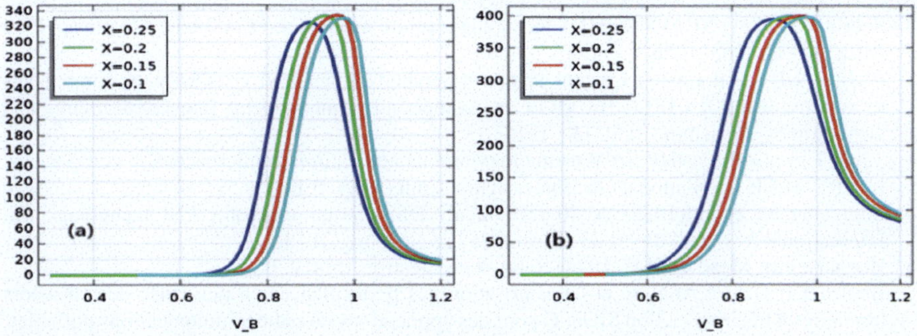

Fig. 5 Cut-off frequencies as a function of germanium fraction rate **a** transition frequency, **b** maximum oscillation frequency

Table 3 Germanium fraction rate effect on frequency performances

Germanium %	10	15	20	25
f_T (GHz)	330.65	332.37	334	325
f_{max} (GHz)	396.3	396.8	397	395

4 Conclusion

The electrical simulations of a SiGe Heterojunction Bipolar Transistor HBT integrated in BiCMOS55nm with DPSA-SEG (Double-Poly silicon, Self-Aligned, Selective Epitaxial Growth) architecture are the subject of this work. In order to examine the electrical features of HBT at high temperatures (T = 300, 350, 400, 450 K), we found that the HBT's electrical performances reduced significantly at high temperatures, when the temperature is 300 K, the cut-off frequencies (ft, fmax) drop from (334 GHz, 397 GHz) to (268 GHz, 343 GHz) respectively when the temperature rises to 450 K.

However, we have also simulated the impact of the germanium profile (10, 15, 20 and 25%) in the SiGe alloy on the HBT's electrical characteristics. As the germanium concentration of the $Si_{1-x}Ge_x$ alloy increases the physical properties of the material change approaching those of germanium and moving away from those of silicon.

We would like to point out that for germanium rate less than 20%, static and dynamic performance improves but these performance deteriorates for germanium concentration over 20%.

References

1. Chevalier P et al (2017) Si/SiGe: C and InP/GaAsSb heterojunction bipolar transistors for THz applications. Proc IEEE 105(6):1035–1050

2. Schröter M et al (2016) SiGe HBT technology: future trends and TCAD-based roadmap. Proc IEEE 105(6):1068–1086
3. Pfeiffer UR et al (2018) Current status of terahertz integrated circuits-from components to systems. In: 2018 IEEE BiCMOS and compound semiconductor integrated circuits and technology symposium (BCICTS). IEEE
4. Takai T et al (2017) IHP SAW technology and its application to microacoustic components. In: 2017 IEEE international ultrasonics symposium (IUS). IEEE
5. Sugiyama M et al (1989) A 40 GHz f/sub T/Si bipolar transistor LSI technology. In: International technical digest on electron devices meeting. IEEE
6. Hoeneisen B, Mead CJSSE (1972) Solid State Electron 15:819
7. Blanchet F (2007) Analyse et caractérisation des performances en puissance de transistors bipolaires à hétéro-jonction SiGe: C pour des applications de radiocommunications portables. Limoges
8. Jiang C et al (2016) A fully integrated 320 GHz coherent imaging transceiver in 130 nm SiGe BiCMOS. IEEE J Solid-State Circuits 51(11):2596–2609
9. Chung S et al (2018) Poly-crystalline silicon waveguide devices on hollow deep trench isolation in standard foundry bulk silicon process. In: Optical fiber communication conference. Optica Publishing Group
10. Rücker H, Heinemann BJSS (2018) High-performance SiGe HBTs for next generation BiCMOS technology. Semicond Sci Technol 33(11):114003
11. Chevalier P et al (2014) A 55 nm triple gate oxide 9 metal layers SiGe BiCMOS technology featuring 320 GHz f T/370 GHz f MAX HBT and high-Q millimeter-wave passives. In: 2014 IEEE international electron devices meeting. IEEE
12. Van-Tuan V et al (2016) TCAD calibration of high-speed Si/SiGe HBTs in 55-nm BiCMOS. Electrochem Soc 75(8):113
13. Chevalier P et al (2015) Nanoscale SiGe BiCMOS technologies: From 55 nm reality to 14 nm opportunities and challenges. In: 2015 IEEE bipolar/BiCMOS circuits and technology meeting-BCTM. IEEE
14. Voinigescu SP et al (2012) Characterization and modeling of an SiGe HBT technology for transceiver applications in the 100–300-GHz range. IEEE Trans Microw Theory Tech 60(12):4024–4034
15. Lakhdara M et al (2014) Mathematical approach and optimisation of nanometric base thickness for a SiGeC HBT dedicated to radiofrequency applications. 259:925–936
16. Selberherr S (1984) Analytical investigations about the basic semiconductor equations. In: Analysis and simulation of semiconductor devices. Springer, pp 127–148
17. Lacave T (2011) Transistor bipolaire Si/SiGe C en nœud CMOS avancé pour applications (sub)-millimétriques. Lille 1
18. Lachkhab CEG et al (2021) COMSOL simulation of a new generation of SiGe heterojunction bipolar transistor (HBT) integrated in a BiCMOS55nm technology. In: AIP conference proceedings. AIP Publishing
19. Bounemeur A, Chemachema M (2023) General fuzzy adaptive fault-tolerant control based on Nussbaum-type function with additive and multiplicative sensor and state-dependent actuator faults. Fuzzy Sets Syst 7:108616
20. Bounemeur A, Chemachema M, Essounbouli N (2018) Indirect adaptive fuzzy fault-tolerant tracking control for MIMO nonlinear systems with actuator and sensor failures. ISA Trans 1(79):45–61
21. Zahaf A, Bououden S, Chadli M, Boulkaibet I, Neji B, Khezami N (2023) Dynamic sensorless control approach for Markovian switching systems applied to PWM DC–DC converters with time-delay and partial input saturation. Sensors 23(15):6936
22. Zahaf A, Bououden S, Chadli M, Chemachema M (2022) Robust fault tolerant optimal predictive control of hybrid actuators with time-varying delay for industrial robot arm. Asian J Control 24(1):1–5

Environmental Protection by the Adsorptive Elimination of Basic Violet 3 Dye from Water

K. Saadallah, C. Ad, M. Djedid, M. Benalia, and S. Saadallah

Abstract Various industries generate effluent that contains persistent pollutants, including dyes, which have detrimental impacts on the environment and human health. Underscore the need for eco-friendly materials to treat dye-containing wastewater and prevent adverse effects on water resources. This study focuses on the potential of using pine tree bark (PTB) as an adsorbent to remove hazardous basic violet 3 dye (BV3) from synthetic water. The study examines factors such as contact time, PTB mass, pH and initial BV3 concentration to understand their impact on batch-mode adsorption. The efficiency of PTB for BV3 dye removal is evaluated through a wide range of basic violet 3 dye (BV3) concentrations (15–150 mg/L), pH values (2–10), and temperatures (25–55 °C), with the kinetics and adsorption isotherms of dyes studied systematically. The kinetic and isotherm study shows that the pseudo-second-order kinetic and Freundlich isotherm models accurately describe the experimental data. Based on these findings, it can be concluded that pine tree bark (PTB) is a cost-effective, eco-friendly, and promising material for removing dye molecules from aqueous solutions. This material can effectively mitigate the adverse effects of dye-containing wastewater on the environment and water resources.

Keywords Adsorption · Pine tree bark · Basic violet 3 · Isotherm · Kinetics

K. Saadallah (✉) · C. Ad · M. Djedid · M. Benalia
Laboratory of Process Engineering, Laghouat University, BP 37G, Ghardaïa Road, Laghouat, Algeria
e-mail: K.saadallah@lagh-univ.dz

S. Saadallah
Laboratory of Thermodynamics and Surface Treatment of Materials, Constantine University, 325 Ain el Bey Road, P.O. Box, Constantine, Algeria

S. Hamamda et al. (eds.), *Proceedings of the 2nd International Conference of Nanotechnology for Environmental Protection and Clean Energy Production*, Springer Proceedings in Materials 45, https://doi.org/10.1007/978-981-97-1916-7_30

1 Introduction

Preserving the environment and securing the future of our planet and humanity has become imperative in the present times. With the advancement of industries and technology, the environment has faced numerous issues, such as pollution and degradation. The rampant industrialization and unregulated use of natural resources, particularly water, have caused a significant decline in water quality worldwide [1, 2]. The excessive use of synthetic dyes and toxic chemicals in industrial activities has further worsened the situation, increasing the influx of pollutants into water bodies [3]. synthetic dyes in industrial wastewater harm aquatic ecosystems by reducing light penetration, increasing biochemical oxygen demand, and ultimately affecting the photosynthetic process of aquatic ecosystem. Consequently, the availability of nutrients to aquatic organisms is significantly affected, leading to potential ecological impacts [4–7]. Dyes are classified into several categories, including acidic, basic, direct, dispersive, reactive, sulfur, and vat dyes [8]. Basic Violet 3 is classified as a cationic triphenylmethane dye, commonly called Gentian violet or Crystal violet. This dye has been utilized for various applications, including biological staining, dermatology, coloring textile fibres and paper printing. However, this dye has been found to possess mutagenic properties and acts as a mitotic poison [9]. This kind of dye is challenging to remove from wastewater due to its complex structures, making the efficient treatment of these pollutants a significant concern. The issue of effluents contaminated with dyes has led to the emergence of a wide range of physicochemical and biological approaches to combat the problem. Adsorption is known as a highly effective technique for decontaminating water from dyes, among other methods. It is favored for its high efficiency, ease of use, adaptability, and economic viability. As a result, it has become the go-to method for removing significant amounts of color from contaminated water sources [10, 11]. The most utilized adsorbent is Active carbon which known of its surface area and small-volume pores. However, the predominant use of coal-based activated carbon raises concerns due to its non-renewable nature and expected depletion of fossil fuels. Additionally, the high expense of activated carbon is a crucial factor to consider in the adsorption process [5, 12]. In light of these challenges, researchers are exploring the development of sustainable adsorbents that are not only cost-effective but also eco-friendly and easily regenerated using agricultural waste [13]. Therefore, this study aims to remove Basic violet 3 dye from water through adsorption using pine tree bark and analyze the kinetics and isotherms of the process.

2 Experimental Procedure

2.1 *Materials*

Pine tree bark (PTB) sourced from Laghouat, Algeria, was selected as the adsorbent. The bark was purified using tap and distilled water to eliminate any traces of dirt and dust. Afterwards, the sample was dried in an oven at 105 °C for 12 h. The dried sample was finely ground, sieved to achieve a uniform particle size and kept in a desiccator, ready for subsequent adsorption experiments. Basic Violet 3 (BV3), a cationic dye with a chemical formula of $C_{25}N_3H_{30}Cl$ and a molar mass of 407.979 g/mol, was selected as the adsorbate and obtained from Sigma Aldrich. To prepare a 1000 mg/L stock solution of BV3, 1 g of adsorbate was dissolved in 1000 mL of distilled water. The resulting solution was covered with aluminum foil and kept in a dark place to prevent UV degradation. BV3 solutions with desired concentrations were prepared by diluting the BV3 stock solution.

2.2 *Methodology*

This research aimed to assess the adsorption process of Basic Violet 3 (BV3) dye by examining its isotherms and kinetics. In this regard, the batch system was utilized to perform the adsorption studies under different conditions. Including Pine tree bark (PTB) mass (0.1–1.00 g), initial basic violet 3 (BV3) concentration (25–150 ppm), pH (ranging from 2 to 10),contact time (1–180 min) at 25 °C. During each experiment, only one adsorption parameter was varied while the others were kept constant. The samples were agitated in a thermostatic water shaker at 150 rpm (rotations per minute) until equilibrium was achieved. Afterwards, the residual concentration of dye was analyzed using UV-Vis spectrophotometer at 588 nm. The formula for calculating the adsorbate uptake is provided in Eq. (1).

$$q_t = \frac{(C_i - C_f) \times V}{m} \tag{1}$$

The quantity of dye adsorbed on the adsorbent surface (mg/g) is represented by qt. The initial and final equilibrium concentrations of the dye (mg/L) are denoted by C_i and C_f, respectively. The mass of the adsorbent (g) is represented by m, and the volume of the solution sample (L) is represented by V) [6]. Following the batch equilibrium results, the study derived the dye adsorption isotherms.

3 Results and Discussions

3.1 Analysis of the Dye Adsorption Isotherms

Assessing the dye adsorption isotherms is important for comprehending the interplay between adsorbate and adsorbent, which could provide helpful insights for enhancing the utilization of adsorbents [13]. In this research, the adsorption isotherm analysis for the removal of basic violet 3 (BV3) onto pine tree bark (PTB). Freundlich and Langmuir isotherm equations were employed to validate the data's suitability.

Equation (2) provides the linear equation for Langmuir isotherm model [14].

$$\frac{1}{q_e} = \frac{1}{q_{\max}.K_L} \frac{1}{C_e} + \frac{1}{q_{\max}} \tag{2}$$

Equation (3) provides the linear equation for Freundlich isotherm model [15].

$$\ln q_e = \ln K_f + \frac{1}{n} \ln C_e \tag{3}$$

The equilibrium amount of dye adsorbed on the adsorbent surface (mg/g) is represented by qe, whereas Ce denotes the equilibrium dye concentration (mg/L) in the solution. The remaining parameters are the isotherms constants that the correlation coefficient of the experimental data can determine.

In the Langmuir equation, the monolayer dye uptake capacity of the adsorbent is represented by qmax (mg/g), whereas the Langmuir constant is represented by KL ($dm^3\ mg^{-1}$) [14].

While for the Freundlich equation incorporates a distribution coefficient, K_F ($L\ g^{-1}$), which is indicative of the adsorption reliability. Along with this, the heterogeneity factor, n, is used to represent the degree of deviation from linearity of adsorption. If the value of n equals 1, the adsorption process is linear. Conversely, if n is less than 1, the adsorption process is chemical in nature and if n is greater than 1, the adsorption process is deemed favorable and physical in nature [16].

To assess the precision of experimental data fitting to isotherm models, the determination coefficients (R^2) of the Langmuir and Freundlich models are taken into consideration. Additionally, the Langmuir isotherm is characterized by the R_L dimensionless separation parameter, which not only indicates the shape of the isotherm but also predicts the feasibility of the adsorption system. R_L is given by: Eq. (4)

$$R_L = \frac{1}{1 + (K_L.C_0)} \tag{4}$$

The R_L value obtained indicates the nature of the isotherm, which can be classified as unfavorable, linear, favorable, or irreversible. An R_L value greater than one suggests an unfavorable isotherm where chemisorption is the dominant process [15].

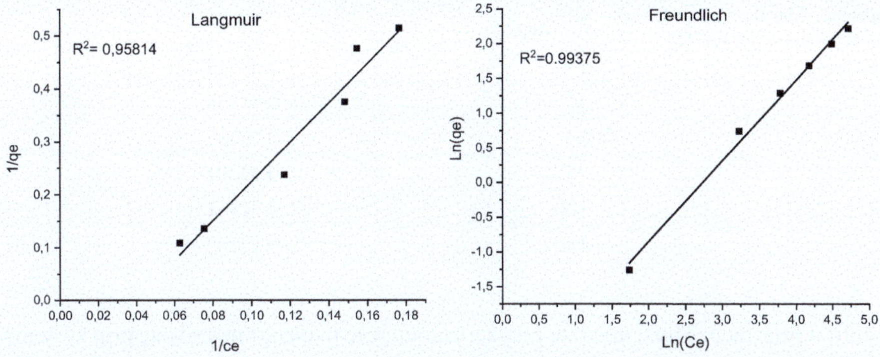

Fig. 1 Langmuir and Freundlich plots: mass of PTB = 0.6 g, initial BV3 dye concentration = 15–150 mg/l, temperature = 25 °C, shaker speed = 150 rpm, time = 60 min

Table 1 The isotherm parameters for BV3 dye adsorption onto pine tree bark (PTB)

Adsorbate	Q_0 (mg/g)	Langmuir		Freundlich		
		K_L	R^2	n	K_F	R^2
BV3	6.7935	0.0396	0.9581	1.1680	0.0522	0.9938

If the $R_L = 1$, the isotherm is said to be linear, indicating that both chemisorption and physisorption occur at similar rates. If $0 < R_L < 1$ implies a favorable isotherm where physisorption is the predominant process. If $R_L = 0$, it suggests an irreversible isotherm where the chemisorption process cannot be reversed [17].

Figure 1 shows the adsorption isotherms for BV3 dye, with the calculated parameters for each isotherm provided in Table 1. The results indicate that the Freundlich model ($R^2 = 0.99375$) is a better fit for the data than the Langmuir model ($R^2 = 0.95814$), suggesting that the adsorption process is heterogeneous and multi-layered [17]. The value of $n > 1$ suggests that physisorption is the predominant process and that adsorption is favorable. These results are consistent with previous research [17–19] which have successfully applied the Freundlich isotherm to the removal of the same adsorbate using different adsorbents. The Langmuir analysis showed a maximum dye uptake capacity (q_{max}) of 6.7953 mg/g and a Langmuir constant of 0.03956 L/mg (refer to Table 1). The values of R_L, as depicted in Table 2, show that the dye uptake process was favorable, within the range of (0–1).

3.2 *Kinetics Study*

To comprehend the involved control mechanisms in adsorption processes, kinetic models serve as an essential tool. Through kinetic studies, the dye uptake rate can be regulated, which consequently influences the residual time of the adsorbate at the

Table 2 The R_L values for the BV3 dye onto (PTB) adsorbent

C_o (mg/L)	R_L
15	0.58512
30	0.41354
50	0.2973
75	0.22001
100	0.17461
150	0.12363

solution interface. Thus, this rate plays a crucial role in designing adsorption systems [20]. In this regard, two sorption kinetic models were applied, the pseudo first order (PFO) and pseudo second-order (PSO) equations, to evaluate the kinetics of BV3 dye uptake onto PTB adsorbent. Equations (5) and (6) provide the linear expressions for the PFO and PSO models, respectively.

$$\ln(q_e - q_t) = \ln(q_e) - k_1 t \quad (5)$$

$$\frac{t}{q_t} = \frac{1}{k_2} \cdot \frac{1}{q_e^2} + \frac{t}{q_e} \quad (6)$$

The BV3 adsorption capacity (mg/g) at time t and at equilibrium is represented by q_t and q_e, respectively. The rate constants for the pseudo-first-order (L/min) and pseudo-second-order (g/mg min) models are denoted by k_1 and k_2, respectively. The study evaluates the applicability of two kinetic models, pseudo-first-order and pseudo-second-order, to the adsorption process of basic violet 3 dye (BV3) onto pine bark adsorbent. The graphs of ln (qe − qt) against t and t/qt against t, respectively, are presented in Fig. 2. The kinetic constants are calculated from the slope and intercept of both models depicted in Table 3. The results reveal that the pseudo-second-order model is a suitable fit for the BV3 dye uptake onto the pine tree bark (PTB) adsorbent, with high R^2 values of 0.99–1. Moreover, the calculated qe values are in agreement with the experimental values, further validating the viability of the pseudo-second-order model for this process. These findings are consistent with previous research [19–23] that also reported the successful use of the pseudo-second-order model for adsorbing the same dye onto different adsorbents.

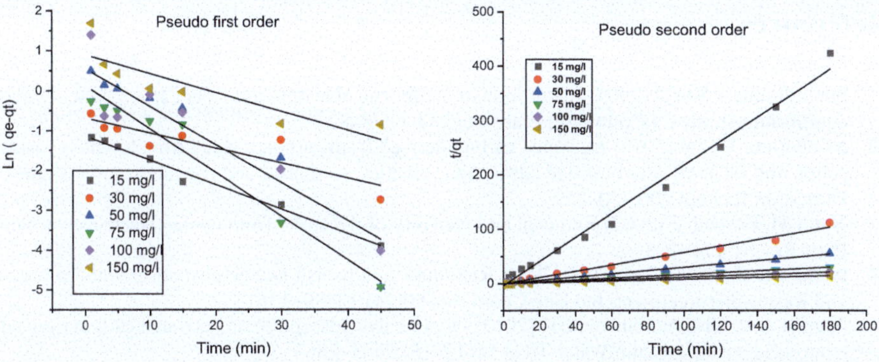

Fig. 2 Plots adsorption Kinetic models: mass of PTB = 0.6 g, initial BV3 dye concentration = 15–150 mg/l, temperature = 25 °C, shaker speed = 150 rpm, time = 60 min

Table 3 Comparative analysis of pseudo-first order and pseudo-second order models for adsorption of BV3 dye

Dye	C_0 (mg/l)	Q_{exp} (mg/g)	First order kinetic model			Second order kinetic model		
			Q_{cal} (mg/g)	K_1	R^2	Q_{cal} (mg/g)	K_2	R^2
BV3	15	0.5512	0.4160	0.070	0.8916	0.4521	1.3340	0.9906
	30	1.8888	0.5901	0.043	0.771	1.7298	0.9891	0.9937
	50	3.5527	0.4161	0.163	0.8916	3.3168	0.7846	0.9979
	75	6.0276	1.4588	0.106	0.905	5.9595	0.6473	0.9999
	100	8.4814	1.5310	0.042	0.8394	8.5034	0.2147	1
	150	12.3114	1.995983	0.081	0.62	12.346	0.1014	0.9999

4 Conclusion

The aim of this study was to investigate the potential of using Pine tree bark (PTB), a local eco-friendly adsorbent, for decontaminating synthetic water from basic violet 3 dye. The study relied on a combination of kinetic and isotherm analysis, which revealed that the experimental data was best, fitted to the Freundlich isotherm equation. At 25 °C, the maximum dye uptake capacity was determined to be 6.7935 mg/g using the Langmuir equation. Furthermore, the kinetics study showed that the pseudo-second-order kinetic model was the most appropriate for the adsorption process. These results indicate that PTB is a promising low-cost adsorbent for effectively removing toxic dye.

References

1. Senthilkumaar S et al (2005) Adsorption of methylene blue onto jute fiber carbon: kinetics and equilibrium studies. J Colloid Interface Sci 284(1):78–82
2. Maneerung T et al (2016) Activated carbon derived from carbon residue from biomass gasification and its application for dye adsorption: kinetics, isotherms and thermodynamic studies. Bioresour Technol 200:350–359
3. Naim M, El Abd Y (2002) Removal and recovery of dyestuffs from dyeing wastewaters. Sep Purif Rev 31:171–228
4. Ahmaruzzaman M, Gupta VK (2011) Rice husk and its ash as low-cost adsorbents in water and wastewater treatment. Ind Eng Chem Res 50(24):13589–13613
5. Ahmad AA, Idris A, Hameed BH (2013) Organic dye adsorption on activated carbon derived from solid waste. Desalin Water Treat 51(13–15):2554–2563
6. Annuar MSM et al (2009) Kinetics and energetics of azo dye decolorization by *Pycnoporus sanguineus*. Water Air Soil Pollut 202(1):179–188
7. Kumar D et al (2020) Chlorophyll a fluorescence kinetics of mung bean (*Vigna radiata* L.) grown under artificial continuous light. Biochem Biophys Rep 24:100813
8. Sharma J, Sharma S, Soni V (2021) Classification and impact of synthetic textile dyes on aquatic flora: a review. Reg Stud Mar Sci 45:101802
9. Ratna, Padhi BS (2012) Pollution due to synthetic dyes toxicity & carcinogenicity studies and remediation. Int J Environ Sci 3:940–955
10. Masel RI (1996) Principles of adsorption and reaction on solid surfaces, vol 3. Wiley
11. Crini G (2006) Non-conventional low-cost adsorbents for dye removal: a review. Bioresour Technol 97(9):1061–1085
12. Grassi MCB et al (2012) Removal of emerging contaminants from water and wastewater by adsorption process
13. Crini G, Badot P-M (2008) Application of chitosan, a natural aminopolysaccharide, for dye removal from aqueous solutions by adsorption processes using batch studies: a review of recent literature. Prog Polym Sci 33(4):399–447
14. Langmuir I (1918) The adsorption of gases on plane surfaces of glass, mica and platinum. J Am Chem Soc 40(9):1361–1403
15. Freundlich H (1906) Over the adsorption in solution. J Phys chem 57(385471):1100–1107
16. Crini G et al (2007) Removal of C.I. basic green 4 (malachite green) from aqueous solutions by adsorption using cyclodextrin-based adsorbent: kinetic and equilibrium studies. Sep Purif Technol 53(1):97–110
17. Crini G (2008) Kinetic and equilibrium studies on the removal of cationic dyes from aqueous solution by adsorption onto a cyclodextrin polymer. Dyes Pigm 77(2):415–426
18. Wang XS et al (2008) Comparison of basic dye crystal violet removal from aqueous solution by low-cost biosorbents. Sep Sci Technol 43(14):3712–3731
19. Muthukumaran C, Sivakumar VM, Thirumarimurugan M (2016) Adsorption isotherms and kinetic studies of crystal violet dye removal from aqueous solution using surfactant modified magnetic nanoadsorbent. J Taiwan Inst Chem Eng 63:354–362
20. Dawood S, Sen TK (2012) Removal of anionic dye Congo red from aqueous solution by raw pine and acid-treated pine cone powder as adsorbent: equilibrium, thermodynamic, kinetics, mechanism and process design. Water Res 46(6):1933–1946
21. Batool M et al (2021) Exploring the usability of *Cedrus deodara* sawdust for decontamination of wastewater containing crystal violet dye. Desalin Water Treat 224:433–448
22. Abdelhamid B, Chemachema M, Bouzina S (2023) Fuzzy fault-tolerant control applied on two inverted pendulums with nonaffine nonlinear actuator failures. Int J Robot Control Syst 3(2)
23. Zahaf A, Bououden S, Chadli M, Chemachema M (2022) Robust fault tolerant optimal predictive control of hybrid actuators with time-varying delay for industrial robot arm. Asian J Control 24(1):1–5

The Effect of Increasing the Thickness of an Annular Space Subjected to a Heat Flux Emitted by a Solar Panel on the Dynamic and Thermal Fields

Sofiane Touahri, Hassen Haithem, and Toufik Boufendi

Abstract The present work is a 3D numerical simulation, in an annular space between two horizontal concentric cylinders in which an incompressible fluid circulates. The wall of the inner cylinder is adiabatic while the wall of the outer cylinder is subjected to a constant thermal heat flux. The model equations of continuity, momenta and energy are numerically solved by the finite volume method with a second order spatiotemporal discretization. The used mesh is $52 \times 44 \times 162$ nodes along the radial, angular and axial directions. The studied hydraulic diameters are: Dh = 0.75, 1.0, and 1.25 cm. The purpose of this work is to improve the heat transfer by increasing the thickness of the annular space, this allowed us to study the dynamic and thermal fields of the heat transfer fluid and make a comparison of average temperatures and Nusselt numbers for three different cases. The results obtained show that despite the qualitative similarity of thermal and dynamic fields for the three studied cases, the heat transfer increases when the thickness of the annular space is increased, this is justified by the increase in the Nusselt number.

Keywords Annular space · Mixed convection · 3D numerical simulation

1 Introduction

Laminar flows in horizontal annular spaces between two concentric cylinders are studied by several researchers, in this context, experimental and numerical works has been carried out for different types of annular space. In their numerical work, Nazrul et al. [1] studied heat transfer by laminar mixed convection of air in a annulus, the inner cylinder is subjected to a constant heat flux, while the outer cylinder is

S. Touahri (✉) · H. Haithem · T. Boufendi
Energy Physics Laboratory, Department of Physics, University of Constantine 1, Constantine, Algeria
e-mail: sofianetouahri04@yahoo.fr

T. Boufendi
University August 20, 1955, Skikda, Algeria

S. Hamamda et al. (eds.), *Proceedings of the 2nd International Conference of Nanotechnology for Environmental Protection and Clean Energy Production*, Springer Proceedings in Materials 45, https://doi.org/10.1007/978-981-97-1916-7_31

adiabatic. The results obtained showed that the velocity and temperature fields are influenced by the transverse Archimedean thrust, the Nusselt number at $Z^* = 0.1$ is higher by 30% and 110% for $Ra = 10^5$ and $Ra = 10^7$ respectively with respect to forced convection. In their numerical work, Touahri et al. [2, 3] studied The effect of variable fluid properties on mixed convection in horizontal pipe and in annular space between two horizontal cylinders, the outer cylinder is heated by an electric current passing its thickness and the inner cylinder is insulated. Thermal convection in the fluid domain is combined with thermal conduction in the solid thickness. The results obtained showed a considerable variation in viscosity and a moderate variation in thermal conductivity of the fluid. The Nusselt number of the mixed convection becomes higher than that of forced convection when the Grashof number increases. With the used parameters, the average Nusselt number is correlated as $Nu_A = 9.9130\ Ri^{0.0816}$.

Gourari et al. [4] numerically studied the heat transfer by natural convection of laminar flow of water in inclined rings, the inner cylinder generates a constant heat source while the outer cylinder is cold, the upper and lower walls are insulated thermally, the results are obtained for the inclination angles: 0°, 45°, 90°, these results showed that the average Nusselt number increases with the increase in the Rayleigh number and that the heat transfer is maximum for $\alpha = 90°$. Similar studies was also treated numerically by Basant et al. [5], Oni [6], Khemici et al. [7], Kotake et al. [8], Kumar [9], Chung et al. [10], Nouar [11] and Chaieb et al. [12]. The Heat transfer by forced and free convection of laminar air flow inside horizontal concentric pipes was studied experementaly by Mohammed et al. [13], their results showed that the axial Nusselt number increases with the increasing of Reynolds number and the average Nusselt can be related with the different dimensionless numbers by the correlation: $Nu_A = 2.964(Gr{\cdot}Pr/Re)^{0.0326}$. A similar study was also treated numerically and experimentally by Yuzhu et al. [14].

2 Mathematical Model

We consider an annular space between two horizontal concentric cylinders having a length $L = 1$ m. The outer cylinder is subjected to a constant thermal heat flux transferred to laminar incompressible flow of distilled water ($Pr = 8.082$) which corresponds to the inlet fluid temperature ($T = 15$ °C), while the inner cylinder is considered adiabatic (Fig. 1).

The objective of our work is to study the effect of increasing the thickness of the annular space: ***e*** on improving heat transfer to the fluid, for this three geometries are studied:

(1) First case: $L = 100$ cm, $D_i = 0.75$ cm, $D_o = 2$ cm, $D_h = 1.25$ cm, $\boldsymbol{e} = 0.625$ cm
(2) Second case: $L = 100$ cm, $D_i = 1.0$ cm, $D_o = 2$ cm, $D_h = 1.0$ cm, $\boldsymbol{e} = 0.5$ cm
(3) Third case: $L = 100$ cm, $D_i = 1.25$ cm, $D_o = 2$ cm, $D_h = 0.75$ cm, $\boldsymbol{e} = 0.375$ cm

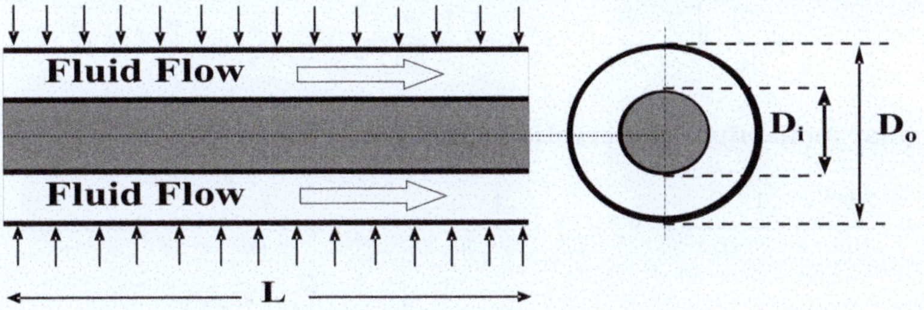

Fig. 1 Geometry of the problem

For the three studied cases, the volume flow of the heat transfer fluid is considered constant and equal to 1.343×10^{-5} m^3/s and the inlet temperature is equal to 15 °C. The thermal conductivity and kinematic viscosity of the fluid at the inlet are equal to: $K_0 = 0.594725$ W/m°K and $\nu_0 = 1.14 \times 10^{-6}$ m^2/s respectively.

To avoid the repetition, the geometry and mathematical modeling presented are those of the hydraulic diameter $D_h = 1.25$ cm. For this case, at the entrance the fluid flows with an average speed equal to 4.97×10^{-2} m/s which gives a Reynolds number equal to 545. The Grashof number is considered equal to 2.44×10^5.

2.1 Modelling Equations

The studied problem is well modeled by the following non-dimensional conservation partial differential equations with their initial and boundary conditions:

$$\text{At } (t^* = 0)\text{: } V_r^* = V_\theta^* = V_z^* = T^* = 0 \tag{1}$$

- **Mass conservation equation at (t^*> 0)**

$$\frac{1}{r^*}\frac{\partial}{\partial r^*}\left(r^*\, V_r^*\right) + \frac{1}{r^*}\frac{\partial V_\theta^*}{\partial \theta} + \frac{\partial V_z^*}{\partial z^*} = 0 \tag{2}$$

- **Radial momentum conservation equation**

$$\begin{aligned}
&\frac{\partial V_r^*}{\partial t^*} + \frac{1}{r^*}\frac{\partial}{\partial r^*}\left(r^* V_r^* V_r^*\right) \\
&\quad + \frac{1}{r^*}\frac{\partial}{\partial \theta}\left(V_\theta^* V_r^*\right) + \frac{\partial}{\partial z^*}\left(V_z^* V_r^*\right) - \frac{V_\theta^{*2}}{r^*} \\
&= -\frac{\partial P^*}{\partial r^*} + \frac{Gr_0^*}{\mathrm{Re}_0^2}\cos\theta\, T^*
\end{aligned}$$

$$+\frac{1}{\mathrm{Re}_0}\left[\frac{1}{r^*}\frac{\partial}{\partial r^*}(r^*\tau_{rr}^*)+\frac{1}{r^*}\frac{\partial}{\partial\theta}(\tau_{r\theta}^*)-\frac{\tau_{\theta\theta}^*}{r^*}+\frac{\partial}{\partial z^*}(\tau_{rz}^*)\right] \tag{3}$$

- **Angular momentum conservation equation**

$$\frac{\partial V_\theta^*}{\partial t^*}+\frac{1}{r^*}\frac{\partial}{\partial r^*}(r^*V_r^*V_\theta^*)+\frac{1}{r^*}\frac{\partial}{\partial\theta}(V_\theta^*V_\theta^*)+\frac{\partial}{\partial z^*}(V_z^*V_\theta^*)+\frac{V_r^*V_\theta^*}{r^*}$$
$$=-\frac{1}{r^*}\frac{\partial P^*}{\partial\theta}-\frac{Gr_0^*}{\mathrm{Re}_0^2}\sin\theta T^*+\frac{1}{\mathrm{Re}_0}\left[\begin{array}{l}\frac{1}{r^{*2}}\frac{\partial}{\partial r^*}(r^{*2}\tau_{\theta r}^*)\\+\frac{1}{r^*}\frac{\partial}{\partial\theta}(\tau_{\theta\theta}^*)+\frac{\partial}{\partial z^*}(\tau_{\theta z}^*)\end{array}\right] \tag{4}$$

- **Axial momentum conservation equation**

$$\frac{\partial V_z^*}{\partial t^*}+\frac{1}{r^*}\frac{\partial}{\partial r^*}(r^*V_r^*V_z^*)+\frac{1}{r^*}\frac{\partial}{\partial\theta}(V_\theta^*V_z^*)+\frac{\partial}{\partial z^*}(V_z^*V_z^*)$$
$$=-\frac{\partial P^*}{\partial z^*}+\frac{1}{\mathrm{Re}_0}\left[\frac{1}{r^*}\frac{\partial}{\partial r^*}(r^*\tau_{rz}^*)+\frac{1}{r^*}\frac{\partial}{\partial\theta}(\tau_{\theta z}^*)+\frac{\partial}{\partial z^*}(\tau_{zz}^*)\right] \tag{5}$$

- **Energy conservation equation**

$$\frac{\partial T^*}{\partial t^*}+\frac{1}{r^*}\frac{\partial}{\partial r^*}(r^*V_r^*T^*)+\frac{1}{r^*}\frac{\partial}{\partial\theta}(V_\theta^*T^*)+\frac{\partial}{\partial z^*}(V_z^*T^*)$$
$$=G^*-\frac{1}{\mathrm{Re}_0\,\mathrm{Pr}_0}\left[\begin{array}{l}\frac{1}{r^*}\frac{\partial}{\partial r^*}(r^*q_r^*)\\+\frac{1}{r^*}\frac{\partial}{\partial\theta}(q_\theta^*)+\frac{\partial}{\partial z^*}(q_z^*)\end{array}\right] \tag{6}$$

2.2 *The Boundary Conditions*

The previous differential equations are solved with the following boundary conditions:

- **At the entrance: (z^* = 0)**

$$V_r^*=V_\theta^*=T^*=0,\ V_z^*=1 \tag{7}$$

- **At the exit: (z^* = 100)**

$$\frac{\partial V_r^*}{\partial z^*}=\frac{\partial V_\theta^*}{\partial z^*}=\frac{\partial V_z^*}{\partial z^*}=\frac{\partial}{\partial z^*}\left(K^*\frac{\partial T^*}{\partial z^*}\right)=0 \tag{8}$$

- **At the internal pipe: (r^* = 0.5)**

$$V_r^* = V_\theta^* = V_z^* = 0, and\left(\frac{\partial T^*}{\partial r^*}\right) = 0 \tag{9}$$

- **At the external pipe: (r^* = 1)**

$$V_r^* = V_\theta^* = V_z^* = 0, and\left(\frac{\partial T^*}{\partial r^*}\right) = 1 \tag{10}$$

2.3 *Nusselt Numbers*

At the solid–fluid interface ($r^* = 1$) the local Nusselt number is defined as:

$$Nu(\theta, z^*) = \frac{h(\theta, z^*)D_i}{K_0} = \left[\frac{\left(K^* \partial T^* / \partial r^*\right)\big|_{r^*=1}}{T^*(1, \theta, z^*) - \partial T_A^*(z^*)}\right] \tag{11}$$

3 Numerical Procedure

The finite volume method described by Patankar [15] is used for the numerical solution of the previous modeling equations. The convective and nonlinear terms were discretized according to the Adams–Bashforth numerical scheme, the spatiotemporal numerical discretization is second order, so the order of temporel numerical error is $(\Delta t^*)^2$ and $(\Delta r^*)^2$, $(\Delta z^*)^2$, $(\Delta \theta)^2$ for radial, axial and azimithal direction respectively. A Fortran code based on the SIMPLER algorithm is used to process the pressure–velocity coupling, the tri-diagonal matrix algorithm (TDMA) is used to obtain the iterative solution of the equations of the discretized model. The meshes used in this work are: 26 × 44 × 162 and 52 × 88 × 162 in the radial, azimuthal and axial directions respectively and the results presented are those of the second mesh. The proposed time step is ($\Delta t^* = 5.10^{-4}$) and the time march is continued until the steady state is obtained which is controlled by the stabilization of the temporal evolution of the hydrodynamic and thermal fields.

The used numerical code has been validated by the comparison of our results with those of Nazrul et al. [1] who studied the laminar mixed convection heat transfer in a horizontal concentric annulus with adiabatic outer wall and a uniform heat flux at the inner wall. The controlling parameters are: $D_2/D_1 = 2$, Re = 200, Pr = 0.7 and Gr = 0, The axial evolution of the circumferentially averaged Nusselt number is illustrated in Fig. 2, it is seen that there is a good agreement between the two results.

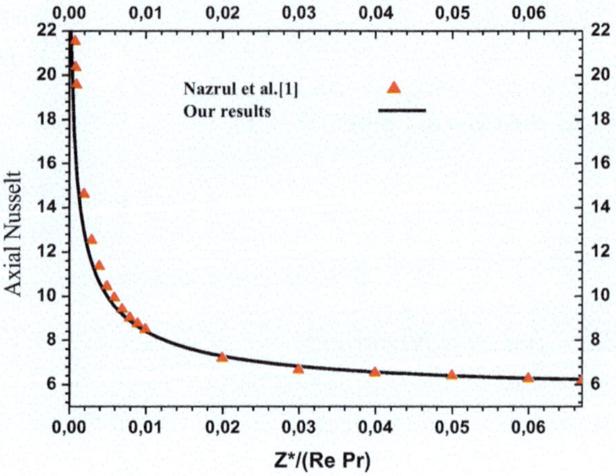

Fig. 2 Axial evolution of the mean axial Nusselt number; a comparison with the results of Nazrul et al. [1]

4 Results and Discussions

Qualitatively the results of the three studied cases are similar but quantitatively there are differences. To avoid repetition, we will present in this section the results of the case $D_h = 1$ cm. The other results are used in the quantitative comparisons of the different studied cases. For this case, the Reynolds number is equal at 545, the Prandtl number equal to 8.082 and the Grashof number is equal to 2.44×10^5. The dynamic field is characterized by a main flow in the axial direction and a transverse flow influenced by the density variation with temperature.

4.1 *Transverse Flow*

At the entrance ($Z = 0$ cm), the transverse flow is nonexistent because the fluid temperature is constant. Just after, the heat transferred to the fluid through a constant heat flux at the wall of the external cylinder creates a temperature gradient oriented inward. This situation induces a transverse movement explained as follows: the fluid moves along the hot wall of the external cylinder from the bottom towards the up and then descends from the top to the bottom near the wall of the internal cylinder (Fig. 3).

4.2 *Axial Flow*

At the entrance to the annulus, the fluid presents itself with a constant speed V_0. At this section, the axial component is constant while the angular and radial components

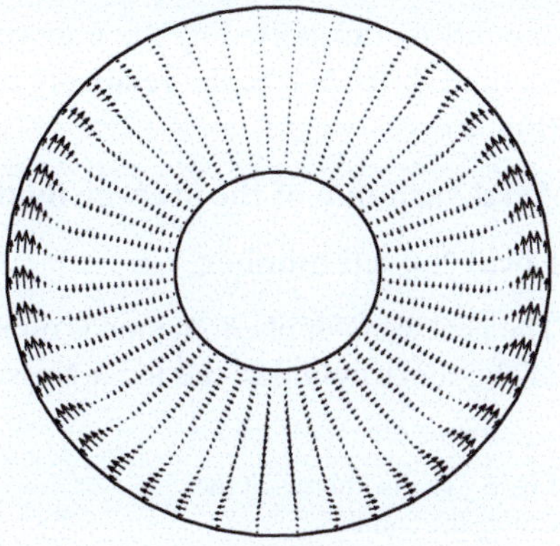

Fig. 3 Transvers flow at the exit of the annulus

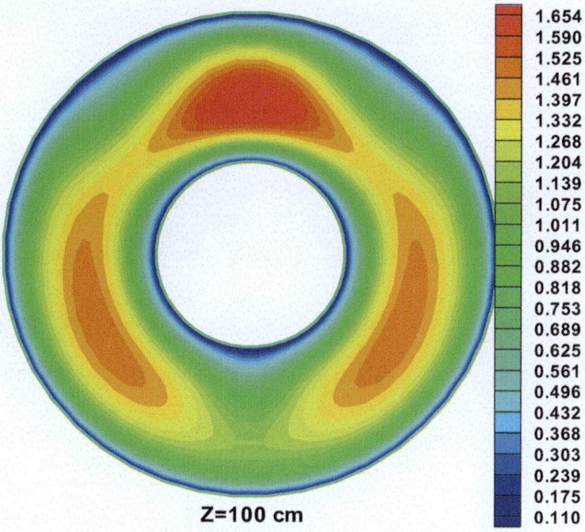

Fig. 4 Axial flow at the exit of the annulus

are zero. After; the axial speed field enters a continuous variation such that the low speeds are located near the walls of the cylinders and the maximum is at the heart of the annular space (Fig. 4).

4.3 Temperature Field and Heat Transfer

At the inlet, the fluid has a constant temperature, after in the presence of the secondary flow, an angular variation of the temperature is imposed. This variation is explained as follows: at a given axial section, the hot fluid near from the outer wall of the annulus is transported by the transverse movement upwards under the effect of the thermal thrust force, the relatively cold fluid descends near from the internal wall (Fig. 5).

Axial variation of the average fluid temperature (see Fig. 6)

Local Nusselt number

The heat transfer phenomenon is quantified by the local Nusselt number. On the wall of the inner cylinder, the local Nusselt number is zero but on the wall of the outer

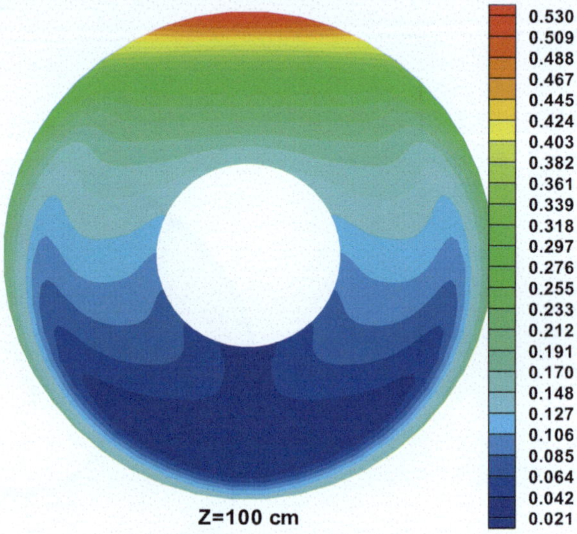

Fig. 5 The isotherms at the exit of the annulus

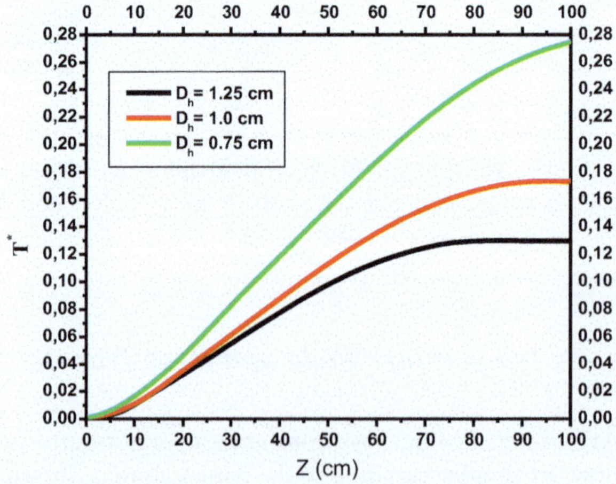

Fig. 6 Axial evolution of the average fluid temperature

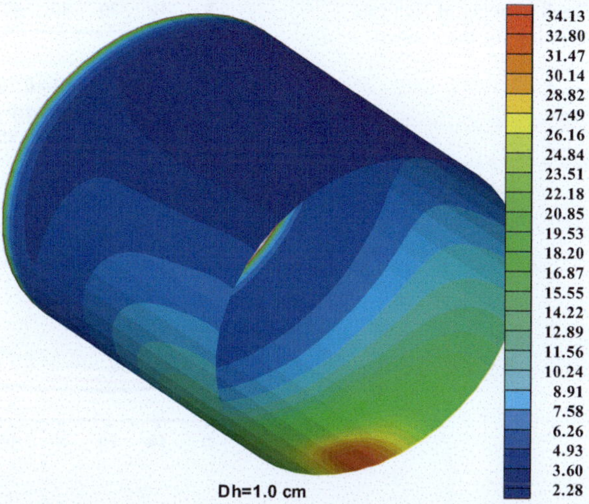

Fig. 7 Variation of the local Nusselt number

cylinder, there is an axial and angular variation in the local Nusselt number from the inlet to the outlet, this variation is shown in Fig. 7.

Axial Nusselt number

The axial Nusselt number is shown in Fig. 8 for all cases studied. Qualitatively there is a similarity in the variation of the axial Nusselt number as a function of Z, in the short entry zone of the annular space, the axial Nusselt number undergoes a very significant decrease followed by an increase up to the exit from the annular space. It is very clear that the axial Nusselt for the case $D_h = 1.25$ cm is the most superior followed by the case $D_h = 1.0$ cm and finally the case $D_h = 1.25$ cm. This is justified as follows: the local Nusselt number defined by Eq. (11) is the ratio between the incident heat flux which is constant and the difference between the local temperature at the wall and the average temperature for the same axial section.

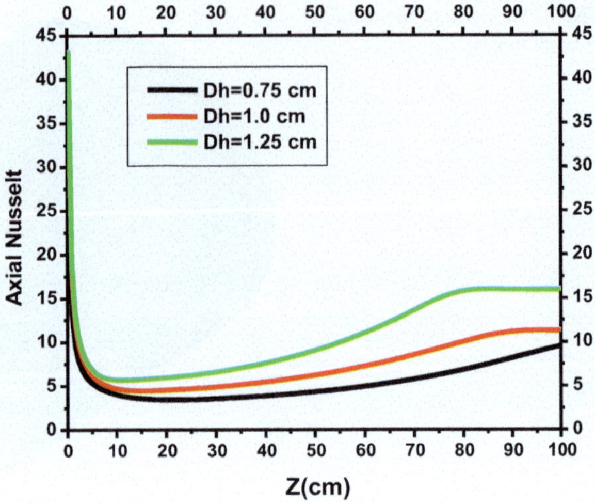

Fig. 8 The axial Nusselt variation for the studied cases

5 Conclusion

In this work, we have numerically studied the heat transfer by mixed convection in a horizontal annulus, the external pipe is heated by a constant thermal flux while the internal pipe is isolated. The objective of this work is to study the effect of increasing the thickness of the annulus on improving heat transfer for the same fluid flow rate.

The obtained results show that the increase in the thickness of the annular space is accompanied by an increase in the Nusselt number and automatically an improvement in heat transfer to the working fluid. The use of a thickness ***e*** equal to 0.625 cm improves the heat transfer by 37.66% and 95.69% compared with thickness equal to 0.5 and 0.375 cm respectively.

Abbreviations

Nomenclature

D	Diameter, m
k	Thermal Conductivity, W/m°C
L	Length, m
Nu	Nusselt number
P	Pressure, N/m^2
Pr	Prandtl number
r	Radial Coordinate, m
Re	Reynolds number

t	Time, s
T	Temperature, °K
U	Radial Velocity Component, m/s
V	Axial Velocity Component, m/s
W	Angular Velocity Component, m/s
z	Axial Coordinate, m

Greek Letters

μ	Dynamic viscosity, kg m/s
θ	Angular coordinate, rad
τ	Viscous stress, N/m^2
ρ	Density, kg/m^3

Subscripts

A	Average at given section
r, θ, z	Reference to the radial, angular and axial direction, respectively

Superscripts

*	Non dimensional

References

1. Nazrul I, Gaitonde UN, Sharma GK (2001) Mixed convection heat transfer in the entrance region of horizontal annuli. Int J Heat Mass Transfer 44(11):2107–2120
2. Touahri S, Boufendi T (2012) Numerical study of the conjugate heat transfer in a horizontal pipe heated by Joulean effect. Therm Sci 16(1):53–67
3. Touahri S, Boufendi T (2015) Conjugate heat transfer with variable fluid properties in a heated horizontal annulus. Heat Transfer Res 64(11):1019–1038
4. Gourari S, Mebarek-Oudina F, Hussein AK, Kolsi L, Hassen W, Younis O (2019) Numerical study of natural convection between two coaxial inclined cylinders. Int J Heat Technol 37(3):779–786
5. Basant KJ, Michael OO, Babatunde A (2018) Steady fully developed mixed convection flow in a vertical micro-concentric-annulus with heat generating/absorbing fluid: an exact solution. Ain Shams Eng J 9(4):1289–1301
6. Oni MO (2017) Combined effect of heat source, porosity and thermal radiation on mixed convection flow in a vertical annulus: an exact solution. Eng Sci Technol Int J 20(1):518–527

7. Khemici M, Boufendi T, Touahri S (2019) Numerical study of developing laminar mixed convection in a heated annular duct with temperature dependent properties. Therm Sci 23(6A):3411–3423
8. Kotake S, Hattori N (1985) Combined forced and free convection heat transfer for fully-developed laminar flow in horizontal annuli. Int J Heat Mass Transfer 28(11):2113–2120
9. Kumar R (1988) Study of natural convection in horizontal annuli. Int J Heat Mass Transfer 31(6):1137–1148
10. Chung SY, Rhee GH, Sung HJ (2002) Direct numerical simulation of concentric annular pipe flow. Int J Heat Fluid Flow 23(4):426–440
11. Nouar C (1999) Numerical solution for laminar mixed convection in a horizontal annular duct: temperature-dependent viscosity effect. Int J Numer Meth Fluids 29(5):849–864
12. Chaieb I, Boufendi T, Nicolas X (2022) Taylor-Couette flow with mixed convection heat transfer and variable properties in a horizontal annular pipe. Therm Sci 26(1A):287–298
13. Mohammed HA, Antonio C, Saidur R (2010) Experimental study of forced and free convective heat transfer in the thermal entry region of horizontal concentric annuli. Int Comm Heat Mass Transfer 37(7):739–747
14. Yuzhu Z, Zhang L, Bu S, Cheng Sun W, Xiao Y, Liu L (2020) Study on heat transfer characteristics of the whole plate fin tube cooler. Int J Thermofluid Sci Technol 7(2):070204
15. Patankar SV (1980) Numerical heat transfer and fluid flow. McGraw-Hill, New-York

Control Systems for Renewable Energy Production

Wind Turbine Energy Production Based on Event-Triggered Speed Control

Abdelmalek Zahaf, Abdelhamid Bounemeur, and Sofiane Bououden

Abstract An improved strategy for energy production is introduced based model predictive control, with a precise control instance using event triggered mechanism. Aiming to optimize the wind turbine speed for energy production while accounting for uncertainties and sensor malfunctions. The proposed strategy leverages model predictive control, a powerful control technique known for its ability to optimize system performance by considering future predictions of power generation. In this context, MPC is utilized to formulate an optimal control sequence for wind turbine operation, accounting for various factors such as wind speed fluctuations, turbine dynamics, and power output constraints. Unlike traditional time-triggered control schemes, our approach employs event-triggered sampling, which means that control updates are triggered only when specific events or significant changes occur in the system, thereby optimal control demands and improving efficiency.

Keywords Wind turbine · Speed control · Sensor fault · Event-triggered

1 Introduction

Clean energy production has been experiencing significant growth in last decades. As known, the wind turbine has a challenging operational conditions, due to the complexity of wind turbine structure several scenarios can have encountered. Usually, occurring faults on actuators/sensors of the system during the processing is often happening. That can result disgrace the performances of the main components. To

A. Zahaf (✉)
Faculty of Technology Sciences, Frères Mentouri University-Constantine 1, Constantine, Algeria
e-mail: zahaf.abdelmalek@umc.edu.dz

A. Bounemeur
National Polytechnic School of Constantine, Constantine, Algeria

S. Bououden
Laboratory of SATIT, Department of Industrial Engineering, Abbes Laghrour University, 40004 Khenchela, Algeria

S. Hamamda et al. (eds.), *Proceedings of the 2nd International Conference of Nanotechnology for Environmental Protection and Clean Energy Production*, Springer Proceedings in Materials 45, https://doi.org/10.1007/978-981-97-1916-7_32

avoid system breakdown due to the control components failures, the diagnosis tasks are significant steps during the working of process [1]. To manage the occurred faults, fault tolerant control (FTC) approach is involved in many applications. To maintain high properties such as stability and performances, FTC based model predictive control is investigated and analyzed by [2–9] to ensure and enhance the stability and the efficient control strategy. Meanwhile, FTC based adaptive control is proposed in [10–19] to improve the effectiveness of the controller. Applying these control strategies on various clean energy sources aims to target the maximum quantity of energy. Thorough that, the nonlinear dynamic state feedback controller is proposed as multivariable control strategy in [20]. To improve the variable speed control, a multimodel optimal control scheme given by [21]. For more precision and reliability of the controlled multimodel based multivariable control strategy, the predictive theory has employed in [22]. To this end, authors in [23] analyses the case of occurring faults for wind turbine during the processing based on fault tolerant predictive control. The using the wind power generation for Load Frequency of Hybrid Power System has discussed by [24] based Model Predictive Control.

In this paper, to improve the wind turbine speed for energy production, uncertainties and sensor malfunctions are taking into consideration for an efficient controller, that employs an observer based predictive control scheme. After given new sight of wind-turbine modeling with uncertainties, a new reformulation of the original system to QP is introduced, which have been used for minimizing the evaluation function to compute the predictive control inputs. The control updates are triggered only when specific events or significant changes occur in the system. As result, the event-triggered based predictive control mechanism improve precise control instance using obtained optimal control. finally, simulation results of wind-turbine are given.

The rest of this paper is presented as next: a new sight for the modelling of wind-turbine systems is given. Control scheme for wind-turbine, and discussions of the simulation results.

2 Model Plant and Modelling of Wind-Turbine

In this work, wind-turbine system is analyzed. Considering the different ways of modelling as [20–22], the wind turbine can be described by the following differential equations:

$$\begin{cases} \dot{\Omega}_T(t) = \widetilde{T}_{aero} - \widetilde{T}_{mec} \\ \dot{\Omega}_{g-ls}(t) = \widetilde{T}_{mec} - \widetilde{T}_{em} \\ \dot{T}_{me}(t) = \kappa\left(\Omega_T - \Omega_{g-ls}\right) + \beta\left(\dot{\Omega}_T(t) - \dot{\Omega}_{g-ls}(t)\right) \end{cases} \tag{1}$$

where, $\widetilde{\mathrm{T}}_{em} = \mathrm{T}_{em} G_g / J_{g-ls}$, $\widetilde{\mathrm{T}}_{aero} = \mathrm{T}_{aero}/J_T$ and $\widetilde{\mathrm{T}}_{mec} = \mathrm{T}_{mec}/J_T$ are the generator torque, the torque caught by the wind turbine and the low-speed shaft torque resulting from the torsion and frictions effects, respectively; while J_{g-ls} and J_T are respectively the generator and the turbine inertia. Ω_{g-ls} is the generator rotational speed reported to the low speed shaft.

Let us define

$$\dot{z}(t) = [\dot{\Omega}_T(t) \quad \dot{\Omega}_{g-ls}(t) \quad \dot{\mathrm{T}}_{me}(t)]^T \tag{2}$$

$$\dot{z}(t) = \begin{bmatrix} 0 & 0 & 1/J_T \\ 0 & 0 & 1/J_T \\ \kappa & -\kappa & 0 \end{bmatrix} z(t) + \begin{bmatrix} 1/J_T \\ -G_g/J_{g-ls} \\ 0 \end{bmatrix} u(t_k) + \begin{bmatrix} \beta \\ -\beta \\ 0 \end{bmatrix} \dot{z}(t) \tag{3}$$

Then, we get

$$\dot{z}(t) = \overline{\mathcal{A}} z(t) + \overline{\mathcal{B}} u(t_k) \tag{4}$$

where, $\mathcal{A}$, and $\mathcal{B}$ are represented the matrix state and matrix of signal control.

$\widetilde{\beta}$ is the event-triggered matrix. $u(t_k) = [\mathrm{T}_{aero}\mathrm{T}_{em}0]$ represent the event-triggered input.

The defining matrices of (4) are describing as follows:

$$\overline{\mathcal{A}} = \tilde{\beta}^{-1}\mathcal{A}, \overline{\mathcal{B}} = \tilde{\beta}^{-1}\mathcal{B} \quad \text{and} \quad \tilde{\beta} = \begin{bmatrix} 1+\beta & 1-\beta & 1 \end{bmatrix},$$

$$\mathcal{A} = \begin{bmatrix} 0 & 0 & 1/J_T \\ 0 & 0 & 1/J_T \\ \kappa & -\kappa & 0 \end{bmatrix}, \mathcal{B} = \begin{bmatrix} 1/J_T \\ -G_g/J_{g-ls} \\ 0 \end{bmatrix}$$

To convert the uncertain non-linear problem (4) to an uncertain linear problem, we use LFT (Fractional Linear Transformation) representations as formalisms.

With appropriate sampling time for the system (4) we get the following discrete-time uncertain system as next:

$$\Psi(k+1) = A\Psi(k) + Bu(k) \tag{5}$$

3 Efficient Dynamic Control Strategy

From Eq. (5) we consider the next closed loop discrete-time system is employed an observer, which is designed as:

$$\widehat{\Psi}(k+1) = (A+B)\widehat{\Psi}(k) + LCe(k) \tag{6}$$

where A, $B\, and\, C$ are state matrices, K are the controller gains matrix and L is the observer gain matrix

The previous uncertain wind-turbine system occurring sensors faults can be written as:

$$\tilde{\Psi}(k+1) = \tilde{A}(k)\,\tilde{\Psi}(k) + \tilde{B}\tilde{u}(k) \tag{7}$$

where, $\tilde{A}_p = \begin{bmatrix} A & LC \\ C & F_s FC \end{bmatrix}$, $\tilde{B}_p = \begin{bmatrix} B \\ 0_n \end{bmatrix}$,

The matrix $\tilde{A}$ and $\tilde{B}$ represent the system and the input matrices, respectively.

Taking into account results in [1], the $\tilde{y}$ is the output of the augmented wind-turbine system, and the control input $u(k)$ represents the desired energy of the current step k. The performance of optimization quadratic function J is given as follows:

$$\begin{cases} \min\limits_{u} J_h(k) = \sum_{j=0}^{h}\left(\| \tilde{\Psi}(k+j/k)\|_Q^2 + \|u(k++j/k)\|_R^2 \right) + \rho\widehat{\Psi}^T(k/k)P_0\widehat{\Psi}(k/k) \\ Subject\ to \begin{cases} \tilde{\Psi}(k+1) \\ Wu \le Z \end{cases} \end{cases} \tag{8}$$

where

$$\begin{cases} \tilde{y}(k) = [y(k+1)y(k+2)\ldots y(k+N)]^T \\ \tilde{u}(k) = [u(k)u(k+1)\ldots u(k+N)]^T \\ \rho = [01]\text{ is the triggering threshold} \end{cases} \tag{9}$$

The h and λ represents the prediction step and weighting factor for input, respectively. The term $[\tilde{y}(k+j/k) - y_d(k+j/k)]^2$ represents the weighting of future error, and $\lambda[\tilde{u}(k+j/k)]^2$ represents the energy of input control.

The predictive model $\tilde{y}(k)$ is described as next:

$$\tilde{y}(k+1) = H(k)\tilde{x}(k) + M\overrightarrow{u}(k) + G\tilde{u}(k) \tag{10}$$

where, G, H and F are respectively the augmented state matrix system, input control and Event-triggered matrices

$$H = \left[\tilde{C}\tilde{A}\tilde{C}\tilde{A}^2\ldots\tilde{C}\tilde{A}^N\right]^T,$$

$$G = \left[\tilde{C}\tilde{B}\tilde{C}\tilde{A}\tilde{B}\ldots\tilde{C}\tilde{A}^{N-1}\tilde{B}\right]^T,$$

$$M = \begin{bmatrix} 0_n & 0_n & \ldots & 0_n \\ \tilde{C}\tilde{B} & 0_n & \ldots & 0_n \\ \vdots & \vdots & \ddots & \vdots \\ \tilde{C}\tilde{A}^{N-1}\tilde{B} & 0_n & \ldots & 0_{n0} \end{bmatrix}$$

Furthermore, in this paper the $\vec{u}$ term indicates the desired energy for the speed control of h step prediction horizon for wind-turbine. for relaxation conditions of the online optimization, we hold the (10) in (8), then, the cost function can be arranged as (11), taking into consideration the constrains of the control input. The performance index J defined to calculate the optimal solution.

$$\min_{\vec{u},\quad i>0} \frac{1}{2}\vec{u}^T P\vec{u} + R^T\vec{u} \quad \text{Subject to} \quad \mathcal{W}u \leq \mathcal{Z} \tag{11}$$

Where W and Z are representing the lower and upper energy limits, respectively.

To deal with actuators and sensors failures the P and R matrices in (11) are defined as follows:

$$P = \lambda\mathcal{V}^T\mathcal{V} + M^T M \tag{12}$$

$$R = M^T(H(k)\tilde{x}(k) + Gu(k)) \tag{13}$$

where $\mathcal{W}$, $\mathcal{V}$ and $\mathcal{Z}$ are appropriate matrices

In this stage to test the convexity and feasibility of (11) for the FDMPC, the definiteness of Matrix P has been evaluated. To compute the optimal control input based on the Quadratic Programming optimization. The necessary condition is that the quadratic function defined in (11) is strictly convex if and only if matrix P is positive definite.

4 Simulation Results

To test and evaluated the effectiveness of the presented approach, the model (6) that reflected the wind –turbine system (1) after modelling and reformulation. The simulation parameters used in this paper are as next (Table 1).

Table 1 Wind-turbine physical parameters.

Parameters	Description	Numerical value
J_T	Turbine inertia	4.75×10^6 N m s^2
J_g	Generator inertia	0.85×10^6 N m s^2
G_g	Gearbox gain	91.3
Ω_{T-nom}	Nominal speed	2.05 MW
P_{nom}	Nominal power	17.75 rpm
β	Mechanical coupling damping coefficient	3.2×10^5 N m^{-1} s
κ	Mechanical coupling stiffness coefficient	109×10^6 N m^{-1} s
ρ	Air density	1.225 kg/m^3

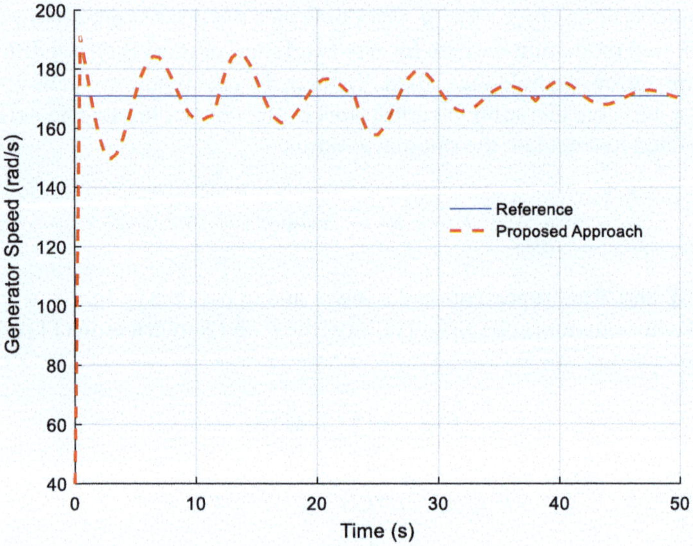

Fig. 1 The estimated generator speed of wind-turbine

The weighting matrices are: $Q_0 = I_n, R_0 = 0.5$.

Figure 1 shows the variation of the predicted generator speed in presence of event-triggered for the scenario of low wind speed. The estimated generator speed is controlled around the desired value (i.e. 171 rad/s), meanwhile, the generator speed has changing between 156 and 184 rad/s, that consider as an acceptable deviation, since it's around of 10% as ratio.

5 Conclusion

The event-triggered based predictive control strategy is studied in this paper. To improve precise control instance using obtained optimal control, an augmented system is reformulated using a state observer based predictive control scheme as a constrained QP. This approach employs event-triggered sampling, mainly, the control updates are triggered only when specific events or significant changes occur in the wind-turbine. The simulations results show good performances for the generator speed of wind-turbine.

References

1. Abdelmalek Z, Abdelhamid B, Sofiane B, Boulkaibet I (2021) Fault diagnosis of uncertain hybrid actuators based model predictive control. In: Proceedings of the 4th international conference on electrical engineering and control applications: ICEECA 2019, 17–19 Dec 2019, Constantine, Algeria 2021. Springer, Singapore, pp 961–971
2. Zahaf A, Bououden S, Chadli M, Chemachema M (2022) Robust fault tolerant optimal predictive control of hybrid actuators with time-varying delay for industrial robot arm. Asian J Control 24(1):1–5
3. Zahaf A, Bououden S, Chadli M, Boulkaibet I, Neji B, Khezami N (2023) Dynamic sensorless control approach for Markovian switching systems applied to PWM DC–DC converters with time-delay and partial input saturation. Sensors 23(15):6936
4. Zahaf A, Bououden S, Chadli M (2017) Constrained fuzzy predictive control design based on the PDC approach. In: Recent advances in electrical engineering and control applications 2017. Springer International Publishing, pp 140–154
5. Abdelmalek Z, Chemachema M, Sofiane B, Boulkaibet I (2021) Fault tolerant predictive control for constrained hybrid systems with sensors failures. In: Proceedings of the 4th international conference on electrical engineering and control applications: ICEECA 2019, 17–19 Dec 2019, Constantine, Algeria. Springer, Singapore, pp 973–984
6. Abdelmalek Z, Bououden S, Chadli M, Zelinka I, Boulkaibet I (2019) Observer based model predictive control of hybrid systems. In: Advanced control engineering methods in electrical engineering systems 2019. Springer International Publishing, pp 198–207
7. Zahaf A, Boutamina B, Bououden S, Filali S (2014) New approach of model based TS fuzzy predictive control using LMI approach. In: 2014 15th international conference on sciences and techniques of automatic control and computer engineering (STA), 21 Dec 2014. IEEE, pp 38–43
8. Zahaf A, Bououden S, Chemachema M. Contribution to robust fault-tolerant predictive control with constraints for hybrid systems (Doctoral dissertation, Université Frères Mentouri-Constantine 1)
9. Bououden S, Boulkaibet I, Chadli M, Abboudi A (2021) A robust fault-tolerant predictive control for discrete-time linear systems subject to sensor and actuator faults. Sensors 21:2307
10. Bounemeur A, Chemachema M, Essounbouli N (2018) Indirect adaptive fuzzy fault-tolerant tracking control for MIMO nonlinear systems with actuator and sensor failures. ISA Trans 79:45–61
11. Bounemeur A, Chemachema M (2021) Adaptive fuzzy fault-tolerant control using Nussbaum-type function with state-dependent actuator failures. Neural Comput Appl 33:191–208
12. Bounemeur A, Chemachema M (2021) Adaptive fuzzy fault-tolerant control for a class of nonlinear systems under actuator faults: application to an inverted pendulum. Int J Robot Control Syst 1(2):102–115
13. Abdelhamid B, Mouhamed C, Najib E (2019) Optimal indirect robust adaptive fuzzy control using PSO for MIMO nonlinear systems. In: Advanced control engineering methods in electrical engineering systems 2019. Springer International Publishing, pp 208–224
14. Abdelhamid B, Mouhamed C, Najib E (2017) Indirect robust adaptive fuzzy control of uncertain two link robot manipulator. In: Recent advances in electrical engineering and control applications 2017. Springer International Publishing, pp 125–139
15. Bounemeur A, Chemachema M (2021) Active adaptive fuzzy fault-tolerant control for a class of nonlinear systems with actuator faults. In: Proceedings of the 4th international conference on electrical engineering and control applications: ICEECA 2019, 17–19 Dec 2019, Constantine, Algeria 2021. Springer, Singapore, pp 985–999
16. Bounemeur A, Chemachema M, Essounbouli N (2014) Robust indirect adaptive fuzzy control using Nussbaum gain for a class of SISO nonlinear systems with unknown directions. In: 2014 15th international conference on sciences and techniques of automatic control and computer engineering (STA), 21 Dec 2014. IEEE, pp 748–754

17. Bounemeur A, chemachema M, Zahaf A, Bououden S (2019) Adaptive fuzzy fault-tolerant control using nussbaum gain for a class of SISO nonlinear systems with unknown directions. In: Proceedings of the 4th international conference on electrical engineering and control applications: ICEECA 2019, 17–19 Dec 2019, Constantine, Algeria 2021. Springer, Singapore, pp 493–510
18. Abdelhamid B, Chemachema M, Bouzina S (2023) Fuzzy fault-tolerant control applied on two inverted pendulums with nonaffine nonlinear actuator failures. Int J Robot Control Syst 3(2)
19. Bounemeur A, Chemachema M (2023) General fuzzy adaptive fault-tolerant control based on Nussbaum-type function with additive and multiplicative sensor and state-dependent actuator faults. Fuzzy Sets Syst 7:108616
20. Boukhezzar B, Lupu L, Siguerdidjane H, Hand M (2007) Multivariable control strategy for variable speed, variable pitch wind turbines. Renew Energy 32(8):1273–1287
21. Khezami N, Braiek NB, Guillaud X (2010) Wind turbine power tracking using an improved multimodel quadratic approach. ISA Trans 49(3):326–334
22. Bououden S, Chadli M, Filali S, El Hajjaji A (2012) Fuzzy model based multivariable predictive control of a variable speed wind turbine: LMI approach. Renew Energy 37:434–439
23. Bououden S, Allouani F, Abboudi A, Chadli M, Boulkaibet I, Barakeh ZA, Neji A, Ghandour R (2023) Observer-based robust fault predictive control for wind turbine time-delay systems with sensor and actuator faults. Energies 16(2):858
24. Liu J, Yao Q, Hu Y (2019) Model predictive control for load frequency of hybrid power system with wind power and thermal power. Energy 172:555–565

PID Controller for Precise Voltage Output of Solid Oxide Fuel Cell

Abdelhamid Bounemeur, Abdelmalek Zahaf, and Mohamed Chemachema

Abstract In this paper, we introduce a novel methodology for precise control of Solid Oxide Fuel Cell (SOFC) systems, aiming to optimize power generation while accounting for uncertainties, disturbances, sensor malfunctions, and nonlinear actuator issues. Our approach leverages PID controller as the main components in our control strategy. PID is known as a powerful tool in control theory due to its ability to handle uncertainties and ensure a precise tracking to any given references. Voltage control is a critical aspect of SOFC operation, directly influencing the cell's efficiency and stability. PID controllers play a pivotal role in maintaining a steady and controlled voltage output by dynamically adjusting operating parameters based on real-time feedback. By employing PID control in voltage regulation, SOFCs can maintain a stable electrical output, preventing overvoltage or underperformance issues. Precise tuning of PID parameters tailored to the specific characteristics of SOFCs is crucial to achieving optimal voltage control. Simulation study will show the advantages behind integrating PID controller on voltage control of a SOFC system.

Keywords Solid oxide fuel cell · PID controller · Voltage control

A. Bounemeur (✉) · A. Zahaf · M. Chemachema
Signla Processing Laboratory, Department of Electronics, Faculty of Engineering, University of Brothers Mentouri Constantine 1, Constantine, Algeria
e-mail: abdelhamid.bounemeur@enp-constantine.dz

A. Zahaf
e-mail: zahaf.abdelmalek@umc.edu.dz

M. Chemachema
e-mail: m_chemachema@yahoo.fr

A. Bounemeur
Polytech Electrical Engineering Laboratory LGEPC, National Polytechnic School of Constantine, Constantine, Algeria

S. Hamamda et al. (eds.), *Proceedings of the 2nd International Conference of Nanotechnology for Environmental Protection and Clean Energy Production*, Springer Proceedings in Materials 45, https://doi.org/10.1007/978-981-97-1916-7_33

1 Introduction

In the last decade, researchers have been widely focused on the study of Solid Oxide Fuel Cell, Solid Oxide Fuel Cells (SOFCs) are innovative energy conversion devices that produce electricity through an electrochemical process. They operate at high temperatures, utilizing a solid ceramic electrolyte that conducts oxygen ions from the cathode to the anode. This unique design allows them to efficiently convert the chemical energy of fuels, such as hydrogen, natural gas, or methane, directly into electricity. Controlling SOFCs is crucial to optimize their performance and ensure their efficiency and durability. Among the various control techniques used, PID (Proportional-Integral-Derivative) control is often employed to regulate and stabilize different aspects of SOFC operation. The PID control system consists of three main components [1–10]. Proportional (P) Control: This component responds proportionally to the error between the desired and actual values. In the context of SOFCs, P control can be used to regulate parameters like temperature, fuel flow rate, or oxygen flow rate. For instance, if the temperature deviates from the setpoint, the P component adjusts the control signal in proportion to the error to bring the temperature back to the desired level. Integral (I) Control: The integral component accumulates the error over time and uses this accumulated error to eliminate any steady-state error [10–15]. In SOFCs, I control can help mitigate long-term deviations from setpoints. For example, it can compensate for slow changes in operating conditions or help maintain consistent fuel or air flow rates. Derivative (D) Control: The derivative component predicts the future trend of the error based on its rate of change. In SOFC applications, D control can anticipate rapid changes in operating conditions and provide preemptive adjustments to prevent overshoots or instability in parameters like temperature or voltage.

In a SOFC system, PID control can be applied to various aspects [7, 16–19]: Temperature Control: PID controllers can regulate the temperature within the SOFC to maintain optimal operating conditions, preventing thermal stresses and ensuring efficient performance. Fuel and Air Flow Control: By utilizing PID control on flow control valves, the flow rates of fuel and air entering the SOFC can be adjusted dynamically to maintain the required stoichiometry and prevent issues like fuel starvation or oxygen depletion.

Compared to adaptive and predictive control scheme, the PID controller has a superiority against modern control theory, due to its cost and execution time [1–24].

Voltage and Current Control: PID control can be employed to regulate the output voltage and current of the SOFC to prevent overvoltage, underperformance, or fluctuations in power output.

In our study, we will focus on the control of the voltage of the SOFC to prevent any overvoltage and fluctuations in the output to the systems. Moreover, PID controller will be integrated in the feed forward path a s a main controller to further enhance the performances in case of faults or uncertainties.

The emphasize of this paper is organized as follows:

Mathematical model of a SOFC; Linear Model; Controller strategy; Simulation; Conclusion.

2 Model Plant and Control Dynamic

The mathematical model of the SOFC plant is defined with the following state-space representation:

$$\begin{cases} \dot{x}_1(t) = \frac{1}{\tau_f}[u(t) - x_1(t)] \\ \dot{x}_2(t) = \frac{1}{\tau_{H_2}}\left[\frac{1}{K_{H_2}}(x_1(t) - 2K_r i(t)) - x_2(t)\right] \\ \dot{x}_3(t) = \frac{1}{\tau_{H_2O}}\left[\frac{2K_r i(t)}{K_{H_2O}} - x_3(t)\right] \\ \dot{x}_4(t) = \frac{1}{\tau_{O_2}}\left[\frac{1}{K_{O_2}}\left(\frac{x_1(t)}{r_{H-O}} - 2K_r i(t)\right) - x_4(t)\right] \\ y(t) = N.\left[E^0 + \frac{RT}{2F}\left(\frac{x_2(x_4)^{\frac{1}{2}}}{x_3}\right)\right] - R_{\text{ohmic}}\, i(t) \end{cases} \tag{1}$$

where the output is considered as the battery voltage V; P_{H_2}, P_{H_2O}, P_{O_2} are respectively the partial pressure of the hydrogen, water, and oxygen. This equation is defined based on the Nernst equation regarding the ohmic losses.

3 Linear Model of SOFC

To make the controller easy to build, a linearization model around an equilibrium point is presented below, with four state matrix A, B, C, D:

$$\begin{cases} \dot{X} = A + Bu \\ Y = CX + Du \end{cases} \tag{2}$$

where

$$V = N.\left[E^0 + \frac{RT}{2F}\ln\left(\frac{P_{H_2}(P_{O_2})^{\frac{1}{2}}}{P_{H_2O}}\right)\right] - R_{\text{ohmic}}\,.I \tag{3}$$

$$\mathbf{x(t)} = \begin{pmatrix} x_1(t) \\ x_2(t) \\ x_3(t) \\ x_4(t) \end{pmatrix} = \begin{pmatrix} q_{\mathrm{H_2}} \\ P_{\mathrm{H_2}} \\ P_{\mathrm{H_2O}} \\ P_{\mathrm{O_2}} \end{pmatrix}, \mathbf{y(t)} = V \tag{4}$$

$$\mathbf{A}=\begin{bmatrix}\frac{\partial f_1}{\partial x_1} & \frac{\partial f_1}{\partial x_2} & \frac{\partial f_1}{\partial x_3} & \frac{\partial f_1}{\partial x_4}\\ \frac{\partial f_2}{\partial x_1} & \frac{\partial f_2}{\partial x_2} & \frac{\partial f_2}{\partial x_3} & \frac{\partial f_2}{\partial x_4}\\ \frac{\partial f_3}{\partial x_1} & \frac{\partial f_3}{\partial x_2} & \frac{\partial f_3}{\partial x_3} & \frac{\partial f_3}{\partial x_4}\\ \frac{\partial f_4}{\partial x_1} & \frac{\partial f_4}{\partial x_2} & \frac{\partial f_4}{\partial x_3} & \frac{\partial f_4}{\partial x_4}\end{bmatrix}_{X=X_{eq}} = \begin{bmatrix}-\frac{1}{\tau_f} & 0 & 0 & 0\\ \frac{1}{\tau_{H_2}K_{H_2}} & -\frac{1}{\tau_{H_2}} & 0 & 0\\ 0 & 0 & -\frac{1}{\tau_{H_2O}} & 0\\ \frac{1}{\tau_{O_2}K_{O_2}r_{H-O}} & 0 & 0 & -\frac{1}{\tau_{O_2}}\end{bmatrix} \quad (5)$$

$$\mathbf{B}=\begin{bmatrix}-\frac{1}{\tau_f}\\ 0\\ 0\\ 0\end{bmatrix}$$

x_{20}, x_{40}: represents the equiliburm points.

$$\mathbf{C}=\begin{bmatrix}0 & \frac{NRT}{2Fx_{20}} & -\frac{NRT}{2F} & \frac{NRT}{4F(x_{40})^2}\end{bmatrix}, \mathbf{D}=0 \quad (6)$$

The real parameters of the SOFC are presented in Table 1.

Table 1 SOFC real parameters

Parameters	Values	Unity	Representation
T	1273	K	Absolute temperature
F	96.487×10^6	C/kmol	Faraday constant
R	8314	J/(kmol K)	Universal gas constant
E^0	1.18	V	Ideal standard voltage
N	450	-	Cellule numbers in the «stack»
R_{ohmic}	3.2813×10^{-4}	Ω	Ohmic losses
K_r	0.996×10^{-6}	kmol/(s A)	Constant
K_{H_2}	8.43×10^{-4}	kmol/(s atm)	Hydrogen constant
K_{H_2O}	2.81×10^{-4}	kmol/(s atm)	Water constant
K_{O_2}	$2.52e \times 10^{-3}$	kmol/(s atm)	Oxygen constant
τ_{H_2}	26.1	s	Response time of hydrogen flux
τ_{H_2O}	78.3	s	Response time of water flux
τ_{O_2}	2.91	s	Response time of oxygen flux
τ_f	5	s	Time-constante for combustion
r_{H-O}	1.145	–	Ratio between H_2 et O_2
i_L	800	A/cm^2	Density of the limited current
a	0.05	–	Constant of Tafel
b	0.11	–	Slope of Tafel

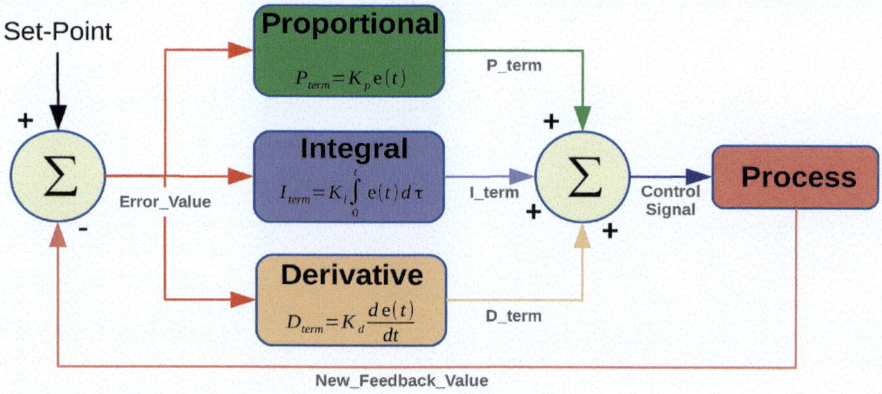

Fig. 1 PID controller scheme

4 Controller Strategy

In this part, a PID controller has been proposed in order to meet the desired performances and deal with parameters uncertainties and disturbances.

PID controller has three actions, the past action (integral), the present action (proportional), and the future action (derivative), the block diagram of this controller in shown Fig. 1.

The process in our case is the SOFC system.

5 Results and Simulation

In this part, simulation result will be presented for the output voltage of the SOFC along with the different pressure and the control input signal (Fig. 2).

The first test is the step input, so that we can assess the behavior of the system for a step testing input.

After choosing the PID gains with tunning, we can further obtain a good result as shown in Figs. 3 and 4.

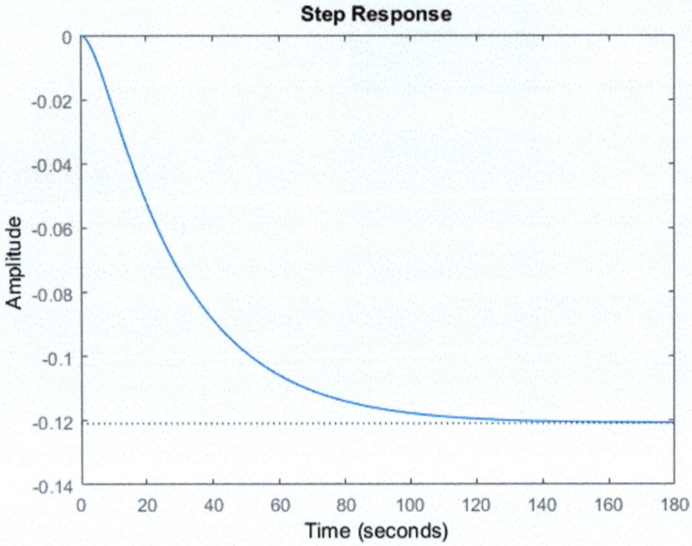

Fig. 2 Step response of the open-loop

Fig. 3 The output voltage in terms of current and temperature

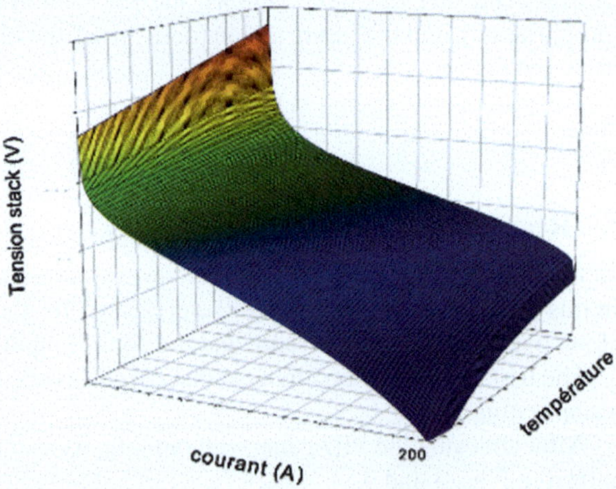

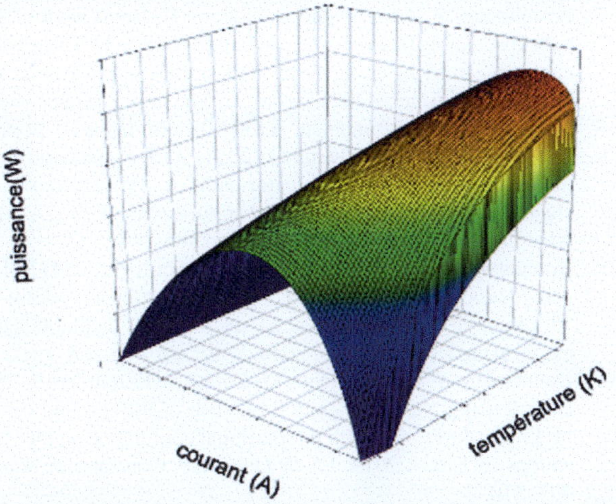

Fig. 4 The output power in terms of current and temperature

6 Conclusion

We can conclude in this paper that the PID controller is a powerful toll for controlling any process with the presence of disturbances and uncertainties. Thus, the SOFC process should have a precise voltage control during its operating mode and also, in other works, authors tried to control also the temperature of the SOFC, which will be our next study. Furthermore, the used PID controller proves the efficiency and the accuracy of the output voltage.

References

1. Bounemeur A, Chemachema M, Essounbouli N (2018) Indirect adaptive fuzzy fault-tolerant tracking control for MIMO nonlinear systems with actuator and sensor failures. ISA Trans 1(79):45–61
2. Bounemeur A, Chemachema M (2021) Adaptive fuzzy fault-tolerant control using Nussbaum-type function with state-dependent actuator failures. Neural Comput Appl 33:191–208
3. Bounemeur A, Chemachema M (2021) Adaptive fuzzy fault-tolerant control for a class of nonlinear systems under actuator faults: application to an inverted pendulum. Int J Robot Control Syst 1(2):102–115
4. Bounemeur A, Chemachema M, Essounbouli N (2014) New approach of robust direct adaptive control of a class of SISO nonlinear systems. In: 2014 15th international conference on sciences and techniques of automatic control and computer engineering (STA), 21 Dec 2014. IEEE, pp 725–730
5. Abdelhamid B, Mouhamed C, Najib E (2019) Optimal indirect robust adaptive fuzzy control using PSO for MIMO nonlinear systems. In: Advanced control engineering methods in electrical engineering systems 2019. Springer International Publishing, pp 208–224
6. Abdelhamid B, Mouhamed C, Najib E (2017) Indirect robust adaptive fuzzy control of uncertain two link robot manipulator. In: Recent advances in electrical engineering and control applications 2017. Springer International Publishing, pp 125–139

7. Abdelmalek Z, Abdelhamid B, Sofiane B, Boulkaibet I (2021) Fault diagnosis of uncertain hybrid actuators based model predictive control. In: Proceedings of the 4th international conference on electrical engineering and control applications: ICEECA 2019, 17–19 Dec 2019, Constantine, Algeria 2021. Springer, Singapore, pp 961–971
8. Chemachema M, Bounemeur A (2017) Commande adaptative floue des systèmes non linéaires
9. Bounemeur A, Chemachema M (2021) Active adaptive fuzzy fault-tolerant control for a class of nonlinear systems with actuator faults. In: Proceedings of the 4th international conference on electrical engineering and control applications: ICEECA 2019, 17–19 Dec 2019, Constantine, Algeria 2021. Springer, Singapore, pp 985–999
10. Bounemeur A, Chemachema M, Essounbouli N (2014) Robust indirect adaptive fuzzy control using Nussbaum gain for a class of SISO nonlinear systems with unknown directions. In: 2014 15th international conference on sciences and techniques of automatic control and computer engineering (STA), 21 Dec 2014. IEEE, pp 748–754
11. Bounemeur A, Chemachema M, Zahaf A, Bououden S (2019) Adaptive fuzzy fault-tolerant control using Nussbaum gain for a class of SISO nonlinear systems with unknown directions. In: Proceedings of the 4th international conference on electrical engineering and control applications: ICEECA 2019, 17–19 Dec 2019, Constantine, Algeria 2021. Springer, Singapore, pp 493–510
12. Abdelhamid B, Chemachema M, Bouzina S (2023) Fuzzy fault-tolerant control applied on two inverted pendulums with nonaffine nonlinear actuator failures. Int J Robot Control Syst 3(2)
13. Bounemeur A, Chemachema M (2019) Contribustion à la commande tolérante aux défauts d'une classe de systèmes non linéaires. Diss. Université Frères Mentouri-Constantine 1
14. Bounemeur A, Chemachema M (2023) General fuzzy adaptive fault-tolerant control based on Nussbaum-type function with additive and multiplicative sensor and state-dependent actuator faults. Fuzzy Sets Syst 7:108616
15. Zahaf A, Bououden S, Chadli M, Chemachema M (2022) Robust fault tolerant optimal predictive control of hybrid actuators with time-varying delay for industrial robot arm. Asian J Control 24(1):1–5
16. Chadli M, Bououden S, Zelinka I (eds). Recent advances in electrical engineering and control applications
17. Zahaf A, Boutamina B, Bououden S, Filali S (2014) New approach of model based TS fuzzy predictive control using LMI approach. In: 2014 15th international conference on sciences and techniques of automatic control and computer engineering (STA), 21 Dec 2014. IEEE, pp 38–43
18. Abdelmalek Z, Chemachema M, Sofiane B, Boulkaibet I (2021) Fault tolerant predictive control for constrained hybrid systems with sensors failures. In: Proceedings of the 4th international conference on electrical engineering and control applications: ICEECA 2019, 17–19 Dec 2019, Constantine, Algeria. Springer, Singapore, pp 973–984
19. Zahaf A, Bououden S, Chadli M (2017) Constrained fuzzy predictive control design based on the PDC approach. In: Recent advances in electrical engineering and control applications 2017. Springer International Publishing, pp 140–154
20. Filali S, Zahaf A. Commande Prédictive Floue Sous Contraintes
21. Zahaf A, Bououden S, Chadli M, Boulkaibet I, Neji B, Khezami N (2023) Dynamic sensorless control approach for Markovian switching systems applied to PWM DC–DC converters with time-delay and partial input saturation. Sensors 23(15):6936
22. Zahaf A, Bououden S, Chemachema M. Contribution to robust fault-tolerant predictive control with constraints for hybrid systems (Doctoral dissertation, Université Frères Mentouri-Constantine 1)
23. Abdelmalek Z, Bououden S, Chadli M, Zelinka I, Boulkaibet I (2019) Observer based model predictive control of hybrid systems. In: Advanced control engineering methods in electrical engineering systems 2019. Springer International Publishing, pp 198–207
24. Bounemeur A, Chemachema M (2023) Optimal adaptive fuzzy fault-tolerant control applied on a quadrotor attitude stabilization based on particle swarm optimization. Proc Inst Mech Eng Part I J Syst Control Eng 5:09596518231199212

Finite-Control-Set Model Predictive Control (FCS-MPC) and Fuzzy Self-Adaptive PI Controller (FSA-PIC) for Wind Turbine System Based on DFIG

Amira Aggoune, Farid Berrezzek, and Khaled Khelil

Abstract This paper aims at conducting a comparative analysis between two control strategies, namely the finite-control-set model predictive control (FCS-MPC) and the fuzzy self-adaptive proportional-integral controller (FSA-PIC), for regulating the rotor side converter (RSC) in a wind turbine system (WTS) based on a doubly-fed induction generator (DFIG). This study seeks to assess the performance of these strategies with regard to precision and stability when faced with random variations in wind speed and changes in system parameters. To achieve this objective, numerous numerical tests have been carried out within the MATLAB/Simulink environment, employing a 1.5 MW WTS model. The results obtained from these tests consistently demonstrate that the FCS-MPC outperforms the FSA-PIC in terms of accuracy, robustness, and the minimization of power fluctuations across all operational scenarios.

Keywords Wind turbine system · Doubly-fed induction generator · Fuzzy self-adaptive PI controller · Finite-control-set model predictive control

1 Introduction

In the last decade, wind energy conversion systems (WECS) equipped with doubly-fed induction generator (DFIG) are becoming the most competitive renewable energy alternative for delivering clean energy to the power grid [1]. Thanks to its primary features of power transmission controllability, operational flexibility at different operation modes and low voltage ride-through capability [2, 3]. However, one of the main issues in WECS based on DFIG is often to lack stability due to the intrinsic randomness of wind energy and the non-linear nature of the DFIG [1]. As a result, many control strategies have been developed including the decoupled vector control with

A. Aggoune (✉) · F. Berrezzek · K. Khelil
LEER Laboratory, Department of Electrical Engineering, Faculty of Sciences and Technologies, University of Souk Ahras, Souk Ahras, Algeria
e-mail: a.aggoune@univ-soukahras.dz

S. Hamamda et al. (eds.), *Proceedings of the 2nd International Conference of Nanotechnology for Environmental Protection and Clean Energy Production*, Springer Proceedings in Materials 45, https://doi.org/10.1007/978-981-97-1916-7_34

standard PI regulator, which have the perfect capability of controlling independently the stator active and reactive power supplied to the power grid [2]. However, this control method still unsuitable for non-linear and uncertain systems, and suffers from poor transient performance resulting from the conventional method of adjusting the PI parameters [4].

Therefore, adopting an adaptive control algorithm is one potential solution to overcome the drawbacks of the PI controllers [3]. The most common adaptive technique presented in the literature is fuzzy self-adaptive PI controller (FSA-PIC) [1]. In this technique the PI gains are updated in real-time according to the system operating conditions [2, 3]. This method exhibits optimal performance addressed to the model uncertainties and external disturbances [1]. However, the major shortcoming of the fuzzy logic control (FLC) resides in the need for expert knowledge of the process.

Recently, the finite-set-control model predictive control (FCS-MPC) has been introduced as an alternative control strategy able to ensure optimal and robust control performance to the WECS [5]. The FCS-MPC is an optimization technique based on the dynamical model of the system [6]. In this technique, the switching states that trigger the rotor side converter (RSC) are generated with respect to the system plant and a cost function. The key feature of the FSC-MPC lies in removing current regulation loop and the modulation block [4]. The control signals are obtained by minimizing a cost function that evaluate the error between the set point and its own prediction value for each possible switching state of the RSC [6].

In this paper, we will conduct a comparative analysis between the FCS-MPC and the FSA-PIC, with the goal of evaluating the introduced control strategies in terms of precision and stability with respect to parametric variations and different operating modes that the DFIG can handle. This paper is intended to be an additional contribution to comparative performance analysis of FSA-PIC and FCS-MPC to control WTS based on DFIG. The rest of the paper is organized as follows: Mathematical description of the major parts of the WECS is presented in Sect. 2. FSA-PIC design is the subject of Sect. 3. The FCS-MBPC description is given in Sect. 4. The comparative analysis between the FSA-PIC and FCS-MPC is conducted in Sect. 5. Finally, Sect. 6 concludes the article.

2 DFIG Based WTS Modeling

2.1 DFIG Modeling

The current state representation of the DFIG in the synchronous d-q reference frame in terms of the stator and rotor resistances R_s, R_r, stator, rotor and mutual inductances L_s, L_r, L_m respectively, the stator angular speed ω_s and generator slip g can be expressed as follows [5, 7]:

$$\frac{d}{dt}\begin{pmatrix} i_{sd} \\ i_{sq} \\ i_{rd} \\ i_{rq} \end{pmatrix} = \frac{L_m}{\sigma}\begin{pmatrix} -\frac{1}{\tau_s L_m} & \omega_s \kappa_m & \frac{1}{\tau_r L_s} & \frac{\omega_s \kappa_s}{L_s} \\ -\omega_s \kappa_m & -\frac{1}{\tau_s L_m} & -\frac{\omega_s \kappa_s}{L_s} & \frac{1}{\tau_r L_s} \\ \frac{1}{\tau_s L_r} & -\frac{\omega_s \kappa_s}{L_r} & -\frac{1}{\tau_r L_m} & \omega_s \kappa_g \\ \frac{\omega_s \kappa_s}{L_r} & \frac{1}{\tau_s L_r} & -\omega_s \kappa_g & -\frac{1}{\tau_r L_m} \end{pmatrix}\begin{pmatrix} i_{sd} \\ i_{sq} \\ i_{rd} \\ i_{rq} \end{pmatrix} + \frac{L_m}{\sigma}\begin{pmatrix} \frac{1}{L_s L_m} & 0 & \frac{-1}{L_s L_r} & 0 \\ 0 & \frac{1}{L_s L_m} & 0 & \frac{-1}{L_s L_r} \\ \frac{-1}{L_s L_r} & 0 & \frac{1}{L_r L_m} & 0 \\ 0 & \frac{-1}{L_s L_r} & 0 & \frac{1}{L_r L_m} \end{pmatrix}\begin{pmatrix} V_{sd} \\ V_{sq} \\ V_{rd} \\ V_{rq} \end{pmatrix} \quad (1)$$

With: $\kappa_m = \left(\frac{1}{L_m} - \frac{L_m g}{L_s L_r}\right)$, $\kappa_g = \left(\frac{g}{L_m} - \frac{L_m}{L_s L_r}\right)$, $\kappa_s = (1 - g)$, $\tau_s = \frac{L_s}{R_s}$, $\tau_r = \frac{L_r}{R_r}$
The stator active and reactive powers of the DFIG are given by [8]:

$$\{P_s, Q_s\} = \left\{\frac{3}{2}\left(V_{sd} i_{sd} + V_{sq} i_{sq}\right), \frac{3}{2}\left(V_{sq} i_{sd} - V_{sd} i_{sq}\right)\right\} \quad (2)$$

2.2 RSC Modeling

The output voltage of a two-level RSC is defined as [4]:

$$\begin{bmatrix} V_{An} \\ V_{Bn} \\ V_{Cn} \end{bmatrix} = \frac{V_{dc}}{3}\begin{bmatrix} 2 & -1 & -1 \\ -1 & 2 & -1 \\ -1 & -1 & 2 \end{bmatrix}\begin{bmatrix} S_a \\ S_b \\ S_c \end{bmatrix} \quad (3)$$

With V_{An}, V_{Bn}, V_{Cn} are the phase voltages, V_{dc} is the DC -link voltage and S_a, S_b, S_c are the RSC switching states. Note that for a two-level RSC the number of switching states is equal to eight. Therefore, eight vector voltages can be obtained (see Fig. 1).

Fig. 1 Possible voltage vectors for the RSC

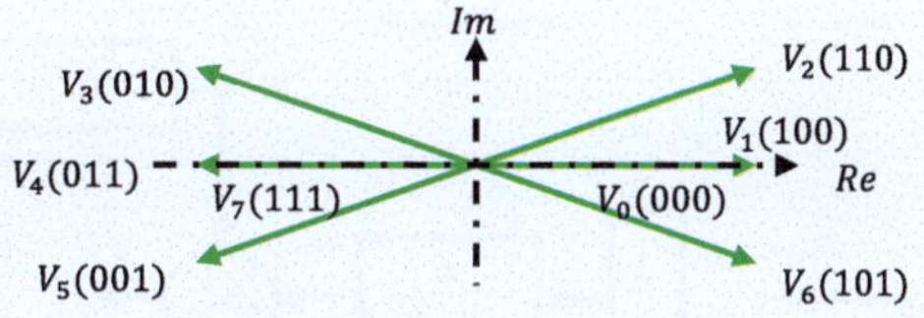

3 Desing of FSA-PIC for DFIG Based WTS

The decoupled d-q vector control is the most frequent control scheme applied to the WECS; however, it suffers from suboptimal performance resulting from the design of the PI regulator gains which is achieved by linearizing the system model at single operating point [3]. Therefore, adopting an intelligent method to adjust PI controller gains is a brilliant solution. For that reason, this section is devoted to the design of a fuzzy self-adaptive PI controller (FSA-PIC) (see Fig. 2).

3.1 Fundamentals on the Decoupled Vector Control

Generally, the stator active and reactive powers of the DFIG are controlled by means of the RSC using the decoupled d-q vector control strategy. In this technique, the d-axis component of the synchronous reference frame is aligned with the stator flux vector as described in (4) [3]:

$$\{\varphi_{sd}, \varphi_{sq}\} = \{L_s i_{sd} + L_m i_{rd}, L_s i_{sq} + L_m i_{rq} = 0\} \tag{4}$$

From (4), the relationship between the DFIG stator and rotor currents is [9]:

$$\{i_{sq}, i_{sq}\} = \left\{\frac{\varphi_{sd}}{L_s} - \frac{L_m}{L_s} i_{rd}, -\frac{L_m}{L_s} i_{rq}\right\} \tag{5}$$

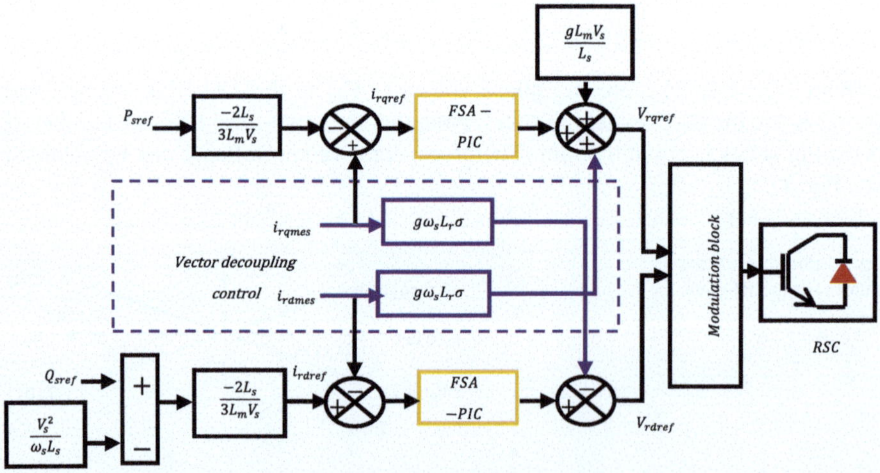

Fig. 2 Decoupled vector control-based FSA-PIC structure

Generally, we assumption that for medium and large-scale wind power generators the stator resistance is neglected, so the stator voltages expression can be written as follows [2]:

$$\{V_{sd}, V_{sq}\} = \{0, \omega_s \varphi_{sd}\} \tag{6}$$

As a result, the rotor voltage components can be deduced as [2, 3]:

$$\{V_{rd}, V_{rq}\} = \left\{R_r i_{rd} + \sigma L_r \frac{di_{rd}}{dt} - \sigma L_r g \omega_s i_{rq}, R_r i_{rq} + \sigma L_r \frac{di_{rq}}{dt} - \sigma L_r g \omega_s i_{rd} + \frac{g L_m V_s}{L_s}\right\} \tag{7}$$

Furthermore, the stator active and reactive powers expressions can be deduced by replacing (4) in (2) as:

$$\{P_s, Q_s\} = \left\{-\frac{3}{2}\frac{V_{sq} L_m}{L_s} i_{rq}, -\frac{3}{2}\left(\frac{V_{sq} L_m}{L_s} i_{rd} - \frac{V_{sq}^2}{\omega_s L_s}\right)\right\} \tag{8}$$

3.2 *FSA-PIC Design*

In this study the fuzzy logic control has been used as a supervisory system that adjusts online the parameters of the PI regulator $k_p(t)$ and $k_i(t)$ as represented in Fig. 3. Consequently, the parameters of the PI can be expressed in function of the PI controller's initial gains k_{p0}, k_{i0} and $\Delta k_p(t)$, $\Delta k_i(t)$ the output variables of the fuzzy inference as follows:

$$\{k_p(t), k_i(t)\} = \{k_{p0} + \Delta k_p(t), k_{i0} + \Delta k_i(t)\} \tag{9}$$

Finally, the expression of the output control system is given by equation:

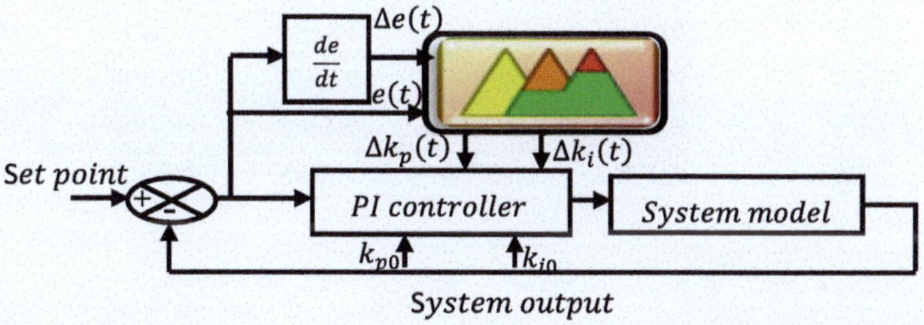

Fig. 3 FSA-PIC structure

$$u(t) = \left(k_{p0} + \Delta k_p(t)\right)e(t) + \left(k_{i0} + \Delta k_i(t)\right)\int_0^t e(t)dt \tag{10}$$

As explained previously, the fuzzy logic technique is used to update in real-time the parameters of the PI regulator in terms of the error $e(t)$ and change of error $\Delta e(t)$, which will serve as inputs for the FSA-PI controller. Each controller variable is labelled.

Each controller variable is labelled with five triangular membership functions (MFs) denoted as follows: NB/NS (negative big/small), ZE (Zero), PS/PB (positive small/big) (see Fig. 4). Additionally, the rule base is composed of 50 rules used to allow the Mamdani fuzzy inference system to proceed the decision-making process, as represented in Table 1. The output signal is generated by applying the centre of gravity (COG) defuzzification technique.

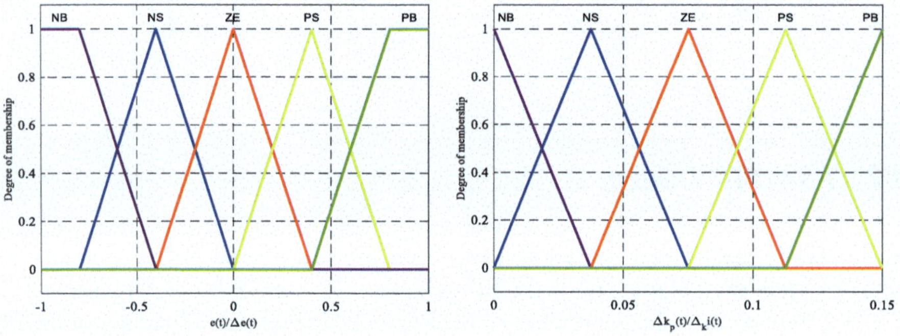

Fig. 4 MFs distribution of inputs and outputs of the FLC

Table 1 Fuzzy rules of the PI controller's gains

$\Delta k_p(t)/\Delta k_i(t)$		$\Delta e(t)$				
		NB	*NS*	*ZE*	*PS*	*PB*
$e(t)$	*NB*	PB/NB	PB/NB	PS/NS	PS/NS	ZE/ZE
	NS	PB/NB	PB/NB	PS/NS	ZE/ZE	NS/PS
	ZE	PB/NB	PS/NS	ZE/ZE	NS/PS	NS/PS
	PS	PS/NS	ZE/ZE	NS/PS	NS/PS	NB/PB
	PB	ZE/ZE	NS/PS	NS/PS	NB/PB	NB/PB

4 FCS-MPC for the Control of DFIG

The FCS-MPC is a control technique that makes an explicit use of the dynamical model of the process to predict its future evolution over a short period of time and choose the optimal control action by minimizing a cost function [6]. In this study, the rotor current regulation loop and the modulation block are replaced by a predictive rotor current control that uses the finite number of switching states that trigger the RSC and the system model to forecast the rotor currents of the DFIG, for each switching state, on a sampling period T_s. For each prediction, the optimization process is repeated using a cost function and the switching state yielding the least error between the predicted variable and the reference is chosen to be implemented during the next sampling period (Fig. 5).

4.1 Predictive Current Rotor Control

Based on the current state representation developed in (1), one can define the DFIG's rotor currents equations as follows:

$$\frac{di_{rd}}{dt} = \frac{L_m}{\sigma}\left(\frac{1}{\tau_s L_r} i_{sd} - \frac{\omega_s \kappa_s}{L_r} i_{sq} - \frac{1}{\tau_r L_m} i_{rd} + \omega_s \kappa_g i_{rq} - \frac{1}{L_s L_r} V_{sd} + \frac{1}{L_m L_r} V_{rd}\right) \tag{11}$$

$$\frac{di_{rq}}{dt} = \frac{L_m}{\sigma}\left(\frac{\omega_s \kappa_s}{L_r} i_{sd} + \frac{1}{\tau_s L_r} i_{sq} - \omega_s \kappa_g i_{rd} - \frac{1}{\tau_r L_m} i_{rq} - \frac{1}{L_s L_r} V_{sq} + \frac{1}{L_m L_r} V_{rq}\right) \tag{12}$$

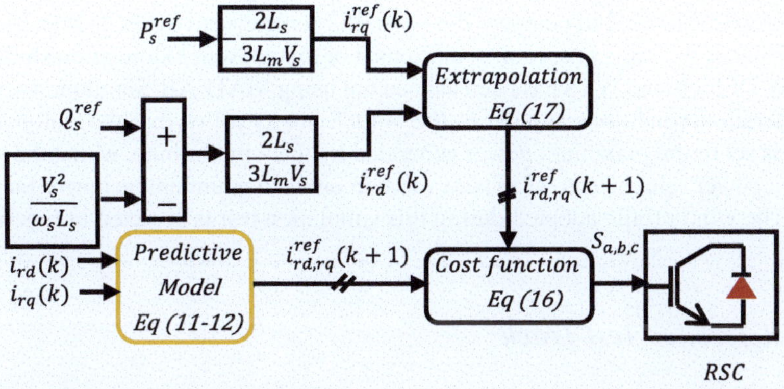

Fig. 5 FCS-MBPC structure

In order to derive the discrete model of the predicted rotor currents, we will adopt the forward difference Euler method as in (13) [10]:

$$\frac{dx}{dt} = \frac{x(k+1) - x(k)}{T_s} \tag{13}$$

Therefore, the discrete model of the predicted rotor currents corresponding to instant time $k+1$ can be written as following:

$$i_{rd}^{k+1} = \frac{L_m T_s}{\sigma}\left(\left(1 - \frac{1}{\tau_r L_m}\right) i_{rd}^k + \omega_s \kappa g i_{rq}^k + \frac{i_{sd}^k}{\tau_s L_r} - \frac{\omega_s \kappa_s}{L_r} i_{sq}^k - \frac{1}{L_s L_r} V_{sd}^k + \frac{1}{L_m L_r} V_{rd}^k\right) \tag{14}$$

$$i_{rq}^{k+1} = \frac{L_m T_s}{\sigma}\left(\left(1 - \frac{1}{\tau_r L_m}\right) i_{rq}^k - \omega_s \kappa g i_{rd}^k + \frac{\omega_s \kappa_s}{L_r} i_{sd}^k + \frac{1}{\tau_s L_r} i_{sq}^k - \frac{1}{L_s L_r} V_{sq}^k + \frac{1}{L_m L_r} V_{rq}^k\right) \tag{15}$$

The cost function can be calculated as follows [11]:

$$g = \left(i_{rdref}^{k+1} - i_{rd}^k\right)^2 + \left(i_{rqref}^{k+1} - i_{rq}^{k+1}\right)^2 \tag{16}$$

where i_{rdref}^{k+1} and i_{rqref}^{k+1} are the reference rotor currents at time $(k+1)$. They can be calculated by utilizing the $2nd$ order Lagrange extrapolation as following:

$$i_{rd,rqref}^{k+1}(k+1) = 3i_{rd,rqref}^{k} - 3i_{rd,rqref}^{k-1} + i_{rd,rqref}^{k-2} \tag{17}$$

5 Results and Discussion

This section is devoted to compare the FSA-PIC controller to the FCS-MPC performances under the same operation conditions. For that reason, a simulation study of a 1.5 MW DFIG based WTS has been carried out using MATLAB/Simulink software. The operation conditions applied to the WECS are as following: the stator active power is set to the maximum power generated by the wind turbine, whilst the stator reactive power reference is set to zero value in order to maintain the power factor at unity. The wind profile adopted during this simulation test is depicted in Fig. 6.

5.1 Reference Test Track

Figure 7 presents the simulation results of the DFIG based WTS using the introduced strategies. The stator active power tracks perfectly its reference value. Meanwhile, the stator reactive power is kept at zero value during the simulation study. It can be

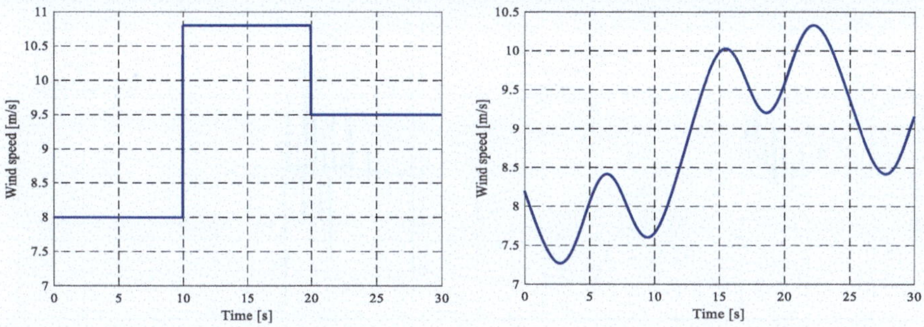

Fig. 6 Applied wind speed profile

clearly seen that the d-q axis rotor currents are the images of the stator reactive and active powers respectively and they follow their imposed references. The comparative analysis of the obtained results shows that the decoupled control is ensured for both control strategies. However, the FCS-MPC shows a faster and more precise system response compared to the FSA-PIC. Moreover, the controlled variables ripples are considerably reduced using the FCS-MPC.

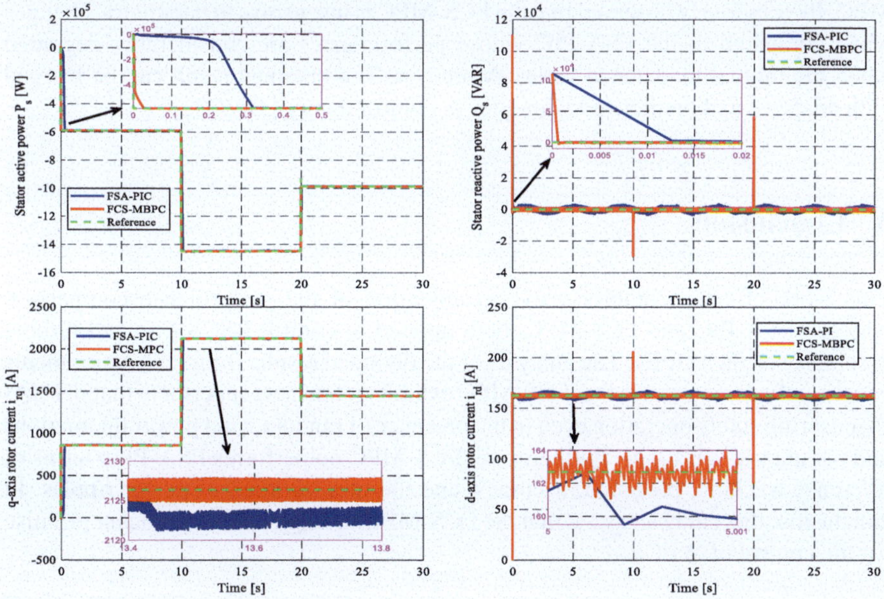

Fig. 7 Reference track test response

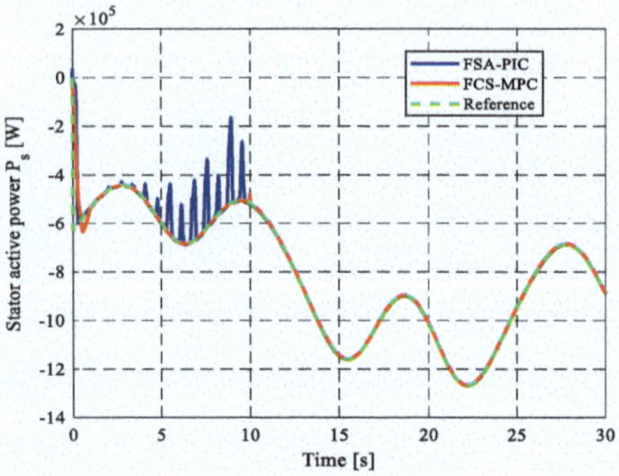

Fig. 8 Robustness test response

5.2 Robustness Test

In this section, the FSA-PIC is compared to the FCS-MPC considering an intermittent wind speed profile and the following parametric variations: 130% of the mutual inductance within the range of [0–10 s]. In the range of [10–20 s]: 200% of stator and rotor resistances. Figure 8 depicts the simulation results of the FSA-PIC and FCS-MPC. It is obviously observed that the FCS-MPC is insensitive to parametric changes when compared to the FSA-PIC whose performances are considerably degraded when the mutual inductance varies. Moreover, The obtained result can be merged with adaptive and predictive techniques in control theory [12–34].

6 Conclusion

This work offers a comprehensive examination of the performance comparison between FSA-PIC and FCS-MPC, both applied to control RSC of a wind turbine system based on a DFIG. The analysis of the obtained results from both experiments consistently demonstrates that FCS-MPC exhibits robustness superior to that of FSA-PI, ensuring stable operation even in the presence of random wind power fluctuations and parametric variations. Furthermore, FCS-MPC outperforms FSA-PI in terms of reference tracking, precision, accuracy, and the minimization of power ripples. In conclusion, this study suggests that the FCS-MPC strategy holds significant promise for the control of WECS.

References

1. Wadawa B, Errami Y, Obbadi A, Sahnoun S, Chetouani E, Aoutoul M (2022) Comparative application of the self-adaptive fuzzy-PI controller for a wind energy conversion system connected to the power grid and based on DFIG. Int J Dyn Control 10(6):2151–2173
2. Kasbi A, Rahali A (2021) Performance optimization of doubly-fed induction generator (DFIG) equipped variable-speed wind energy turbines by using three-level converter with adaptive fuzzy PI control system. Mater Today Proc 47:2648–2656
3. Tamaarat A (2019) Active and reactive power control for DFIG using PI, fuzzy logic and self-tuning PI fuzzy controllers. Adv Model Anal C 74(2–4):95–102
4. Kou P, Liang D, Li J, Gao L, Ze Q (2018) Finite-control-set model predictive control for DFIG wind turbines. IEEE Trans Autom Sci Eng 15(3):1004–1013
5. Gontijo GF, Tricarico TC, Franca BW, Da Silva LF, Van Emmerik EL, Aredes M (2019) Robust model predictive rotor current control of a DFIG connected to a distorted and unbalanced grid driven by a direct matrix converter. IEEE Trans Sustain Energy 10(3):1380–1392
6. Nagy ZK, Braatz RD (2018) Nonlinear model predictive control for batch processes. Control Handb Second Ed: 15-1–15-30
7. Kaloi GS, Wang J, Baloch MH (2016) Active and reactive power control of the doubly fed induction generator based on wind energy conversion system. Energy Rep 2:194–200
8. Nguyen TA et al (2021) An efficient model predictive control based on Lyapunov function for doubly fed induction generator fed by a T-type inverter. Electr Eng 103(1):663–676
9. Zamzoum O, El Mourabit Y, Errouha M, Derouich A, El Ghzizal A (2018) Power control of variable speed wind turbine based on doubly fed induction generator using indirect field-oriented control with fuzzy logic controllers for performance optimization. Energy Sci Eng 6(5):408–423
10. Zhang Y, Jiao J (2018) Model predictive rotor current control for doubly fed induction generators under unbalanced grid voltages. Proc IEEE Int Power Electron Appl Conf Expo PEAC 2018:1–5
11. Yaramasu V, Wu B (2016) Model predictive control of wind energy conversion systems
12. Bounemeur A, Chemachema M, Essounbouli N (2018) Indirect adaptive fuzzy fault-tolerant tracking control for MIMO nonlinear systems with actuator and sensor failures. ISA Trans 1(79):45–61
13. Bounemeur A, Chemachema M (2021) Adaptive fuzzy fault-tolerant control using Nussbaum-type function with state-dependent actuator failures. Neural Comput Appl 33:191–208
14. Bounemeur A, Chemachema M (2021) Adaptive fuzzy fault-tolerant control for a class of nonlinear systems under actuator faults: application to an inverted pendulum. Int J Robot Control Syst 1(2):102–115
15. Bounemeur A, Chemachema M, Essounbouli N (2014) New approach of robust direct adaptive control of a class of SISO nonlinear systems. In: 2014 15th international conference on sciences and techniques of automatic control and computer engineering (STA), 21 Dec 2014. IEEE, pp 725–730
16. Abdelhamid B, Mouhamed C, Najib E (2019) Optimal indirect robust adaptive fuzzy control using PSO for MIMO nonlinear systems. In: Advanced control engineering methods in electrical engineering systems 2019. Springer International Publishing, pp 208–224
17. Abdelhamid B, Mouhamed C, Najib E (2017) Indirect robust adaptive fuzzy control of uncertain two link robot manipulator. In: Recent advances in electrical engineering and control applications 2017. Springer International Publishing, pp 125–139
18. Abdelmalek Z, Abdelhamid B, Sofiane B, Boulkaibet I (2021) Fault diagnosis of uncertain hybrid actuators based model predictive control. In: Proceedings of the 4th international conference on electrical engineering and control applications: ICEECA 2019, 17–19 Dec 2019, Constantine, Algeria 2021. Springer, Singapore, pp 961–971
19. Chemachema M, Bounemeur A (2017) Commande adaptative floue des systèmes non linéaires

20. Bounemeur A, Chemachema M (2021) Active adaptive fuzzy fault-tolerant control for a class of nonlinear systems with actuator faults. In: Proceedings of the 4th international conference on electrical engineering and control applications: ICEECA 2019, 17–19 Dec 2019, Constantine, Algeria 2021. Springer, Singapore, pp 985–999
21. Bounemeur A, Chemachema M, Essounbouli N (2014) Robust indirect adaptive fuzzy control using Nussbaum gain for a class of SISO nonlinear systems with unknown directions. In: 2014 15th international conference on sciences and techniques of automatic control and computer engineering (STA), 21 Dec 2014. IEEE, pp 748–754
22. Bounemeur A, chemachema M, Zahaf A, Bououden S (2019) Adaptive fuzzy fault-tolerant control using Nussbaum gain for a class of SISO nonlinear systems with unknown directions. In: Proceedings of the 4th international conference on electrical engineering and control applications: ICEECA 2019, 17–19 Dec 2019, Constantine, Algeria 2021. Springer, Singapore, pp 493–510
23. Abdelhamid B, Chemachema M, Bouzina S (2023) Fuzzy fault-tolerant control applied on two inverted pendulums with nonaffine nonlinear actuator failures. Int J Robot Control Syst 3(2)
24. Bounemeur A, Chemachema M (2019) Contribustion à la commande tolérante aux défauts d'une classe de systèmes non linéaires. Diss. Université Frères Mentouri-Constantine 1
25. Bounemeur A, Chemachema M (2023) General fuzzy adaptive fault-tolerant control based on Nussbaum-type function with additive and multiplicative sensor and state-dependent actuator faults. Fuzzy Sets Syst 7:108616
26. Zahaf A, Bououden S, Chadli M, Chemachema M (2022) Robust fault tolerant optimal predictive control of hybrid actuators with time-varying delay for industrial robot arm. Asian J Control 24(1):1–5
27. Chadli M, Bououden S, Zelinka I (eds). Recent advances in electrical engineering and control applications
28. Zahaf A, Boutamina B, Bououden S, Filali S (2014) New approach of model based TS fuzzy predictive control using LMI approach. In: 2014 15th international conference on sciences and techniques of automatic control and computer engineering (STA), 21 Dec 2014. IEEE, pp 38–43
29. Abdelmalek Z, Chemachema M, Sofiane B, Boulkaibet I (2021) Fault tolerant predictive control for constrained hybrid systems with sensors failures. In: Proceedings of the 4th international conference on electrical engineering and control applications: ICEECA 2019, 17–19 Dec 2019, Constantine, Algeria. Springer, Singapore, pp 973–984
30. Zahaf A, Bououden S, Chadli M (2017) Constrained fuzzy predictive control design based on the PDC approach. In: Recent advances in electrical engineering and control applications 2017. Springer International Publishing, pp 140–154
31. Filali S, Zahaf A. Commande Prédictive Floue Sous Contraintes
32. Zahaf A, Bououden S, Chadli M, Boulkaibet I, Neji B, Khezami N (2023) Dynamic sensorless control approach for Markovian switching systems applied to PWM DC–DC converters with time-delay and partial input saturation. Sensors 23(15):6936
33. Zahaf A, Bououden S, Chemachema M. Contribution to robust fault-tolerant predictive control with constraints for hybrid systems (Doctoral dissertation, Université Frères Mentouri-Constantine 1).
34. Abdelmalek Z, Bououden S, Chadli M, Zelinka I, Boulkaibet I (2019) Observer based model predictive control of hybrid systems. In: Advanced control engineering methods in electrical engineering systems 2019. Springer International Publishing, pp 198–207

SPRINGER NATURE

GPSR Compliance

The European Union's (EU) General Product Safety Regulation (GPSR) is a set of rules that requires consumer products to be safe and our obligations to ensure this.

If you have any concerns about our products, you can contact us on ProductSafety@springernature.com

In case Publisher is established outside the EU, the EU authorized representative is:

Springer Nature Customer Service Center GmbH
Europaplatz 3
69115 Heidelberg, Germany

The manufacturer's authorised representative in the EU is Springer Nature Customer Service Centre GmbH, Europaplatz 3, 69115 Heidelberg, Germany. If you have any concerns regarding our products, please contact ProductSafety@springernature.com

Printed and bound by CPI Group (UK) Ltd, Croydon, CR0 4YY
01/10/2026
02225855-0010